Design-Build

Planning through Development

Jeffrey L. Beard

Design-Build Institute of America
Washington, D.C.

Michael C. Loulakis Sr.

Wickwire Gavin P.C.
Vienna, Virginia

Edward C. Wundram

The Design Build Consulting Group
Beaverton, Oregon

Boston, Massachusetts   Burr Ridge, Illinois
Dubuque, Iowa   Madison, Wisconsin   New York, New York
San Francisco, California   St. Louis, Missouri

Library of Congress Cataloging-in-Publication Data

Beard, Jeffrey L.
  Design-build : planning through development / Jeffrey L. Beard, Michael C. Loulakis
Sr., Edward C. Wundram.
    p.   cm.
  ISBN 0-07-006311-9
  1. Engineering design. 2. Building. 3. Industrial project management. I. Loulakis,
Michael C. II. Wundram, Edward C. III. Title.
TA174.B42   2001
  658.4'04—dc21   00-069936

McGraw-Hill

A Division of The McGraw·Hill Companies

  6 7 8 9 BKM BKM   0 9 8 7 6

ISBN 0-07-006311-7

*The sponsoring editor for this book was Larry S. Hager, the editing supervisor
was David E. Fogarty, and the production supervisor was Pamela A. Pelton. It
was set in Century Schoolbook by Joanne Morbit of McGraw-Hill Professional's
composition unit, Hightstown, N.J.*

McGraw-Hill books are available at special quantity discounts to use as premiums and sales promotions, or for use in corporate training programs. For
more information, please write to the Director of Special Sales, Professional
Publishing, McGraw-Hill, Two Penn PLaza, New York, NY 10121-2298. Or
contact your local bookstore.

To my beautiful and understanding wife,
June Natalie Beard, who endured many sunny
weekends mostly devoid of adult companionship while
I hunched over a laptop computer surrounded by books
and papers. Once again, it is "our" time.

JEFF BEARD

To Karen, my wife, best friend and most ardent
supporter, for giving me the encouragement and
freedom to participate in this project; and to our
children, Christina, Andrea and Charlie, who will
someday read this book and finally understand (and
hopefully be proud of) what dad was doing in front of
the computer for all those hours.

MIKE LOULAKIS

To Vicky, wife, friend, partner, reader, and critic, who
encouraged me to think and write beyond the
conventions of a technical manual; and to the rest of
my family who never quite understood "design-build,"
but who nevertheless smiled and urged me to continue.

ED WUNDRAM

Contents

Preface

In the design and construction industry today, there is probably no term that has spawned more discussion and debate than "design-build." Experience has shown that this project delivery method is a broadly encompassing and flexible concept, not a narrow, inflexible or rigid methodology. It is a strategic instrument that is more akin to a multimotional robot assembling multiple objects in a CAD-CAM environment than a hammer swinging in a single motion at a metal spike.

Within a capitalistic economy, business processes that achieve greater efficiencies emerge out of competitive pressures. The object is to create an instrument that motivates marketplace providers to offer greater value for the facility owner. Fundamentally, facility or infrastructure asset value is *performance over time*. Design-build's ability to offer innovation and increased value makes it one of the most important tools in the toolbox of capital facilities owners worldwide. However, like any tool, design-build cannot be applied to every solution. The key is to clearly understand what design-build (or any other delivery system) can do or what it cannot do for a given project.

Design-Build: Planning through Development is intended to give readers a grasp of this ambidextrous tool. Discrete sections in the chapters cover the many new roles, phases, responsibilities, risks, and challenges that accompany an innovative business model, and the text follows the process from inception to operation. As the integrated process matures, design-build naturally moves upstream to garner feasibility studies, site evaluation and selection, financing, and programming, and downstream to commissioning, operation, maintenance, and asset disposition. The process is metamorphosing into design-build-finance, design-build-operate-maintain, and integrated total facilities services. The reader who works within the core business of design and construction will be stimulated by the possibilities of a maturing design-build marketplace.

This book will serve as a resource to both owners and A/E/C practitioners. Among the practical topics treated in the chapters are

- Aligning goals and incentives on multidisciplinary projects
- Writing project criteria in performance terms

- Identifying the key responsibilities of owners in design-build
- Determining the important components of a conceptual estimate
- Drafting requests for proposals (RFPs)
- Analyzing the differences between acquisition/project delivery and procurement/purchasing
- Identifying risks and liabilities in design-build

Our intent in writing the book was to bring varying perspectives to integrated design and construction. Jeffrey Beard is trained as a facilities planner and construction project manager. Edward Wundram is a registered architect. Michael Loulakis is an engineer and attorney. As we wrote our respective chapters, we were reminded of how a comprehensive grounding in design-build philosophies will lead to a fundamental change in the design and construction industry. A true integration with manufacturers, suppliers and specialty contractors is possible with the design-build approach, just as America's automotive and aviation industries have evolved in their design and production consolidation over the past generation.

The past focus of the construction industry was upon rules rather than results, and on means and methods rather than performance. Design-build shifts the emphasis from a fragmented approach to a combined effort in support of the owner's objectives. Achievement of the owner's ultimate "drivers" (whether they be innovative technical solutions, world-class aesthetic design, safety and security, environmental sustainability, or price) becomes an overarching goal of the combined team.

The opinions expressed in the writings are our own and not tied to any single professional organization, trade group, or policy. We welcome inquiries and commentaries from readers as we each continue to discover the potential for greater collaboration within our chosen industry.

JEFFREY BEARD, *Washington, D.C.*
MICHAEL LOULAKIS, *Vienna, Virginia*
EDWARD WUNDRAM, *Beaverton, Oregon*

Acknowledgments

Inspiration comes in many forms. Teachers, mentors, parents, competitors, and compatriots have inspired us. We have also been inspired by creations of people in the built environment, from buildings to bridges; from process plants to piscineries. And we have been inspired by the collaborative process of design-build, in all of its many permutations: from the simple combination of design and construction to the total integrated life cycle of a capital asset, including planning, financing, design/construction, operation, maintenance, and sale/transfer/renovation.

For all of us, the greatest source of inspiration and commitment for this book came from the founding Chairman of the Design-Build Institute of America, Preston Haskell, Chairman of the Haskell Company, Jacksonville, Florida. His articulate and impassioned leadership placed DBIA on sound and principled footing, and propelled the nonprofit organization to codify best practices and to create practical tools for owners and practitioners who are implementing design-build delivery. He motivated each of us to be proficient in what we do and to be ever aware that design-build was a "work-in-progress" that needed to be constantly challenged and improved upon.

We have been heavily influenced by other key leaders of the design-build movement in the 1980s and 1990s, including Kraig Kreikemeier (Sverdrup Corporation), Rik Kunnath (Pankow Builders), Don Warren (Suitt Construction Company) and Grant McCullagh (McClier Corporation). Each played pivotal roles in the fledgling years of the Design-Build Institute and provided phenomenal service to the industry in their nonprofit leadership roles. We also salute some of the more recent Chairpersons of DBIA, including Tanya Matthews (Parsons-Brinckerhoff); Jeff Raday (McShane Construction Corp.) and Steve Halverson (The Haskell Company). With their energy and intellect, the welcoming of new integrated services concepts and the flourishing of better ideas will continue to shape the entire industry.

We are grateful to those opinion leaders and seasoned spokespersons within the design-build movement within academia, including Victor Sanvido (Penn State), James Rowings (Iowa State) and Saied Sadri (Georgia Tech). Pioneering research has also been done by Mark Konchar (Centex Construction), Tony Songer (Virginia Tech), Keith Molenaar (University of Colorado), Barbara

Jackson (Cal Poly), Doug Gransberg (University of Oklahoma); among many others. For the design-build movement to truly extend roots, the industry will need strong support from universities and colleges, and the above individuals have been leading the way.

Tireless workers for the collaborative process also deserve mention. Fundamentally, the members and committee activists within the Design-Build Institute breathed life into the new business model. DBIA staff also contributed to the cause: Dave Johnston, who left the American Institute of Architects to work for DBIA, devised strategies to attain pan-government federal design-build legislation. His strong and sound arguments for procurement flexibility imperiled his career as those who clung to rigid applications of the old order sought to overturn his accomplishments for "best practices." More progressive industry thinking has validated his pioneering work on design and construction procurement. Don Jessup, employing his meeting-planning experience based on years with the Associated General Contractors of America, built the world's finest annual conclave addressing design-build: DBIA's Professional Design-Build Conference. Don's meticulous work habits have placed the organization on a firmer financial footing and have given the Institute international renown.

There would be no tome entitled *Design-Build: Planning through Development* without the initial recruitment of Jeff Beard by M.D. "Doc" Morris, F.ASCE of Ithaca, N.Y., who serves as the Construction Series editor for McGraw-Hill Books. His periodic "this is your conscience calling" phone messages kept the process from miring in the mud of incompletion.

Finally, we take this opportunity to give a nod to each of the co-authors. Jeff Beard was the "brains" and "glue" behind this project. As one of the nation's leading thinkers about design-build, Jeff provided the outline for the overall scope and organization of the book, and recruited Mike Loulakis and Ed Wundram for those areas where their expertise would help. Jeff Beard drafted Chapters 3, 7, 8, 9, 11, 12, 18, and 19. Mike Loulakis, who has been writing and speaking about the legal aspects of design-build for almost 15 years, penned Chapters 6, 13, 14, 15, 16, and 17. He also fulfilled the important role as surrogate for Doc Morris among the authors ("I've got my stuff done; when can I see yours?").

Finally, Jeff Beard rounded out the group by asking Ed Wundram to be a co-author. Ed Wundram became a design criteria professional in the 1970s, well before it was fashionable. He was responsible for some of the most conceptually challenging areas of the book (Chapters 1, 2, 4, 5, and 10) and ensured that the book would reach beyond the ordinary literature on the design-build process.

Together, we hope that this book will be part of a growing body of work about *integrated services delivery* that has design-build at its core. We salute those who have preceded us and look forward to reading the publications of those who will expand our concepts and continue to push the bounds of integrated services.

An Introduction to Design-Build Project Delivery

1.1 Need to Restructure

The design and construction industry in North America, and those parts of the industry focused on buildings and civil infrastructure projects in particular, is a very fragmented industry. In traditional procedures used to design and construct facilities, the designers are required by professional ethics and by the nature of the competitive methods used to select general contractors, to maintain an arm's-length relationship with builders, trade subcontractors, material suppliers, and others on the construction side of the industry. Owners must hire architectural and engineering designers separately from builders. They, in response to the owners' programs, prepare 100 percent complete construction documents, also referred to as *plans* and *specifications*. Builders, in turn, are forced by price competition to select subcontractors, equipment, and materials on the basis of the lowest price for the work or item that meets the detailed requirements of the owners' bidding documents. Because builders or general contractors are most often selected on the basis of price, they have little opportunity to offer the owners alternative design solutions other than limited material choices or superficial design options. The three separate sequential steps required by this process are referred to as the *design-bid-build project delivery method.*

With the advent of the post–World War II building boom in North America, owners sought alternatives to the split responsibilities of the design-bid-build method of project delivery. One of the first alternative procedures employed by owners was the engagement of a construction manager whose responsibilities included the transfer of knowledge between the design and the construction sides of the industry. This method had some success and is still employed in the industry on about 10 to 15 percent of projects. However, from the owner's perspective, it still did not offer a single point of responsibility. Construction managers complained that their lack of control over the designers prevented them from achieving more positive results for owner's objectives. By the early 1980s, it became evident to many facility owners that a "restructuring" of the design and construction industry would be necessary to improve the efficiency and accountability of project delivery.

1.2 Design-Build and Singular Responsibility

Some private sector owners whose facilities were used to produce measurable quantities of products, such as electric power, chemical, and food processing industries, found a contractual method to make a single entity responsible for all aspects of the facility design, equipment selection, and construction necessary to produce a specified output. This procedure evolved into the design-build project delivery method.

In the late 1980s and 1990s, there was a trend for the managements of both private-sector firms and public agencies to restructure their organizations and reduce or eliminate staff not directly associated with their core goals. This has led to a procurement policy that transfers many of the risks associated with facilities acquisition to the single entity or design-builder. First and foremost of the "risks" are design management and its corresponding design control. With both design and construction in the hands of one entity, there is a single point of responsibility for coordination, quality, cost control, and schedule adherence, which avoids "buck passing" and blaming others for errors and shortcomings. Owners are able to avoid the role of referee between designers and builders and can focus on needs and scope definitions and timely decisionmaking.

1.3 Quality

The greater responsibilities and accountabilities implicit in the design-build process serve as motivation for high quality and proper performance of building systems. Once the owner's requirements and expectations are documented in terms of program and performance, it becomes the design-builder's responsibility to produce a facility that meets or exceeds those criteria. Prior to construction, the design-builder warrants to the owner that the design documents are complete and free from error. (By contrast, with "traditional" design-bid-build, the owner warrants to the contractor that the drawings and specifications are complete and free from error. Because of the owner's warranty for the construction documents, the traditional process must rely on restrictive contract language, extensive audit and inspection, and occasionally the courts, to ensure final project quality.) Under the terms of most design-build contracts, the design-builder is responsible, at least through the postoccupancy warranty period, for the performance of the facility and its component parts, as defined in the owner's performance specifications. This guarantee of performance, which the design-build entity can now offer because it controls both design and construction, motivates the design-builder to assure the quality of both design and construction in order to mitigate the risk of performance failure. Effectively, the design-build process requires the design-builder to accept and adopt the owner's quality objectives.

1.4   Cost Savings and Value

Design professionals and construction personnel, working and communicating as a design-build team, evaluate alternative materials, building systems, and methods efficiently, accurately, and creatively. Value engineering and constructability reviews are utilized more effectively when the designers and builders work as one body during the design process. Cost saving, however, is not a goal in itself, but rather is part of the design-builder's broader objective of creating value. A comprehensive knowledge of labor and material costs coupled with an awareness of the cost relationships between the various project components, and the ability to control design, allow the design-build team to increase a project's value while reducing its overall cost.

1.5   Time Savings

Because design and construction can be overlapped, and because general contract bidding periods and redesign time are eliminated, total design and construction time can be significantly reduced. Design-build is ideal for the application of fast-track construction techniques, without the corresponding costs and budget risks to the owner. Under this method, construction work is allowed to begin in advance of the completion of the construction documents (plans and specifications). The time savings translate into lower costs and earlier utilization of the completed facility.

Design-build and its associated fast-track methodology reduce the overall project delivery time in other ways. First, in order to prudently enter a design-build contract, an owner must clearly and accurately document its facility needs and project objectives. This discipline requires the owner's organization to analyze their function needs, document them in a way to effectively communicate them to senior management or governing body, and justify them in relation to other needs and available resources, all abstractly and in the absence of any particular facility design. Second, using the fast-track scheduling procedure, the design process parallels procurement and construction, and any delays in the owner's design review and approval process creates additional costs in construction. This cost penalty encourages the owner's organization to conduct their reviews and provide decisions in a timely manner.

1.6   Reduced Administration

Under this streamlined procurement process the potential exists to reduce the owner's administrative tasks. Initially, documenting the program of requirements, preparing requests for proposals (RFP), and conducting evaluations will require intensive involvement by senior managers, consultants, and others; however, this effort will be rewarded by a facility that meets their specific functional and performance needs. After award of the design-build contract, the owner will not be required to spend time and effort coordinating and arbitrating between separate design and construction contracts. While the process does require the owner to provide "prudent oversight" of the design and construction progress, this responsibility is considerably less time-consuming and exposes the owner to far fewer risks than would the traditional separate contracts for design and construction.

1.7   Early Knowledge of Firm Costs

Because the entity responsible for design is simultaneously estimating construction costs and can accurately conceptualize the completed project at an early stage in design development, guaranteed construction cost proposals can be delivered sooner than is otherwise possible. This permits early establishment of financing and reduced exposure to cost escalation, and avoids the risk of committing substantial time and money for architectural and engineering services, only to learn later that the cost of the project is prohibitive.

1.8   Risk Management

Project performance aspects of cost, schedule, and quality can be clearly defined and appropriately balanced (individual risks are managed by the party best able and positioned to manage that risk). Change orders due to errors and omissions in the construction documents are eliminated because the correction of such is the responsibility of the design-builder, not the owner. After the owner outlines its needs in the RFP, it will receive different design solutions and cost proposals representing the best thinking of several design-builders (actually, teams of architects, engineers, contractors, and specialty consultants). These alternative designs will enable the owner to better weigh the risks and benefits of several competing proposals before committing to any single design solution.

This aspect of assigning risks to those best capitalized, staffed, and experienced to assume and manage them is the raison d'être for the design-build project delivery method. Recently, owners have added additional responsibilities to teams of designers, builders, developers, financiers, and facility operators. These new risks have included the responsibilities for site acquisition, land-use variances, permitting, environmental remediation, project financing, and facility maintenance and operations. Almost any real estate responsibility that can be clearly defined can be assigned to others who are more familiar with the associated risks and who may, therefore, assume these responsibilities for less cost than would an inexperienced facility owner.

1.9   Best-Value Selection

In no other aspect of public or private procurement, except low-bid design-bid-build construction, are the choices limited to initial cost. In the extreme examples of an individual buying an apple from a grocer's fruit stand or an airline's purchase of a fleet of new passenger airplanes, quality and quantity for price, or "value," are the basis of the buyer's decision to purchase. Common sense tells us that a similar analysis is appropriate for the purchase of buildings and infrastructure facilities.

Design-build project delivery methods allow the purchaser to compare the quality and scope of a proposed facility with the price offered. Most buildings and civil engineering structures are unique, one-of-a-kind facilities and will likely require a technical and often an aesthetic evaluation to determine their value to the owner. By giving credit in the evaluation and award mechanism for design excellence, materials and systems quality, function efficiency of the plan, the design and construction team's experience, and other intangible factors beyond price, an owner can select a proposal that offers the greatest benefit, and not simply the lowest first cost.

1.10   Caveats to Consider to Ensure a Successful Design-Build Process

1.10.1   Complexity

Design-build project delivery requires careful planning and professional execution to be successful. The owner should choose a design-build process variation on the basis of factors such as the project's complexity, funding, design intent, schedule, risk allocation, and other important issues. For owners who do not have in-house staff with expertise in preparing and administering design-build RFPs and contracts, a design-build consultant (sometimes called a *design criteria professional*) should be retained to prepare scope definition and RFP documents. This consultant can also assist the owner in the evaluation of proposals and monitoring the subsequent design development and construction.

There is often a misconception that very complex facilities are not suitable candidates for competitive design-build procurement. This is not true; in fact, the opposite may be more applicable. Within the design and construction industry, teams of designers, specialty consultants, builders, and design-build subcontractors specialize in very specific and often complex facility types. The key to the appropriateness of a competitive design-build selection is the ability of the owners and their consultants to clearly define the project programs and performance requirements in objective terms. Those requirements should be included in the RFP and must be written in terms that are contractually enforceable and are likely to be interpreted equally by all proposers.

Even complex projects with undefined requirements may still be excellent candidates for other types of design-build selection, such as direct or negotiated selection. The advantage of involving the design-builder in the early program definition stages of a project is that the design-build team will be able to provide realistic cost estimates, alternatives, and practical advice on building systems selection, scheduling, site selection, and other preconstruction services. Complex projects, regardless of the selection methods employed, require the expert counsel of a design criteria consultant and the guidance of an expanded owner-designer-builder team.

1.10.2 Converting owner needs into a statement of needs

Preparing a statement of facility requirements (sometimes called the *design criteria package*) that is sufficiently comprehensive to assure compliance by the offerors, but avoids overly restrictive requirements or details that would inhibit creative solutions, is the most challenging aspect of preparing a design-build RFP. When facility requirements are stated in performance terms and are related to recognized industry standards, the approach not only provides flexibility to the offerors to meet the desired objectives but also fixes responsibility on the design-builder in clearly understood objective and measurable terms.

A facility's performance criteria are generally divided into two separate parts: program requirement and performance requirements. The program requirements or program describes the functional and quantitative needs of the project, such as a proposed building's net floor area, the number of vehicles to be accommodated in a parking garage, or the number and width of vehicle lanes on a bridge. The performance requirements are described in the specifications and include the owner's expectations for the performance of the facility and its component parts, building assemblies, and materials. These specifications should include design standards, quality levels, performance criteria, and methods to substantiate that these requirements have been met.

1.10.3 Insurance and bonds

Certain insurance carriers and bond sureties may not be familiar with the design-build process. This may lead to some hesitation on their part to provide professional liability insurance and contract bonds on design-build projects. However, this situation is changing as sureties and insurance companies gain experience with the design-build project delivery method. In view of the claims to date, their fears about the design-build process have been unfounded. The number and magnitude of owners' claims on design-build projects have been fewer and smaller on a comparison of similar projects. One reason appears to be that the early involvement of builders on the design-build teams lessens the potential for design error and "buildability" issues. Another reason is because both the design and the construction responsibilities lie with a single entity—the design-builder—and the parties to a dispute have no choice when confronted with design errors or construction cost overruns but to seek creative solutions (or recourse) from their team members.

The parties to a design-build contract must assure themselves that adequate coverage exists in both the professional liability and surety areas. Design liability insurance and performance-payment bonds, both specifically written for design-build projects, are now available from some providers.

1.11 Overview of Some of the Steps in a Typical Design-Build Process

1.11.1 Strategic facility planning

This function includes an analysis of current and projected facility requirements to determine the appropriate, long-range development plan for the owner. Facility goals and objectives are established, which, in turn, assist in the selection of appropriate project delivery methods.

1.11.2 Program definition

The owner establishes the project "needs" of functional areas, their occupancy (staff and visitors), equipment to be furnished or accommodated (with utility information), the indoor environment standards for each area, relative locations within the facility, materials, finishes, special conditions, and applicable codes and regulatory conditions. These requirements are summarized in the request for qualifications (RFQ), and defined and articulated in more detail in a request for proposals (RFP).

1.11.3 Request for qualifications (RFQ)

Requirements for offerors are defined and articulated in an RFQ, either by in-house staff or by outside consultants. Teams of designers and builders are organized in response to the requirements of the owner, as stated in the RFQ document. Other particulars influencing team formation are established, such as the need for local or regional designers or for specialized technical expertise, as well as the criteria the owner will utilize to evaluate the capabilities of the design-build organizations applying.

1.11.4 Qualification statements

The project is advertised, qualification statements are received (in response to RFQ), and preferably three, but not more than five, of the best-qualified design-builders (or teams of designers and builders) are selected or "shortlisted." The qualification statement is a description of the design-build team's composition and organization. The composition of the team may not be changed later without the owner's permission.

1.11.5 Request for proposals (RFP)

Design and cost proposals are solicited from the shortlisted design-builders in an RFP document. Among the items found in a typical RFP are project design criteria, program requirements, performance specifications, site information, contract requirements, selection procedures, and proposal requirements or deliverables. The initial RFP will often have precedence over all subsequent submittals and contract documents, except mutually acceptable change orders.

1.11.6   Proposal preparation

Each shortlisted design-build team is furnished copies of the owner's program, budget, schedule, and other requirements, together with a description of the site, utilities, soils, and other existing conditions. Working together, the team of designers, builders, and estimators develops a preliminary or schematic design of the proposed facility for the owner's consideration. This design, described in drawings, in written specifications, and sometimes by a physical model, and accompanied by a formal price proposal, is submitted to the owner. Some owners offer to pay an honorarium or stipend to the proposers in consideration of their efforts to prepare a proposal.

1.11.7   Proposal submission and evaluation

Once received, proposals are evaluated on the basis of quantity, quality, functional efficiency, aesthetics, price, and other factors. The owner may request additional clarification of specific aspects of the proposals from respective design-build teams. Before making a final contract award, the design-build teams may be asked to make in-person presentations to the selection panel and answer the panel's questions concerning their proposals.

1.11.8   Contract award

The selected proposer enters into a contract with the owner and is issued a notice to proceed with the design work after the submittal of required bonds and certificates of insurance.

1.11.9   Design development and construction documents

In accordance with the design-builder's proposal and the owner's comments, the team proceeds to develop more detailed architectural and engineering documents, often in close coordination with representatives of the owner and local building code officials. This latter phase of design work, referred to as the *design development phase,* precedes work on the final construction documents. The design development documents are submitted to the owner for review, comment, and approval.

Construction documents are specific and detailed instructions to the builders, trade subcontractors, building material and equipment suppliers, and others for the construction of the facility. These plans and prescriptive specifications are also used to apply for construction permits from the local building officials. They will incorporate the owner's comments and instructions for change, if any, from the reviews of the design development documents. The owner must review and approve the construction documents before the design-builder can start work on the site.

1.11.10   Construction

On completion and approval of the construction documents for all elements or for specific parts of the work, construction commences. Some contracts, particularly those employing fast-track methods, require that construction proceed after logical phases of design are completed, but prior to completion of the entire body of construction documents. During construction, the owner's representative will monitor the work for quality and degree of completion. The owner will then make progress (partial) payments to the design-builder in a manner consistent with the contract requirements. After the completion of construction, the owner's representatives will examine the facility for compliance with the initial program and performance requirements, as well as with the design-builder's proposal and its construction documents. The design-builder warrants the quality, completeness, and performance of the facility for a period of time thereafter, in accordance with the contract.

1.12   Proposal Evaluation and Selection Methods

1.12.1   Purpose

The purpose of the evaluation and selection process should be to determine which proposal provides the greatest value to the owner. A variety of evaluation and selection techniques are available to private- and public-sector owners. Each has been successfully used, and each has its merits. No single process is appropriate for every situation. The owner must determine which procedures are most appropriate to the project conditions, and if the owner is a public agency, which procedures are permitted under applicable laws and regulations.

1.12.2   Weighted criteria

The RFP lists the final and detailed requirements for the submittal of a qualitative proposal and a firm price. The owner establishes in the RFP a point rating for qualitative factors and for price. The owner receives the qualitative proposals. Price proposals are submitted simultaneously in a separate envelope. The owner reviews each proposal, and may hear oral presentations from each proposer. The owner then assigns points on a scoring matrix for the proposers' responses to each evaluation factor.

After the design and qualitative factors are evaluated, the price proposals are opened. Maximum price points are assigned to the lowest dollar bid, and all others are scaled inversely proportional to that amount. High total points then determine the successful proposal.

1.12.3   Adjusted low bid

A variant of the weighted criteria selection process is the adjusted low bid. The process follows the same steps through receipt of qualitative proposals. Following the oral presentations, qualitative aspects are scored and totaled on a scale of 0 to 100, expressed as a decimal. After all proposals have been reviewed and scored, the price proposals are opened. Prices are then divided by the decimal score to yield an *adjusted bid*. The lowest adjusted bid is recommended for contract award. The adjustment to the bid price is for selection only. The firm's initial proposal price is the actual contract amount.

1.12.4   Equivalent design/low bid

This evaluation procedure parallels the two previous processes up to receipt of design proposals. All steps of qualifications, shortlisting, and design and price submittal are the same. Design proposal reviews, however, are followed by a separate critique session with the individual design-build teams. Proposers are given a deadline to respond to the critique with specific design changes. The revised design proposals are accompanied by a corresponding price adjustment (either add or deduct). The revised design proposals are evaluated by the owner, and price envelopes, both base bid and revised bid, are opened. Award can be made with heavy or exclusive emphasis on price because the proposal critique should have created relatively equivalent designs.

1.12.5   Stipulated sum/best design

Contract price is established by the owner and is stated in the RFP. This process also uses the same initial prequalification and shortlisting methods described earlier. The shortlisted proposers submit their qualitative proposals. Oral presentations are made (optional), and the owner uses its evaluation criteria to score the proposals. Recommendations for contract award are made for the firm offering the best design solution for the fixed budget amount.

1.12.6   Meets criteria/low bid

This method of evaluation most closely resembles the traditional design-bid-build process. Typically, the RFP criteria document allows very little creativity in the proposed design solutions. Very specific outline or conceptual designs are issued as part of the design criteria package. Proposals are solicited from qualified firms. The proposals are evaluated and deemed to meet the base criteria, and award is made to the lowest bidder. The selected firm's role is simply to complete the construction documents, not to change or vary the initial design concept.

1.12.7 Emergency

As implied, when public safety and welfare are threatened, the owner may authorize negotiations with the best-qualified design-builder available at the time. The owner may utilize references, capability, and/or previous experience with the firm to justify the selection.

1.13   Elements of a Successful Design-Build Project

- Set aside traditional procedures and relationships.
- Prequalify proposers.
- Conduct balanced evaluations.
- Establish reasonable submission requirements.
- Develop succinct program and performance criteria.
- Pay stipend or honorarium proportional to the effort required to prepare a priced proposal.
- Limit early design submittals in the RFP proposal phase to the level of detail necessary for evaluation and definition of scope.
- Balance responsibility and risk between owner and design-builder in contract.
- Conduct separate evaluations of qualitative proposals and price proposals.
- Appoint learned and independent judges.
- List selection criteria and weighting in the RFP.
- Use lump-sum contracts when selection is competitive.
- Include requirements for financial guarantees.
- Promptly award contract.
- Prohibit the use of design concepts from unsuccessful proposers.

2

Design-Build: A Brief History

2.1 Origins of the Master Builder

2.1.1 Master Builders in History

As students of architectural history, we recall the ancient *master builders* or *master masons:* Ictinus and Callicrates, builders of the Parthenon in Athens; Abbé Suger for his twelfth century Gothic Royal Abbey Church of Saint Denis, outside Paris; and Filippo Brunelleschi for the Dome of the Florence Cathedral (Fig. 2.1) in the early fifteenth century. They each provided a seamless service that included what we now refer to as *design and construction,* or more recently as *design-build.* It never occurred to these masters to consider one without the other. Such specialization would have been foreign to these design-builders. The Greeks had a word for the role of master builder. It was *arkhitektōn,* which literally translated means first or principal builder or craftsperson, and from which we have derived the modern word *architect.*

2.1.2 Code of Hammurabi

But even before these master builders, there was reference to the singular responsibility for design and construction in the Code of Hammurabi. Hammurabi (1795–1750 B.C.) was the ruler who established the greatness of Babylon, the world's first metropolis. By far the most remarkable of the Hammurabi records is his code of laws, the earliest known example of a ruler proclaiming publicly to his people an entire body of laws. The code regulates in clear and definite terms the organization of society. The judge who blunders in a law case is to be expelled from his judgeship forever, and heavily fined. The witness who testifies falsely is to be slain. Indeed, all the serious crimes are made punishable with death, others by mutilation, on the strict principal of an eye for an eye or a tooth for a tooth.

Figure 2.1   Florence Cathedral.

The parts of the code that address the work of the builder do not distinguish between the responsibility for faulty design and the liability for improper construction. Design was not mentioned very likely because most utilitarian structures of the day were "designed" and built according to traditional trial-and-error means and methods. Uniqueness and originality were not considered necessary for any but the most public structures, such as palaces, temples, and tombs for royalty.

Six of Hammurabi's numerous codes specifically address the work of the builder[1]:

228. If a builder build a house for someone and complete it, he shall give him (the builder) a fee of two shekels in money for each sar of surface.

229. If a builder build a house for someone, and does not construct it properly, and the house, which he built, fall in and kills its owner, then that builder shall be put to death.

230. If it kill the son of the owner, the son of that builder shall be put to death.

231. If it kill a slave of the owner, then he shall pay slave for slave to the owner of the house.

232. If it ruin goods, he shall make compensation for all that has been ruined, and inasmuch as he did not construct properly this house which he built and it fell, he shall re-erect the house from his own means.

233. If a builder build a house for someone, even though he has not yet completed it; if then the walls seem toppling, the builder must make the walls solid from his own means.

These codes clearly imply that the builder must know the appropriate design for the required structure, and must then build it according to those traditionally accepted materials and forms. Any deviations from these "designs" would literally risk the builder's life and limbs.

2.1.3   Vitruvius: *De architectura libri decem*

For the practice of architectural and engineering design to be first defined, and then to make the transition from innovative to routine, design knowledge must be captured, organized, and disseminated. Since the time of Roman Emperor Augustus, handbooks were the popular means of doing this. The creator of the original design handbook was a Roman writer, engineer, and architect, Vitruvius, who lived in the first century B.C. His *De architectura libri decem* (10 books on architecture) written about 40 B.C., carefully describe existing practices, not only in the design and construction of buildings but also in what are today thought of as engineering disciplines. His books include such varied topics as the manufacturer of building materials and dyes, machines for heating water for public baths, amplification in amphitheaters, and design of roads and bridges. His writing is prescriptive and gives direct advice: "I have drawn up definite rules to enable you, by observing them, to have personal knowledge of the quality both of existing buildings and those which are yet to be constructed."[2] As a handbook, *De architectura libri decem*

---

[1]From http://www.yale.edu/lawweb/avalon/hammint.htm. The original paper was entitled *The Code of Hammurabi: Introduction,* written by Charles F. Horne, Ph.D., 1915 and published by The Avalon Project at the Yale Law School.

[2]Preface, Book I, Morgan's translation.

was successful in establishing design and construction management as learned professions. Vitruvius' advice was followed for centuries. His writings assumed that the responsibilities for what we now call design and construction was vested in a single individual or *arkhitektōn*.

2.1.4 Filippo Brunelleschi, quintessential design-builder

Filippo Brunelleschi (1377–1446) was a Florentine artist-architect-builder and one of the initiators of the Italian Renaissance. His revival of classical forms and his championing of an architecture based on mathematics, proportion, and perspective made him a key artistic figure in the transition from the Middle Ages to the modern area. He is credited with discovering the mathematical laws of perspective and producing the first demonstrations of these principles around 1425.

Brunelleschi was born in Florence and received his early training as an artisan in silver and gold. In 1401 he entered, and lost, the famous design competition for the bronze doors of the Florence Baptistery. He then turned to architecture and construction and in 1420 received the commission to execute the dome of the unfinished Gothic Cathedral of Florence, also called the Duomo.

> The construction of the dome of the Florence Cathedral (was) one of the germinal events of Renaissance architecture.... The problem had been posed in the middle of the fourteenth century when the definitive plan for the octagonal crossing had been laid down. The diameter of the dome at 39.5 meters (130 feet) precluded the traditional use of wooden structuring to support the construction of the vault, while the use of buttresses as in northern Gothic cathedrals was ruled out by the building's design.[3]

The dome, a great innovation both artistically and technically, consists of two octagonal vaults, one inside the other. Its shape was dictated by its structural needs — one of the first examples of architectural functionalism. Brunelleschi made a design feature of the necessary eight ribs of the vault, carrying them over to the exterior of the dome, where they provided the framework for the dome's decorative elements. This was the first time that a dome created the same strong effect on the exterior as it did in the interior. Many historians consider Brunelleschi's double-shelled dome to be the most impressive architectural work of the fifteenth century.

The design and construction of the dome presented considerable design and construction challenges that were beyond the technology available at the time. Architects and builders of the day were convinced that they would never find a way to vault the cupola or beams to make a bridge strong enough to support the framework and mass of so tremendous an edifice. In May 1417, Brunelleschi proposed a design to the wardens of Santa Maria del Fiore (Florence Cathedral) that required no scaffolding or falsework. Intrigued, but

---

[3]M. Raeburn, ed., *Architecture of the Western World,* p. 130.

not satisfied that his solution was practical, or even possible, the church officials, at Brunelleschi's own suggestion, invited artists and builders from throughout Europe to submit competing designs.

In 1420, the competitors assembled in Florence to present their schemes for the dome. They assembled in the Office of Works of Santa Maria del Fiore, in the presence of the wardens and consuls and a number of the most able citizens, all of whom were to listen to each builder's suggestions and then reach a decision on how to vault the cupola. The proponents were called into the audience and everyone spoke their minds in turn, all master builders explaining their own plans. Filippo alone said that the cupola could be raised without a great deal of woodwork, without piers or earth, at far less expense than arches would entail, and very easily without any temporary framework to support the dome during construction.

To the consuls, expecting Brunelleschi to expound some beautiful scheme, it seemed that he was talking nonsense. They mocked and laughed at him and turned away, saying that his ideas were as mad as he was. Filippo took offense at this and later prepared a detailed written description of his proposal for the design and construction of the cupola. After considerable lobbying by Brunelleschi with individual consuls, wardens, and citizens, they were encouraged to call another meeting and the builders once again disputed the matter.

The wardens were interested in his scheme and suggestions and urged him to make a model which they could study. He refused. When urged by the wardens and the competing experts to explain his design and construction methods, he suggested that whoever could make an egg stand on end on a flat piece of marble could build the cupola, since this would show how intelligent each person was. So an egg was produced and each builder tried in turn to make it stand on end, but they were unsuccessful. When Filippo was asked to do so, he cracked the bottom of the egg on the marble and made it stand upright. The others complained that they could have done as much; Filippo retorted that they would also have known how to vault the cupola if they had seen his model and plans. Therefore, in 1420, the wardens resolved that Brunelleschi should be given the commission to design and construct the cathedral dome.[4]

The construction of the dome took almost 16 years, with Brunelleschi supervising every aspect of its design and construction. It is recorded that he personally selected the individual stones for the structural ribs of the church's cupola. He even set up a foundry on site to forge the large iron chain that surrounds the base of the dome to keep the ribs from spreading outward. He gained much renown among his fellow engineer-builders for his structural innovations, and for the construction machines he invented to build the dome, such as the revolving crane. The drawings and details of Brunelleschi's crane can be found in the notebooks of Leonardo da Vinci (1452–1519).

---

[4]Giorgio Vasari, *Lives of the Artists,* translated by G. Bull, Penguin Books. Reprint by Viking Press, New York, 1988.

It is a measure of his success and genius that contemporary artists, architects, engineers, and builders each claim Filippo Brunelleschi as one of their own. By their collective definitions, he is a true master builder. Brunelleschi died in Florence in 1446.

2.2   The Rise of Professionalism

2.2.1   Leone Battista Alberti, first architect

The first known record of the intentional separation of the art of architecture from the craft of building was in Italy in the mid-fifteenth century. In 1456, Leon Battista Alberti (1404–1472), a poet, philosopher, and papal secretary to Pope Eugene IV, convinced the Pope that, by way of drawings and models, he could direct a master mason to build a new facade on the Gothic church, Santa Maria Novella in Florence. Although his buildings rank among the best architecture of the Renaissance, he was a theoretical architect rather than a practical architect-builder. He furnished the plans of his buildings but never supervised their construction.

Alberti was born in Genoa in 1404, the son of a Florentine noble. He received the best education available in the fifteenth century. He was proficient in Greek, mathematics, and the natural sciences. As a talented writer, and one of the first organists of his day, Alberti greatly influenced his contemporaries. In 1432, he was appointed a papal secretary by Pope Eugene IV. Alberti's architectural training began with the study of antique monuments in Rome from 1432 to 1434. Later, he joined the papal court in Florence, where he became intensely involved in the cultural life of the city. Among his friends and associates were the great architect-builder Filippo Brunelleschi.

In 1452, Alberti returned to Rome, where he was secretary to six popes. Under Pope Nicholas V, he was in charge of the projects for rebuilding Saint Peter's Basilica and the Vatican. His *De Re Ædificatoria* (1485) was the first printed work on architecture of the Renaissance. This work and writings did much to establish the art of architecture as a profession distinct from the science of engineering and the craft of building. Alberti died in Rome in 1472.

2.2.2   Sir Christopher Wren, London's master builder

After the fifteenth century and well into the nineteenth century, architects continued to retain responsibility for both design and construction, administratively if not always physically. The best-known example of designer-builder during that period was Sir Christopher Wren (1632–1723).

Wren, who is considered his country's foremost architect, was also an astronomer and a mathematician. After becoming famous as a scientist and mathematician, he started his career as an architect at age 29. He was appointed assistant to the surveyor general in charge of the repair and upkeep of public buildings. In 1667, Wren was named deputy surveyor general for the reconstruction of Saint Paul's Cathedral, numerous parish churches, and other buildings destroyed by the London fire of 1666. From 1669 until about 1720, he was surveyor of the royal works and had control of all government construction in Britain.

Wren's designs for Saint Paul's Cathedral were accepted in 1675, and he superintended the building of the baroque structure until its completion in 1710.

2.3   Industrial Revolution and Its Effects on Design and Construction

The Industrial Revolution began in Great Britain during the last half of the eighteenth century and spread throughout regions of Europe and to the United States during the following century. The Industrial Revolution had a profound effect on the manner in which design and construction were organized. The principal factors for change that are directly attributable to the Industrial Revolution include those described in Secs. 2.3.1 to 2.3.5.

2.3.1   Task specialization

Productivity and technical efficiency grew dramatically, in part through the systematic application of scientific knowledge to the manufacturing process. These new facilities of mass production included different machines and the factories to house them, fresh sources of power and energy, and the infrastructure to distribute them—all of which put new and increased demands on designers and builders. Because of the relative complexity of the new industrial facilities, design expertise and specialization were required of the designers, but not to the same degree from the builders.

2.3.2   Ability to communicate design intent

The Industrial Revolution led to the growth of cities as people moved from rural areas into urban communities in search of industrial work, thereby creating needs for new housing, governmental buildings, and support facilities. The expanding market for design services, which did not have to be performed locally, encouraged its separation from the builders, who had to work on or near the building sites. The designers' requirements for construction could now easily be communicated to remote builders by a standardized system of drawings and written instructions (specifications).

2.3.3   Division of labor

The nature of work changed because of the division of labor, an idea important to the Industrial Revolution that called for dividing the production process into basic, individual tasks. The dramatic difference between the intellectual process of design and the physical act of construction made the design and construction industry a likely target for work segregation.

2.3.4 Entrepreneurship

The Industrial Revolution encouraged and rewarded entrepreneurship in the formation of businesses and industries. Architectural and engineering design, by definition, is the science of reducing risk in the construction and use of physical facilities. Design practitioners were, by nature and training, risk-adverse. Builders, because the new factory owners required some degree of advance assurance of performance and cost, had to routinely take considerable contractual risks.

2.3.5 Need for capital

Because the Industrial Revolution was caused by, and relied on, new and powerful machines and factories, it created a need for large amounts of capital. The money markets of Europe and North America responded to this need by organizing collective ownership by nonparticipating owners (stockholders). The construction industry, with its new machinery and large number of laborers, needed this capital, whereas the design professions, with no machinery and small numbers of skilled workers, did not. Additionally, the design professions, with their ethic of individual professional responsibility, could not consider nonparticipating partners.

2.4 Professional Societies

It was not until the middle of the nineteenth century that we find a clear distinction being made between the design professions and builders. British architects organized the Royal Institute of British Architects in 1835. Perhaps a dozen inspired American architects met in New York in late 1836 and formed the American Institution of Architects. However, they were too few in number and from far-flung points, and that organization failed.

In 1852, civil engineers organized the American Society of Civil Engineers and Architects (ASCE). Initially, there was only one architect member, Edward Gardiner, who felt that an independent organization of architects was needed, and soon resigned as the society's vice president to help form a separate professional society. The ASCE dropped "and Architects" from its name in 1869.

Gardiner and 12 other architects met in New York City in early 1857 to organize the American Institute of Architects (AIA). Their objective was "to promote the scientific and practical perfection of its members and elevate the standing of the Profession." The architects took this action, in part, because they believed that the ownership by civil engineers of building material companies (primarily foundries for the manufacture of cast-iron structural members and storefronts) was a conflict of interest with their professional responsibilities.

These organizations were symptomatic of the attempts by the design professionals to separate themselves from the sometimes corrupt builders of the era. This cultural difference between professional designers and builders in 1857 was illustrated by a statement issued by the AIA's secretary, Glen Brown, in 1907, concerning the prevailing attitudes at the time of the Institute's founding 50 years earlier stated[5]: "The profession was not appreciated by the public, the fine arts were thought of little importance, and constructors and carpenters, as practical men, were thought far superior to the architect." Until very recently, the AIA's Code of Ethics forbade a member from having any financial relationship with a building contractor, or with a building material manufacturer or supplier.

2.5   Legal Separation

2.5.1   The Miller Act

A significant cause of the absolute separation of the design professions from the construction trades in America was, and remains, the Miller Act of 1935. This law requires a contractor on a federal project exceeding $100,000 to post two bonds: a performance bond and a labor and material payment bond. The surety company issuing these bonds must be listed as a qualified surety on the Treasury List, which the U.S. Treasury issues each year. Additionally, each of the 50 states, the District of Columbia, Puerto Rico, and almost all local jurisdictions have enacted legislation requiring surety bonds on public work projects. Private owners often adopt similar policies and require partial or full bonding in their contracts for construction.

Surety bonds, unlike other forms of insurance, are actually a form of credit. The surety does not lend the contractor money, but instead allows the surety's financial resources to be used to back the commitment of the contractor; if the contractor defaults, the surety pays for completion of the contractor's work. To issue a surety bond, the surety must be confident that the contractor, among other things, has the capital assets to reimburse the surety in the event of a default.

It is this need for capital, or rather the lack of capital, that has prevented or discouraged professional design firms from acting as construction contractors. The surety requirement is still today a significant barrier to design professionals offering construction services and cost guarantees.

2.5.2   Public contract laws

The reform movement in the design and construction industry in North America continued with the development of public contract laws mandating separation of design from construction, and requiring the selection of building contractors solely on the basis of lowest responsible bid. Except for some industrial sectors, most private owners, at the encouragement of the AIA and

---

[5]From an article entitled "Our Institute Founders, The AIA's First Steps Toward Leadership" by Tony P. Wrenn, Hon. AIA, and published in the newsletter, *AIArchitect,* July 1999.

other professional societies, adopted similar procedures for the procurement of buildings and infrastructure facilities.

Most public procurement statutes require that professional design services be purchased on the basis of qualifications suitable for the design task, and that construction contracts be awarded on the basis of "the lowest responsible bidder." The term "responsible" generally meant financially responsible as demonstrated by the ability to provide performance and payment bonds to cover the contract. These same laws also specified that public agencies must prepare detailed plans and specifications before soliciting competitive cost proposals from builders. Implied in these statutes is the role of the design professional as the agent of the owner, and the prohibition of the designer from having any financial relationship with the builder.

2.5.3   Professional licensing

The first architectural licensing laws were passed in the United States in 1897. Now, in each of the states, the professions of architect and engineer are regulated for the protection of the public. In nearly every state, architects, engineers, and contractors are obligated to register, to become licensed, or to obtain a certificate before engaging in business. In most instances, professional licensing laws do not require that design and construction be separate functions, but like the public procurement regulations, they reflect the prevailing practice at the time they were first enacted.

2.6   The Advent of Modern Design-Builders

As long as buildings remained relatively simple, the separation of design professionals from the balance of the construction industry posed no serious impediment to owners attempting to acquire cost-effective facilities. With the advent of technically more demanding building systems (e.g., air conditioning, elevators, curtain walls) and systems for the concealed distribution of power, lighting, and telephones, responsible design required architects and engineers to coordinate their efforts with manufacturers, vendors, and builders. The post–World War II building boom put a severe strain on this voluntary participation within the design and construction industry. This led to the creation of the discipline of construction management (CM), both CM for fee and CM at risk. The CM model offered building owners additional assurances that the designs developed by their architects and engineers were, for the most part, practical and cost-effective. However, the CM process still lacked the single point of responsibility that owners sought. Private owners could resort to commercial building developers, and industrial clients could avail themselves of the products of the "package builders" to provide that single source. However, public owners were locked into a rigid framework of public contract laws that demanded absolute separation of design from construction, with the owner responsible for the coordination between the two, where the owner guarantees to the builder the accuracy of the architect's plans and specifications.

The first use of public funds involving the design-build process in the United States probably occurred in Indiana. In 1968, an assistant superintendent of public instruction convinced a small community in the southern part of the state to issue performance specifications that had been written by his department, and to purchase a school building by the design-build method. The performance specifications were marginal and the bid was awarded to a Wisconsin firm that was not the low bidder. The decision was based on an evaluation of the educational environment and what was called "the best buy for the dollar." The experience in southern Indiana was repeated by an adjoining county in 1970 and again in 1971, when several elementary schools and a junior high school were bid using the design-build-bid process. The early 1970s saw competitive design-build procurement utilized by several public agencies, primarily in the area of educational facilities and university dormitories.

In 1973, Fairfax County, Virginia received a grant from Educational Facilities Laboratories, Inc. (a division of the Ford Foundation) to study design-build procurement methods. A combined project consisting of one elementary school and the remodeling and enlargement of an existing school were bid by the recommended process, and resulted in not only a control of costs but also a generally accepted good design solution.

The AIA, in response to a continuing concern on the part of its members with issues of professionalism and liability related to participation by architects on design-build teams, appointed a number of investigative task forces on turnkey and package building procedures, beginning in 1968. Key among these efforts was the Design-Build-Bid Task Force, which published their final report in 1975. That report recognized the desire of owners to establish a single point of responsibility for the entire design and construction process and made specific recommendations to clarify the role of the architect in competitive design-build procurements. In response to those recommendations, the AIA published its first version of design-build contract documents in 1985, which were subsequently revised and republished in 1996.

In 1980, the City of Portland (Oregon) accepted a design-build proposal from a team of designers and builders that included architects Michael Graves, FAIA, Architect; and Emery Roth & Sons, Associate Architects, with the joint-venture design-builders Pavarini/Hoffman. The design, the first postmodern public building of any significance, was extremely controversial, but was selected by the city council, in part because of the cost guarantees offered by the design-builder and the results of the best-value analysis performed by the city's construction managers, Morse/Diesel, Inc. In 1983, the building received an Honor Award from the AIA in its annual design awards program.

In an action that brings public contracting for buildings and infrastructure project full circle, the U.S. federal government, in 1997, modified its Federal Acquisition Regulations to include new regulations for design-build procurement.[6]

The Design-Build Institute of America (DBIA), an organization of builders, designers, owners, and others founded in 1993, and whose stated purpose is the promotion of design-build project delivery in the United States, predicts that by the year 2010, half of all nonresidential construction in America will be by the design-build method.

---

[6]Subpart 36.3, "Two-Phase, Design-Build Selection Procedures."

The Facilities Acquisition Process: Business Process, Planning, and Programming

3.1 Investment Allocation and Facilities Acquisition

During the 1970s and 1980s, American business and government managers were stirred to action by external pressures, including foreign competition, deregulation, increasing costs of materials and services, and shareholder or taxpayer demands for better accountability. Shedding the organizational complacency that had set in during the early postwar years, leaders in business and government began to explore new alternatives; to seek long-term rather than short-term solutions; to weed out unproductive procedures and divisions in favor of productive work; and to innovate with ideas that spanned across divisional boundaries.

New attitudes about corporate and government agency management seem to underscore some of the basic tenets of capital economies: (1) management decisions are made about the levels of current consumption, savings or long-term investment; and (2) if an individual decision is made for long-term investment, then management must decide how much should be placed in capital goods for the further production of goods and services, and what types of capital formation, in combination with other capital goods, will generate the greatest future value. True net investment (or net capital formation) consists of net additions to a company's or a government's stock of equipment, facilities, or inventories.

3.2 Business Process Decisions for Industry and Government

Changing attitudes toward management and business processes by business and government are bringing a new focus to organizational strategy. The greater emphasis on the core business (or businesses) of these organizations stimulates a rethinking of how to modify internal procedures, of alternatives for employing assets and of ways of concentrating on the "customer," whether internal or external. As businesses and government agencies go through reengineering to better define their mission and reorient their energies, these organizations typically concentrate on four issues:

Determining how to serve a targeted market. A business can't be all things to all people, so it may choose to concentrate on low cost service (such as Southwest Airlines) or service to upscale business travelers (such as Midwest Express).

Servicing the market faster (to get there sooner). The ability of technology companies to service customers with cellular telephone technology before their competitors left the traditional phone companies scrambling to catch up.

Adopting a culture that supports continuous improvement. The continuous improvement model recognizes that change is a constant for modern business and government, and it reflects the advice of W. Edwards Deming about never being satisfied for reaching a goal unless there are further goals already identified.

Embracing a systems approach, instead of linear approaches. Companies and government agencies are slowly abandoning linear processes, and are adopting concurrent or multiphased processes that integrate procedures and people. Emphasis is placed on the multidisciplinary team, with the more complex the products or services needed, the more complex the team required to populate the project.

Three of these four issues have the effect of encouraging owners to seriously consider the use of integrated design and construction services for their facility needs. During a 1997 National Academy of Sciences symposium entitled "Federal Facilities Beyond the 1990s," a Department of Defense official remarked that the design and construction industry was the only major industry in the United States that continued to segregate engineering and production. The ongoing search for better quality and higher efficiency, coupled with technological tools that are accessible to all disciplines, will allow these two related bastions to join on common ground.

3.3   Acquisition and Procurement

3.3.1   Deciding to acquire or develop a facility

The strategic business plan drives the direction of modern organizations. This strategic plan may have a horizon of 3, 5, or even 10 years. Naturally, the strategic plan captures the goals and aspirations of the organization, and it will frequently outline the business culture that the organization embraces (or seeks to embrace) to accomplish its mission. The annual operating plan and budget are management tools that implement the strategic business plan.

A parallel but supporting strategic plan formulated by many organizations is the long-range facilities plan, which is, in turn, supported by a capital investment plan, a leasing plan, and an operation and maintenance plan. Each of these plans has a corollary annual forecast that can support the organization's annual operations. The long-range facilities plan must mirror the changing organizational objectives outlined in the strategic business plan, and should anticipate requirements for space allocation, expansion or contraction and resource needs (see Fig 3.1). The beginning of each agency's (government) or division's (corporate) long-range facilities plan should begin with a statement of an objective that is tied to the overall organizations strategic plan. For example, a division of the Department of the Navy may want to "reduce the cost of refueling coastwise vessels" by improving land-side pumping facilities and performing maintenance dredging on harbors, approaches, and berths. A division of General Motors may want to "increase utilization of engineering and manufacturing space in South America." A museum in Hampton Roads may have the objective of "improve visitor

Relationship of Long-Range Corporate Planning and Facility Planning

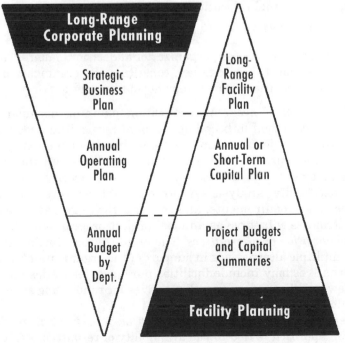

Figure 3.1   Corporate planning and facility planning.

impact by designing and constructing a new visitor center as a frontispiece to the existing Mariners' Museum."

Each facility owner or manager will have a multiyear forecast of the required facilities and sites for year-end dates, which is substantiated by forecasts of how the facility contributes to the objective(s). The Navy division plan will include how much savings are expected in fueling operations, the South American General Motors plants will forecast improved utilization rates for engineering and manufacturing, and the Mariners' Museum will show projected head counts as a result of their new visitor facilities.

3.3.2 Steps in acquisition

Organizations are well served by having a checklist (see steps outlined in Fig. 3.2) that considers all aspects of capital facilities decisionmaking. Influencers on the process include reorganization, consolidation, downsizing, outsourcing, expansion through growth or acquisition, and changes in mission and objectives. Information technology and finance, along with business management trends, also have enormous influence on facilities and real property decisions. The goals of a facilities acquisition plan are

Have the ability to adjust for new programs or new products

Maximize use and return on existing facilities

Minimize facility operating costs consistent with occupancy needs

Anticipate personnel relocations and facility closures

Allow for cost-effective future growth or expansion

To fulfill the acquisition plan, organizational decisionmakers will undertake a series of steps that, taken together, constitute the organization's strategy for real estate and facilities. Those steps normally include:

1. *Identification of need.* An acquisition planning document begins with owner needs expressed in broad operational terms. The needs are translated into performance objectives and eventually into system-specific requirements.

2. *User requirements study.* The plan will address all the technical, business, management, and user/occupant needs as ascertained through a useability/functionality analysis. A process that captures functional needs, addresses space requirements, and converts the aggregate needs into performance attributes will serve as an ideal project programming tool.

3. *Consideration of alternatives.* The organization's facility strategy should consider multiple alternatives in support of the organization's overall mission. The alternatives may include facilities purchase, lease, design, and construction of new facilities, as well as alternatives for renovating space, selling facilities, or doing nothing at all.

4. *Determining the acquisition approach or delivery strategy.* Selection of an acquisition approach is tied to the availability of resources. Early identification

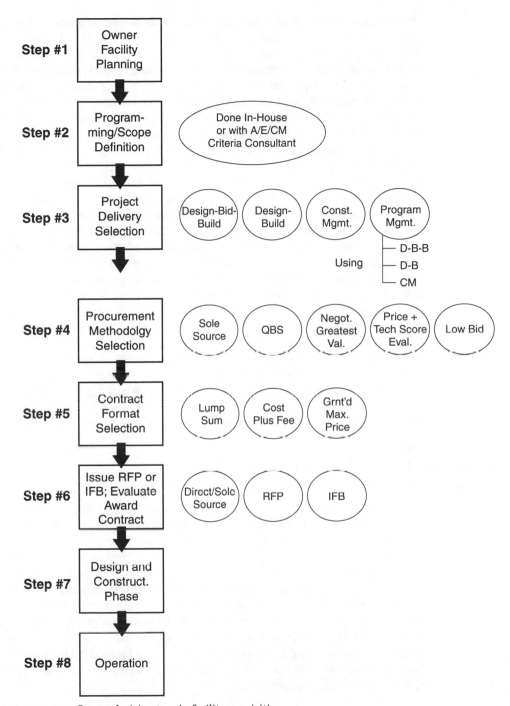

Figure 3.2 Owner decision tree in facility acquisition.

of such constraints, and communication of these constraints to the acquisition team, can save later embarrassment. Section 7 of the Federal Acquisition Regulation instructs government acquisition teams to consider the costs and risks related to each procurement. Cost goals are divided into three components: *should cost* (the first cost of the procurement), *design to cost* (a combination of management goals that balance acceptable performance, cost, and schedule), and *life-cycle cost* (a cost that incorporates acquisition cost as well as operating, maintenance, and other costs of ownership over the useful life of the asset). In the private sector, companies are guided by sound business practice. The firm will first decide on the corporate direction or need, and then proceed with a supporting lease, buy, or other acquisition decision. A company will employ market research, comparative market analysis, and financial feasibility to determine opportunity costs and potential return on investment (ROI). The final decision to acquire a facility, by whatever means, will be predicated on consideration of costs, performance and schedule; underlying drivers will be market presence or share, future corporate growth, response to competition, and other strategic considerations.

5. *Execution of the plan.* Private-sector acquisition experts tend to place a higher premium on schedule, whereas government procurement officials concentrate more on cost. In addition, management emphasis in the private sector is largely on results, contrasted with government traditions that tend to place greater emphasis on process. Either way, the purchasing of services and goods by well-managed organizations is a mechanism for supporting the mission and strategic goals. Commitment to change should cross divisional lines, and result in adoption of acquisition policies that maximize quality measures, provide cost discipline, increase time savings, improve technology, inspire innovation in process or product, and attain overall efficiency of asset acquisition.

3.4   Acquisition Feasibility and Physical Development Policies

The decision to acquire a capital asset is inextricably tied to the overall goals of the organization, as was discussed at the beginning of this chapter. Acquisition is the act of gaining control or ownership over tangible objects or property. Government entities have the ability to gain control or ownership through means other than purchasing, using their powers under eminent domain or condemnation or to mitigate or prevent a public safety hazard. Private organizations, on the other hand, typically rely on purchasing or leasing through the business mechanism of a contract. Single or multiple purchasing methods may be required depending on the *project delivery* system that is employed.

Private-sector acquisition practices are characterized by business approaches that facilitate commercial transactions to the satisfaction of both parties. Among those acquisition approaches are

Management focusing on outcome rather than process. Private-sector asset managers are usually given significant responsibility with minimal oversight.

Financial rewards, remuneration, and advancement are tied to mission success. The private-sector organization's overriding interest is in performance and schedule while maintaining agreed-on quality levels.

Avoiding early specification of design details. By avoiding the tendency to prescribe design solutions, commercial organizations can allow broad latitude in design variation as the concept is maturing. The need for formal change orders and accompanying paperwork at predesign and early design stages is eliminated, and the ability to conduct meaningful tradeoffs of performance and costs is heightened.

Using functional specifications to describe need and finished product. Commercial customers in the automotive and aircraft industries often rely on suppliers and subcontractors to interpret the functional description of a product. The supplier has the flexibility to produce the product in a cost-effective manner while exploring innovative solutions and maintaining high quality. Minimum product documentation is required by the end user except for basic operating instructions.

Developing relationships with services and products suppliers. Private-sector buyers of goods and services rely on long-term relationships that produce reliable, competitively priced items. Relationships endure as long as there is a mutually beneficial arrangement, with consistent records of performance, price, and delivery. Suppliers who want the stability of a long-term business relationship are likely to invest time and energy in assessing their customers' needs and adapting their products to those needs.

Instituting best-value evaluation processes for acquisition. Best value is a method of evaluating the differences among various acquisition choices by looking at multiple attributes of the product or service, rather than looking at only a single attribute (such as low first cost). Best value is the application of common sense to the acquisition process, allowing the purchaser to trade off quality, cost, delivery time, and other issues to arrive at the decision that provides suitability, usefulness, and life-cycle performance at a fair cost.

Since the late 1980s or so, traditional governmental acquisition has been under scrutiny by reform-minded elected officials. Procurement reform movements in the federal government and the states have resulted in a number of new practices. Among the new directions is a slow but inexorable move away from a cost-based purchasing culture. For decades, government procurement practices were based on low-bid procedures instituted during the first half of the nineteenth century to remove cronyism and graft from public purchasing. Nearly all products deemed to be commodities-oriented were placed in a formal government purchasing system that made the use of low first cost as the required selection attribute.

Since the Competition in Contracting Act of 1984, government purchasers have become more bold in seeking out the potential benefits of best value. The new objective of progressive purchasing officials is no longer simply low price or maximum performance, but a commonsense balance among what are often unquantifiable or subjective criteria, such as reliability of the product, strict dependence on delivery schedule, and additional functionality of certain products. The performance of products or systems over time was a major factor in the government's decision to adopt multiattribute value-based purchasing methods; recognizing that while development and acquisition costs can be considerable, it is the performance of the purchases over their life cycle that has the greatest effect on the operations costs (and revenue needs) of the government.

Government agencies have also made strides toward relaxing the rigid formality of their purchasing practices. The government approach had been characterized by being extremely detail-oriented and procedure-bound. Formal rules are necessary to protect the expenditure of public funds, but purchasing agents were being hampered by having to use multipage contracts with numerous sign-offs simply to buy office supplies. The use of government credit cards for small purchases and adoption of negotiating procedures for many items from computer systems to combined design and construction contracts has helped to speed up the public purchasing process.

Additional contracting procedures that have been incorporated into government purchasing include reductions in the use of voluminous prescriptive specifications in favor of performance-based specifications; experimental forms of quality assurance and quality control, such as self-certification and independent testing; and the growing importance of past performance, both to weed out poor performers and to encourage excellent contractors to stay involved with their government customer over the long term.

One relatively recent development in government purchasing has been the attention to market conditions, including surveys and research. Market research can be used to refine buyers' requirements, allowing purchase descriptions that accurately describe their needs without excluding available goods or services that could conceivably satisfy those needs. Market surveys often examine both public and private markets. Some observers contend that former British Prime Minister Margaret Thatcher's extensive use of design-build-operate for infrastructure projects in the United Kingdom during the 1980s spurred the Reagan Administration's experimental use with design-build for federal projects in the United States.

As training of government contracting and procurement personnel is expanded to include best value and integrated services, cultural barriers to design-build will begin to disappear. The implementation of design-build procurement within public agencies has frequently been top down, rather than bottom up, with reform-minded appointees imposing commercial practices on sometimes unwilling bureaucrats. Pilot design-build projects build confidence and help to meet organizational and personal goals, resulting in an increasing number of rank-and-file government employees becoming converts.

The facilities development policy of an organization consists of guidelines for determining how to support the mission and goals of the parent entity through facility expansion or contraction; by renovating; adding; constructing, or disposing of physical assets. A comprehensive facilities plan implements organizational policy by including a current inventory of physical assets, a capital budget, a facilities management plan, and an operation-maintenance plan. Of special importance to the facilities delivery process are the capital budgeting and programming procedures.

Capital budgeting is a formalized forecast of expenditures for major renovations, new construction, and equipment replacement. Annual funding and cash flow requirements are listed for a period of years (depending on the organization's overall budgeting horizon or on the economic life of the facility itself). A year-by-year capital budget plan will include cost summaries for each project, with schedules, method(s) of funding, impact of the project in the organization's operating budget, and cost estimates for planning, executing, and operating the facility.

Programming is a systemized process for recording missions, functions, and user requirements of an organization (or a subset of that organization) that will determine the scope of the facility. Project needs and requirements, including information obtained from the owner and future users of the facility, define what is expected from the facility solution rather than the solution itself. Facilities last many years, and the time taken to solicit and record owner or user needs will usually provide benefits that outweigh the cost of doing the programming. A thorough project program can result in a better fit between the facility and its users, fewer changes during design and construction, savings in construction costs, savings in time to construct, and a longer lasting or more flexible facility solution. Facility programming is mentioned here as an important supporting document for the multiyear capital budget; the actual process of programming is covered in more detail later in this chapter.

3.5   Methods of Acquisition

The decision to gain possession and use of a facility is the culmination of the acquisition planning process. Once the decision has been made to acquire, the organization (private or public entity) must decide where to locate the facility and how the facility should be obtained. Under feudal justice in the middle ages, a ruler could acquire facilities or land for the crown by edict or force, but today's laws offer other acquisition alternatives to the sovereign government.

To acquire a sensitive environmental site as a wildlife refuge, a state government may use its power of eminent domain, using the process of condemnation. The property owner will be compensated with an amount that is determined as market value by the state's appraiser. Right-of-way acquisition for roads and bridges depends on similar statutes and public powers.

An organization's decision to purchase or lease facilities is often tied to the goal of providing the desired utility at the least cost. The decision may involve a series of tradeoffs, including quality level sought, availability (or lack of availability) of suitable sites, and performance requirements for the facility or land. Alternatives also exist in the area of funding: Is the facility being paid for out of operating funds or current year appropriations? Is the decision based on an investment model that must provide a revenue stream or tax advantage? Are financing types, even private and public sources, being combined to allow the otherwise undoable project to move forward?

For reasons of (1) flexibility, (2) ways of treating facility expenses (off balance sheet), (3) availability of funding, and (4) lack of time, the decision to lease or rent rather than buy can sometimes be the wisest approach. Corporate real estate managers have become adept at leveraging their firm's facility requirements to obtain additional services or favored financial treatment from leasing companies. Among the concessions may be free or lower rent during the initial lease period, low-voltage connections or telecommunications services, tenant fitup/design and construction allowances, relocation reimbursement, and additional amenities such as dedicated parking.

For many reasons, the options of acquiring facilities by outright purchase, leasing or other financial transaction may be less attractive than the ability to design, construct or renovate a facility. Project delivery systems consist of the procedures used in the business of design and construction to manage the costs, schedule, and quality elements of creating a new or remodeled facility. The major project delivery systems being used in North America today are described in Secs. 3.5.1 to 3.5.4.

3.5.1 Design-bid-build

Design-bid-build, the delivery process in which an owner hires a designer to design a project to 100 percent, then puts the plans and specifications out for competitive bids or negotiation to a separate contractor. Design-bid-build was the project delivery process of choice for most of the twentieth century. By the beginning of 2000, the so-called traditional design-bid-build process was still used on nearly two-thirds of the projects in the United States.

Benefits of design-bid-build

Established way of doing projects. The architect-engineer (A/E) prepares a complete set of plans and specifications, followed by a low dollar bid or competitive negotiation from construction contractor.

Appropriateness for competitive bidding. Many public owners and some private owners are familiar with low first cost based on 100 percent design documents.

A/E working directly for the owner. The architect-engineer is engaged by the owner in a familiar agency (not at-risk) relationship as a professional giving advice.

Established legal precedents. Some history of litigation has provided established legal findings for allocating risk and responsibility.

No legal barriers to procurement and licensing. Two key issues for the practice of architecture, engineering, construction, and related services under state laws are the processes by which designers and constructors are hired and the required licensing of individuals and firms. The procedures for licensing A/Es and construction firms in all states are firmly established.

Availability of insurance and bonding. Risks and exposures that are covered by insurance are widely available from the underwriting community. Many surety firms also write bonds for the constructor community on the basis of established review procedures to provide a pledge against default.

Problems with design-bid-build

Owner is the arbiter between designer and constructor. A separate contract is made between the owner and the constructor, which results in the owner bearing the risk for adequacy of design, and any disagreements between the designer and constructor have to be resolved by the owner.

The designer and the constructor have differing goals. Rather than focusing on the ultimate goals of the project and the owner, the designers' goals typically have more to do with quality of physical product, and the constructor is focused on schedule and cost management. The output of a design firm, in economists' terms, consists of drawings, specifications, and advice. And legally, the output of a construction firm is accomplishment of "the work" represented by the drawing and specifications. Neither party under this system is totally focused in an economic or legal sense on the ultimate goals of the project or the owner.

An initial low bid does not necessarily result in final best value. The preoccupation with low first cost as the sole indicator of the successful contractor ignores the importance of other selection criteria. By contrast, a final best-value selection would include other important criteria such as past performance of the team, use of good environmental practices, concern for life cycle performance of facility systems, and other measures of importance to the owner.

Bids over budget present problems for owners. Under the design-bid-build system, the market is not *officially* asked for any pricing until the design is 100 percent complete. If the bids come in over the owner's budget, the designer must redesign the project (or the owner will be forced to conduct a value engineering negotiation with the contractor) to arrive at a scheme that fits the budget.

Constructor is not involved in the design. There is a traditional segregation between the work of the designer and the input of constructors in the design-bid-build process. Some owners say that constructability of the project is adversely affected by the lack of construction input into design.

Design-bid-build is slower than other delivery systems. Because of its linear structure, with the production of substantial design required before at-risk construction pricing and construction work begins, design-bid-build tends to be slower than CM at risk with fast tracking and design-build.

History of litigation. During the 1970s and 1980s, design-bid-build began to experience an increasing number of disputes leading to litigation. The system, when working well, has delivered projects to the satisfaction of all parties; however, when there are disagreements, design-bid-build pits the designer against the constructor over design clarity, errors, omissions, in-place construction quality, time delays, and other project-related issues.

3.5.2   Construction management at risk

With *CM at risk,* the owner hires a designer under a separate contract to create the design for a facility, but rather than waiting until the entire design is completed to gain the input from a construction entity, the owner instead brings in a construction manager at some point before design decisions have been finalized. The CM reviews the design and provides constructability input and estimating services, and takes on the construction portion of the project as an at-risk contractor (contrast *CM at risk* with *CM at fee,* wherein the construction manager is only in a consulting role; i.e., giving advice, and is not at risk for the amount of the construction contract).

Benefits of CM at risk

Earlier involvement of the constructor. At-risk construction management can solve some of the constructability problems of traditional design-bid-build by allowing the CM to have input into design at some point during the schematic or design development phases.

Earlier knowledge of costs. The meaningful involvement of the constructor earlier in the process allows access to all aspects of the design, regardless of level of completeness, and permits the constructor to produce realistic and evolving cost estimates.

More professional relationship with the constructor. The nature of construction management is to recognize, early in the design and construction process, the value that a constructor can bring to the project. Rather than being in a vendor (commodities provider) relationship as under traditional design-bid-build, the constructor also assumes a partial role as a professional consultant.

Owner has familiarity with the process. Contractually, CM at risk retains the separate contract system of owner-designer/owner-constructor, as in traditional design-bid-build. This format represents less change for an owner, instead of going directly to a single contract for both design and construction as would be done under an integrated services contract.

CM at risk is faster than traditional design-bid-build. The CM-at-risk delivery process produces projects more quickly than the traditional method, according to university studies. By engaging the constructor during the design process, the work can begin on sequential packages of work, thereby compressing the overall delivery schedule.

Problems with CM at risk

There are still two contracts for the owner to manage. Under a CM-at-risk process, the owner is purchasing the design from the A/E and it is the *owner* who provides an implied warranty of design adequacy to the constructor. Any major disagreements on the project go through the owner for resolution.

Contractually, the parties have different agendas and objectives. As in traditional design-bid-build, the architect-engineer's responsibility is to produce drawings and specifications, and it is the constructor's duty to perform the work represented on those documents. These objectives, and the very nature of the parties' relationship with the owner (that is partially determined by the commercial terms of the contract), place the A/E and constructor on different pages.

Constructor input may not be included by the designer. Although the CM-at-risk process sets up a relationship wherein the designer has input from the constructor during design, it is seldom mandatory for the designer to accept any of the constructor's ideas. Under the contract, the designer may stay with materials and systems with which it is familiar, avoiding the use of any experimental or innovative process or materials that the constructor may propose.

Firm costs for the project are seldom known until later. CM at risk rarely allows for firm costs to be combined with a concept design in a proposal to the owner. Final project cost may not be known until the project is completed under many CM contractual arrangements.

CM-at-risk project delivery is slower than design-build. With the owner issuing separate contracts for the designer and constructor, the process retains some of the linearity of the traditional process. University studies (University of Colorado; Penn State) have shown that CM at risk, while swifter than traditional, is not as time-efficient as a combined design-construction contract.

3.5.3 Design-build

The single-source approach of design-build consists of a firm or team of architect, engineer, and constructor professionals who are at risk for the cost, schedule, quality, and management of the project. The total cost of the construction plus the cost of design are gathered into the design-build contract. By contrast, under traditional design-bid-build, the constructor assumes the cost and schedule risk, the designer is responsible for communicating quality, and the owner manages the two contracts or diffuses the management responsibility among the three parties. Simply stated, the design-builder is both the A/E of record and the at-risk constructor.

Benefits of design-build

Single point of responsibility. The owner who employs design-build delivery has a single point of contact for all questions regarding the design and delivery of the facility. The design-build entity is responsible for quality, budget, schedule, and performance of the completed facility. With the single point of contact, owners can concentrate on definition of needs (scope of work) and timely decisionmaking, rather than on coordination between designer and builder.

All parties treated as professionals. There is an old workplace adage that says that if you treat individuals as professionals, they will respond as professionals. Design-build places the designer and constructor on equal professional footing (from a business standpoint) so that they can provide unified recommendations and jointly developed solutions to the owner.

Time savings. Design-build is considered to be the fastest project delivery system because it encourages overlapping of design and construction phases. Bidding periods and redesign, two events that can occur with traditional design-bid-build, are eliminated. Materials and equipment procurement, and advance construction work, may progress before construction documents are completed. The resulting time savings provides the owner with lower costs and earlier use or occupancy of the facility.

Early knowledge of firm costs. Guaranteed project costs can be known much earlier in the design-build process. The same entity that is responsible for the design is at the same time conducting cost estimates as the design evolves, providing accurate costs for the completed project while the design is only at the conceptual stage. The owner can make an informed decision to proceed with the project with reassurance about scope and final project cost.

Higher quality. Singular responsibility for design and construction motivates the design-builder. The design-build entity has total responsibility for the finished product, and cannot shift design errors or construction defects to another party. Traditional design-bid-build contracts rely on restrictive wording, adversarial audit and inspection requirements, and the legal system to attain project quality.

Cost-effectiveness of design-build. Design and construction personnel, who are working and communicating as a unit, can evaluate alternatives, choosing systems, methods, and materials that enhance the project. Value engineering and constructability are part of the ongoing process and work most effectively when the designers, constructors, specialty contractors, material suppliers, and manufacturers are in constant and concurrent communication during the project. Third-party university research has also shown that cost growth on design-build projects is lower than with design-bid-build or CM at risk, resulting in the most cost-effective final project for the owner.

Encourages innovation. Design-build is the one project delivery methodology that elicits creative responses from the project teams. Normally, the ability to innovate in design and construction is severely curtailed by the use of prescriptive specifications. With design-build, performance requirements are stated, and the design-builder(s) may use different solutions to meet the owner's ultimate project goals. In fact, the selection process often encourages apples to oranges comparisons to see which proposal provides the most value to the owner.

Risks are allocated to the party best able to manage the risk. The design-build process allows the contract to assign risks in a way that produces the most efficient agreement among the parties. Risks can be assigned, as appropriate, to the owner; to the design-builder; shared between the two principal parties; or mitigated by the securing of insurance coverage. All risks can be accounted for, discussed, and dealt with in a manner that is at once more clear and comprehensive than with other delivery methods.

Lower claims and litigation. Data from some of the largest professional liability insurers in the country say that the number of claims in design-build is far fewer than with traditional project delivery. Claims for errors or omissions or for time delays tend to disappear, because the design-build team would have no one to blame for these shortcomings but itself.

Some problems with design-build

Unfamiliarity with the process. Both owners and practitioners may not have used design-build delivery in the past. Changing corporate cultures to a collaborative rather than an adversarial process can take some time. Also with design-build, owners can be pushed for earlier and timely decisions about the project.

Communicating owner's needs in design-build is different. The front-end definition process of defining user needs and translating those needs to a facility program and technical performance requirements is a new wrinkle for those who have been comfortable going through traditional design phases. The design-builder wants to receive *criteria for design* from the owner, not the design itself. Further, most design-builders want to submit proposals that define the project as the team proposes to design and construct the facility. They do not want to be relegated to submitting only a "number" signifying the low first cost for constructing the facility.

Barriers in procurement and licensing laws. Some states and localities have not kept up with changes in project delivery. Procurement laws in these jurisdictions mandate the use of separate design and construction contracts, with different procurement procedures required for design and construction. Similarly, licensing laws have been used to prevent the use of combined design and construction contracts by making it impossible for other than a licensed professional to hold a contract under which professional design services will be furnished. The constituencies that formerly supported these anachronistic laws and regulations are slowly but inexorably changing their views and removing the barriers.

Availability of insurance and bonding products for design-build. During the mid-1990s, a number of the major insurance companies began developing products for the design-build market. Among these were project-based professional liability insurance, a design-build rider for A/E professional liability policies, and new products providing professional liability for constructors in contractor-led design-build contracts. Despite these additions, the market is still skittish about providing ready coverage at the same premiums as traditional delivery. Surety bonding has the same spotty record: outstanding availability for specific markets and large practitioners, but low availability for other markets and entities. Training of agents and brokers about the intricacies of the process should help eliminate any difficulties or misunderstandings in the coming years.

3.5.4   The differences between project delivery and procurement

There has been much confusion about the use of the term *procurement* in the design and development field. As a direct analogy, acquisition is to project delivery as procurement is to purchasing. Acquisition and project delivery are comprehensive processes that an organization may use to gain possession of a new facility. Procurement (or purchasing) involves the use of money or equivalent consideration (value transfer) that executes the exchange of one asset for another. If an organization decides to acquire a facility through CM at risk (a form of project delivery), it will procure (purchase) architectural services on one hand and it will procure (purchase) construction management services on the other hand to have the services and goods necessary to create the facility.

Project delivery options within an organization's acquisition strategy have been reviewed in this chapter. Procurement and purchasing variations, used to implement a project delivery method, are discussed in Chapter 8.

3.6   Professionalism and Project Delivery

Implementation of an acquisition strategy requires action on the part of professionals who have specific fiduciary, legal, and ethical responsibilities to owners, users, and the public at large. A design-build team that is executing a contract to deliver a commercial building (or a public bridge) will be required to follow numerous licensing statutes, safety codes, procurement laws, and ethical guidelines. Buildings or other structures that are open for public use, have specific types of occupancies, or are constructed of special types of materials or assemblies usually require the seal of a registered professional architect or engineer. The skills and judgment of these designers are an integral part of the professional design-build process, as applied to nonresidential facilities, civil infrastructure, and process/power projects.

A careful review of the standards of conduct of the professional design organizations reveals common tenets to which these professionals subscribe:

1. The design professional (whether architect, civil engineer, landscape architect, mechanical engineer, or other design professional) promises to protect the health, safety, and welfare of the public. The duty to protect the safety of the public is an overriding tenet. This duty or responsibility is one of the significant items that separates a professional from a nonprofessional, according to the professional societies that promulgate these standards.

2. The professional assumes a responsibility to the project. A design professional has a duty of professional care, which means that the services provided on the facility will meet a standard of care that is comparable to others practicing in the same field. The built facility, whether it is designed by an A/E in Minneapolis or an A/E in St. Paul, will benefit from professional knowledge and judgment that can be expected from these disciplines.

3. The professional accepts a responsibility to the owner. The owner, despite any personal proclivities at the time of the design, will own and operate the facility, or will transfer the ownership and operation to a new owner. It is this "office" of owner, rather than the individual person, to whom the professional owes a duty.

4. The professional owes the duty of full disclosure on all matters relating to the project to the owner. If the owner is unaware of certain permitting requirements, it is the duty of the professional to inform the owner of such requirements. If the owner has specific needs, it is the design professional's duty to help the owner answer those needs. For example, an owner may be under court order to build a prison because of overcrowding, and the designer or facility advisor may proceed under traditional design-bid-build (a delivery system not known for its speed). The project causes the local government to pay hefty fines for nontimely compliance with a court order. The designer or advisor failed to tell the owner about alternative delivery systems, such as CM at risk using fast-track techniques. Under the duty of full disclosure to the owner, the professional has violated his or her code of conduct.

In an age of multiple project delivery approaches, it is incumbent on the professional to become knowledgeable about acquisition variations. The reality of legitimate project delivery options means that there can no longer be a formulaic or standard delivery approach to every facility. It also means that professionals, if they are to follow their published standards of conduct, must become *project delivery system neutral.*

3.7 Identifying User Requirements

Gathering and gauging user requirements for a facility is a critical first step in the facility development process. A definition of user needs should be done through an orderly procedure that asks those who use or occupy a facility to think through their operations. The process should also create a format for capturing facility "demographics" so that these data can be communicated to those who write (and respond to) technical performance requirements.

When a facility planner is developing user requirements, there is an assumption that the "users" are goal-oriented. The questioner will ask the users to describe the way features of the facility "should be" or "ought to be." In most cases, there are multiple users with differing views of what should or ought to be, a condition that makes the composite results of the user survey of critical importance to facility design solutions.

Frequently, a user requirements study can be used to compare a proposed design to determine whether it meets the user's needs and how well it will meet those needs. Any areas where the design falls short of expectations (or exceeds expectations) can be discussed and dealt with by the facility owner. The expectations can be quality-oriented, that is, expressions of concern about aesthetics, inadequate or nonexistent amenities, or types or sizes of space in a building. Expectations can also be quantitative in nature; including items that can be measured for size or capacity (such as lighting, heat, and water). Some user expectations are a combination of qualitative and quantitative requirements, and the surveyor must carefully assess the relative importance of each part of the combined requirement.

Facility providers are beginning to apply systematic user requirements procedures as a way to attain quality. Quality is the *fitness for purpose* of facilities and their services. Owners want to understand their options in terms of the facility's value to users and value to the organization's core business. An organizational methodology that will compare spaces against their functional requirements within the organization, and that will allow comparison against outside users employing the same standard, is extremely beneficial. Additionally, the same methodology should be configured to ease the translation between user and facility manager communication to technical performance language used by facility designers and constructors.

Collection of user requirements information is often the first step of defining a design-build project. Implementation of the collection process will involve a number of steps:

What is the owner's basic objective for the project? The owner and its consultants should be able to clearly state why the facility project is needed and how gathering of user information will support the need. Is there an overall functional need for the project, such as for a manufacturer's new product line, or because of overcrowding in a hotel? Are there technical reasons for doing the new project, such as the existing facilities do not meet today's environmental standards.

Who are the users, and what is the most efficient way of gaining their input? People who will be using the facility should be represented in proportion to the size of their cohorts or population of their units. It may be advisable to overestimate the influence of specific groups, while at the same time being wary of political influences that are not consistent with overall organizational goals. A programming faux pas would surely be to underestimate the number of users—resulting in an inadequate budget, or worse, a facility that does not have the capacity needed.

How should information about the facility activities and operations be organized and tabulated? Although facility planners have many different methods of organizing project planning information, most include the following elements: mission of the organization; function of the organizational unit (people) or facility element (physical system); activities of the unit or element (subfunctions); current problems, trends, and developments; operating systems and equipment; and task/space needs and adjacencies.

3.8 Performance and Systems

During the 1990s, there were three breakthroughs in standardization to assist the facility planner in producing performance-based asset analyses:

1. The International Organization for Standardization (ISO) issued guide-lines[1] stating that quality and functionality of facilities directly affect ser-vice performance. According to the standard, "characteristics that [are] specified in requirement documents are facilities, capacity, number of per-sonnel and quality of materials."

2. The American Society for Testing and Materials (ASTM) added a family of standards that enabled a facility serviceability rating process in 1995. The *ASTM Compendium on Standards for Whole Building Functionality and Serviceability* allows facility planners, managers, or programmers to bench-mark facility performance using scales embedded in the standard.

3. The ASTM serviceability standard (presently limited to buildings, but the concept is adaptable to other types of facilities) has been accepted by the American National Standards Institute (ANSI) and is being introduced in the form of working papers to ISO for international adoption.

The ASTM serviceability standard is written in nontechnical language for clear understanding by users to determine the effectiveness of the physical asset or space to support its intended purpose. For example, the standard is conceived in a way that recognizes (for workplaces in buildings) how workplace quality is tied to the value of core business activities, not just the raw real estate value of the facility. The data of serviceability are tied to facility options, permitting facility owners and users to balance real estate costs and staff retention, time to market and product/service competitiveness based on price or quality.

Use of the ASTM standards on whole-building serviceability and functionality can be applied to ratings of plans or proposals for new or existing facilities. ASTM

---

[1] *ISO 9000-2 Guidelines for Services,* Section 4.1, 1991.

E1679-95 provides an outline of the standard practice of rating a facility to perform to a given set of user requirements. According to the standard, the

> practice can be used to ascertain the requirements of a group or organization at the time when the group (1) needs to determine the serviceability of the facility it occupies; (2) is contemplating a move and needs to assess the relative capacity of several existing facilities to perform as required, before deciding to rent, lease or buy; (3) needs to compare its requirements to the serviceability of a facility that is being planned, or is designed but not yet built; and (4) is planning to rehabilitate the space it occupies and needs to establish the required level of serviceability that the remodeled facility will have to meet.

The actual requirements scales are a set of descriptions for each topic of serviceability that indicate various levels from the lowest to the highest. For example, the "aspects of serviceability" of buildings include

Support for office work

Meetings and group effectiveness

Sound and visual environment

Typical office information technology

Change and churn by occupants

Layout and building factors

Protection of occupant assets

Facility protection

Image to the public and occupants

Amenities to attract and retain staff

Special facilities and technologies

Location, access, and wayfinding

Structure and building envelope

Manageability

Management of operations and maintenance

Cleanliness

To expand further, within the "aspect" of typical office information technology, topics and features may include those listed in the following table.

Typical Office Information Technology (Aspect)

Building power (topic)

   Present capacity (feature)

   Potential increase (feature)

   Reliability and quality of supply (feature)

Typical Office Information Technology (Aspect) (*Continued*)

Data and telephone systems (topic)

   Distribution (feature)

   Future capacity (feature)

   Shielding of data cables (feature)

   Local area network (feature)

   Rooms for data and telephone connections (feature)

Cable plant (topic)

   Unshielded twisted pair (feature)

   Distance to cable connection (feature)

   Coaxial cable (feature)

   Fiberoptic cable (feature)

Figure 3.3 shows an occupant requirement scale and a facility rating scale as applied to a new or proposed facility.

Performance levels of a whole facility are determined by the entire building system, which consists of subsystems, components, and materials and their relative performance measured against various criteria such as acoustical, thermal, air purity, economic return, and other requirements. Using BOMA or similar guidelines, whole facilities may also be scored against their durability, serviceability, and level of finishes to determine whether the facility can be designated "class A" or another (lesser) alphanumeric score.

Whole-facility performance can be described in a facility project brief (statement of work) that is transmitted to the designer or design-builder. The brief will include services to be provided by the architect, engineer, or other design professional for the project such as functional, technical, and design requirements as distilled from the user requirements. The brief often contains a time plan, cost plan, and technical design data. A design program, discussed in the next section, specifies what facilities will be provided to the owner and/or users and confirms to the owner the facility requirements.

3.9 Facility Programming

The difference between the process of programming and the process of design was succinctly stated in the late 1960s in the CRS classic *Problem Seeking*. *Programming is problem seeking,* affirmed the authors in their 1969 book, and *design is problem solving*. According to Edward T. White, an architect and prolific writer about the art of programming, "The two activities demand different skills and should probably be performed by different people." The content of the program can include owner or user preferences, image and movement, space adjacencies and facility demographics; and, when coupled with facility and systems performance requirements, are the ingredients for a design-build RFQ or RFP.

E 1663

A.5. Typical Office Information Technology

Scale A.5.4.   Data and telephone systems

Occupant Requirement Scale	Facility Rating Scale
9 ☐ Operations require very extensive cabling for data and communications to any location on the floor. Require excellent access to cable distribution systems that frequently need to be altered and upgraded without disruption to the office. Can use existing LAN. Need up to 75% spare capacity in cable routes, e.g. risers, ducts and cable trays. Data cables must not be affected by electromagnetic fields.	**9** ☐ O **Distribution:** Horizontal distribution of data and phone cables is by lay-in trays in ceiling, in raised access floor, or a combination of screens and fixed-position walls. Distribution is possible to any **8** ☐ location on the floor. O **Future capacity:** Data risers and ducts connecting office floors have 75% spare capacity and can be increased. Cable routes in the ceiling or access floor have 75% spare capacity. O **Shielding of data cables:** There is power wiring in grounded steel conduit or duct to shield data cabling from electromagnetic fields due to power distribution, or data cabling self-shielding. O **Local area network:** LAN cabling is installed as part of the building, with a patching board on each floor. O **Rooms for data and telephone connections:** Rooms for data and telephone connections are generous, e.g. floor area is 2% of the office area served. Access is from the public circulation area or in the service core.
7 ☐ Operations require above average amounts of cabling for data and communications to any location on the floor. Require very good access to cable distribution systems that need to be regularly altered with minimum disruption to the office. Will install own LAN. Need up to 25% spare capacity in risers and ducts, and 50% spare capacity in horizontal cable routes. Data cables must not be affected by electro-magnetic fields.	**7** ☐ O **Distribution:** Horizontal distribution of data and phone cables is by conduit, or by lay-in cable trays in the ceiling. Distribution via power poles is possible to any location without interfering with ceiling **6** ☐ fixtures. O **Future capacity:** Data risers and ducts connecting office floors have 25% spare capacity and can be increased. Cable routes in the ceiling or floor have 50% spare capacity. O **Shielding of data cables:** There is power wiring in grounded steel conduit or duct to shield data cabling from electromagnetic fields due to power distribution. O **Local area network:** LAN cabling is installed as part of the building. There is space for a patching board on each floor. O **Rooms for data and telephone connections:** Rooms for data and telephone connections are adequate, e.g. floor area is 1% of the office area served. Access is from the public circulation area or in the service core.
5 ☐ Operations require average amounts of cabling for data and communica-tions. Can tolerate minor limitations on where workstations can be placed on the floor. Require average access to cable distribution systems for occasional changes to cables. Can tolerate some disruption to the office. Will install own LAN. Need minimal spare capacity in risers and ducts, and in horizontal cable routes. Data cables must not be affected by electromagnetic fields.	**5** ☐ O **Distribution:** Horizontal distribution of data and phone cables is by conduit, or by lay-in cable trays in the ceiling, or floor ducts. Distribution is from the ceiling by power pole, with positions governed by the ceiling grid and fixtures. **4** ☐ O **Future capacity:** Data risers and ducts connecting office floors have little spare capacity, but could be increased at minimum cost. Cable trays or ducts in the ceiling or floor have little spare capacity. O **Shielding of data cables:** Floor ducts and above-ceiling conduit are shielded to separate electromagnetic fields adequately from data cables. O **Local area network:** No LAN cabling is installed as part of the building. There is a patching board on each floor. O **Rooms for data and telephone connections:** Rooms for data and telephone connections are just adequate, e.g. floor area is less than 1% of the office area served. Access is from the public circulation area.

Scale A.5.4. continued on next page

Figure 3.3   ASTM chart on occupant rating.

ⒶⓈⓉⓂ E 1663

A.5. Typical Office Information Technology

Scale A.5.4.   Data and telephone systems (continued)

Occupant Requirement Scale	Facility Rating Scale
3 ☐ Can operate satisfactorily with minimum amounts of cabling for data and communications and use existing cabling. Can tolerate limitations on locations of workstations. Require minimal access to cable distribution systems for occasional changes to cables. Can tolerate some disruption to the office. Will install own LAN. Can tolerate no spare capacity in risers and ducts, and in horizontal cable routes. No requirement that data cables not be affected by electro-magnetic fields.	**3** ☐ O <u>Distribution:</u> Horizontal distribution of data and phone cables is in overfilled floor ducts or in ceiling without a cable management system, e.g. no lay-in cable trays. Distribution is from the ceiling by surface conduit or power pole, with positions governed by the ceiling grid dimension and fixtures. **2** ☐ O <u>Future capacity:</u> Data risers and ducts connecting office floors have no spare capacity, but capacity can be increased at moderate cost, e.g. by increasing throughput on cables, use of networks, etc. O <u>Shielding of data cables:</u> Data cables are not adequately shielded from electromagnetic fields caused by electric power. O <u>Local area network:</u> No LAN cabling is installed as part of the building. A patching board on each floor could be located in occupant space near risers. O <u>Rooms for data and telephone connections:</u> Rooms for data and telephone connections are barely adequate, e.g. floor area is less than 1/2% of the office area served. Access is from the public circulation area. Additional closet space, if needed, must be in occupant space.
1 ☐ Minimal need for cabling for data or phone. Can tolerate surface mounted cables. No LAN required, and no foreseeable need for much expansion or changes to cabling. Can tolerate disruption to office in the event that cables must be provided or altered	**1** ☐ O <u>Distribution:</u> Distribution to individual workplaces is by poke-through from the ceiling below. Floor ducts are full. There is no usable ceiling space. Data risers and ducts connecting office floors have no spare capacity. O <u>Future capacity:</u> It is difficult and a major cost to increase the capacity of the distribution system, e.g. by backbone to on-floor distribution centres, and enhanced networks. O <u>Shielding of data cables:</u> Data cables are not shielded from electromagnetic fields caused by electric power, e.g. no shielding in floor ducts. O <u>Local area network:</u> No LAN cabling is installed as part of the building. A patching board on each floor would have to be located in occupant space, remote from the best place for risers. O <u>Rooms for data and telephone connections:</u> Rooms for data and telephone connections have inadequate access, e.g. from public circulation area. Additional space, if needed, must be in occupant space.

☐ Exceptionally important. ☐ Important. ☐ Minor importance.	
☐ Mandatory minimum level (threshold) =	☐ NA or NR

NOTES
Space for handwritten notes on Requirements or Ratings

Figure 3.3 *(Continued).*

Owners, architects, engineers, and builders tend to confuse programming ideas and design ideas. Programming zeros in on the owner's and user's performance problems. Design generates the physical solutions to the owner's problems. A programming issue may be office flexibility due to churn; a design response may be demountable partitions. A programming issue for a building is protecting the occupants from the elements; the design response would be to create the type of wall and roof that would be suitable for the climate given the user comfort requirements.

The evolution of a design-construction project may proceed through user requirements and programming in the following way:

Time ↓	Owner's reference	Design-builder's terminology
Ideas or thoughts	Asset need	Early project lead
Exploration	Feasibility study	Scoping/budgeting
Outlining constraints	Written project definition	Programming/problem definition
Market options issuing	RFQ or RFP	Reviewing RFQ/RFP
Graphics	Concepts	Conceptual design
Project package	Responses from proposers	Submissions/promissory solutions
Facility realization	Award and execution	Design and construction
Facility life	Occupancy/use	Operation and maintenance

The facility programmer can start the amorphous gleam in the owner's eye and translate the wants, needs, and desires into a verbal and diagrammatic representation. Depending on the type of design-build delivery that is being employed, the programming can be accomplished by the design-builder (direct design-build), or can be part of the solicitation package that is provided to the design-build entity by the owner (design criteria design-build). Under the third form of design-build (preliminary design design-build), the owner or the owner's consultant will likely have moved beyond the program to concept design. Nevertheless, it would be useful for the design-build team (under preliminary design design-build) to have access to the original facility program to better understand the owner's design intent.

It is important to note that in many cases (particularly in the private sector for low- to medium-complexity structures), facility programming is not a discrete event conducted by an independent programmer. Rather, programming is a process that is part of the proposal phase iterations that are conducted by the design-build team. The work and communication leading up to the design-builder's proposal to the owner constitute the programming phase, and the proposal itself can embody, narratively and graphically, the essence of the facility program information. User requirements research, analysis, and facility programming are not prevalent in traditional design-bid-build delivery either. However, as the use of integrated services and design-build increases, the demand for these services will also increase.

A facility program for a design-criteria-based design-build project may contain the following information:

Project description and background. A basic facility description is probably outlined in the organization's master plan or strategic plan. This brief project history captures the essence of the proposed project by including the owner identification and contact information, location of the project, gross size of the project in square footage, volume and/or capacity, length, or other pertinent measure of project magnitude. The project description may also contain a budget range, schedule, and elements from the project feasibility study (if one has been done). This overview allows the user of the program to transition into the physical facts, owner goals, project needs, and programming concepts that lead up to clear problem definition.

Facility functions and user activities. Drawing on the ASTM standards referenced earlier in this chapter, or by using a user requirements method as described in Roger Brauer's *Facilities Planning* (developed during his tenure at the U.S. Army Construction Engineering Research Laboratory), a facility programmer can build a compendium of owner/user requirements data. When programming an office building, the organization's activities, staffing, mission, and functions should be explored and understood before venturing into technical requirements for the facility. User requirements information for a building may include organizational charts, position descriptions, operational unit inputs and outputs, operations guidelines, process charts, concentration or distribution of workers, necessity to work independently or in teams, contact with outside personnel, need for communication tools or specialized equipment, and other related operations data.

Space requirements. After the gathering of owner or user requirements data for a building, the programmer will assign space names to locations, work areas, and rooms both within and without the proposed structure. Spaces may be internal or external, bounded by walls, or simply a volume in which an activity takes place. The facility programmer must decide how extensive the list of spaces should be to define the building, and will have to determine whether the space is dedicated exclusively to a particular function or group of users or is shared by multiple groups and functions. For example, a boat manufacturing facility may have departments of warehousing (raw materials, finished hulls), manufacturing (assembly line and finishing), and administration (management, sales, and marketing). An airport facility would consist of a terminal, maintenance hangars, parking structure, administration, air traffic control, runways, and lighting/navigation at a minimum.

A further magnification of the boat manufacturing facilities sales and marketing department would lead the programmer to functional areas within sales and marketing, such as outside meeting, management offices, support staff, and storage. Spaces to support the function of outside meetings may include conference rooms, offices, administrative assistants and schedulers, catering kitchen, and restrooms. The main conference room would be the scene of activities that require careful detail, such as reading, videoteleconferencing, budgeting, writing, discussing, decisionmaking, eating, recording, and other individual and collective actions. The naming of spaces is critical in the program, because spaces that are inadvertently or intentionally omitted may be skipped by the design-builder during the design process.

Physical and environmental requirements of each programmed area. All the spaces within the project site will be influenced by physical and environmental factors. Physical factors may include site limitations such as rock or water bodies, hurricane or seismic zones, expected snow or wind loadings, and site security issues. Physical issues for individual interior spaces will encompass function of the space, movement of people and material, work areas, and communication requirements. The space requirements are sometimes identified using standards that provide baseline sizes, dimensions, and shapes for various types of spaces. A parallel area for defining spaces is the consideration of access and circulation for the communication between spaces, for moving of people and equipment, and for establishing access and critical distances between functional areas.

Environmental conditions may include levels of sound, lighting, thermal conditions, air quality, and privacy. The interest in environmental factors has exploded since 1980, as the economics of external environmental costs and awareness of sustainability have risen. Owner and user expectations about comfort and acceptable auditory, sight, smell, taste, and touch variables have been heightened. Potential facility health hazards and risks [e.g., possible chemical odor or volatile organic compounds (VOCs) from synthetic carpet and glue] will influence material selection, costs, and design applications.

Codes and regulations analysis. A comprehensive programming process will include a site analysis, zoning analysis, building code survey, and historic or critical areas survey. Code language is crafted to provide minimum standards for life safety, fire separation, emergency equipment access, efficient land use, density [such as floor area ratios (FAR) ratios], and soil and water conservation. The uncovering of these "givens" will establish the constraints and boundaries of the project. These limitations, when fully illuminated through a programming exercise, impact engineering options and facility return on investment. If the codes and regulations analysis is issued as part of an RFP, the design-builder will rely heavily on information in the owner's code analysis when developing a proposal.

Enforcement of fire codes is a fundamental role for local governments, who apply regional building codes dictating occupancy allowed (for the type of materials used in construction), exits required, setbacks, stairs (type, number, size, fire rating, access), sprinkler and other fire suppression systems, and alarm systems. Codes also include provisions for wind loading, earthquake resistance, and flooding, depending on the location of the facility. Zoning regulations will stipulate present uses that are allowed, and will provide directions for access points to the facility or property, density allowed on the property, height of buildings, and minimum parking required. Urban zoning statutes will address further issues such as exterior lighting, circulation, geometry and materials landscaping, and structure orientation.

Financial opportunities and constraints. Cost control begins with financial feasibility studies, which often precede programming. An owner establishes budgets on the basis of a realistic needs scope, and on what funds can be leveraged according to credit, reputation, expected revenue stream, or available resources. Facility funding can include appropriations, bonds, loans, fund raising, or other sources. Elements of facility cost are the scope of the requirements (expressed in a building program as spaces), available financial resources (money in the budget), quality of the facility (whether the building is bare-bones-basic, sumptuously grand, or somewhere in between), and time to complete (compressed schedule, drawn-out schedule, or somewhere in-between). The facility acquisition plan must make an allowance for efficiency of the structure, such as for a building that has a net rentable area in relation to the gross area that must be designed and constructed.

The owner's financial goals may be predicated on achieving the lowest predicted life-cycle cost or most efficient energy/operation and maintenance schedule for a multiyear period. Other owners may desire maximum return regardless of other factors. In determining financial opportunities and constraints, if the owner has arrived at a conclusion as to the facility requirements, quality of the facility, and time for completion, there may be a need to adjust the resources required. Conversely, if the owner's decisions are based on facility quality, budget, and time for completion, the facility requirements in terms of size and type of spaces may require adjustment. At least one of the four variables of cost (size requirements, financial resources, quality, and time) must float until the owner arrives at an acceptable balance.

Schedule. Especially in cases where the facility program is being used to communicate the owner's needs to design-build teams, the program should reinforce the rationale for the stipulated design and construction schedule. Time to market is a mantra for today's facility owners, and it frequently considers design-build to gain the time savings (schedule acceleration of approximately 5 to 30 percent over traditional methods) that have been documented by independent studies. Time is often established by an estimate of realistic beneficial use or occupancy dates.

Space adjacencies analysis using functional flow diagrams and other charting methods. The organizational scheme of a space adjacency analysis is a representation of the operational and design problems uncovered during the facility programming exercise. The programmer is responsible for relating the facility performance factors in a manner that will inform the designer (the A/E of record is part of the design-build team) about the areas or components of a facility that are functionally important. Within the proposed structure, there will be areas, spaces or components that are absolutely critical to the function of the facility, and certain elements that are much less important can be used to transition or adjust to the more critical areas.

There are many formats for compiling and presenting facility programming information, and the four described in the following section are applicable to design-build projects:

Tabular bar charts. One useful method that has been developed for determining owner/user functional needs is the ASTM (American Society for Testing and Materials) series of standards (described earlier in this chapter) on serviceability and functionality of facilities. The survey process embedded in the standard leads to a ranking of expectations and fulfillments for each functional category. A review of the facility *aspects* and *topics* (ASTM terminology) reveals the degree of the owner/user's concern about issues of image, information technology requirements, change and churn within the facility, location and access needs, building envelope, sound and visual, and other priorities. A comparison of the scaled aspects and topics can be modulated into a overall profile of the proposed facility, with weighting toward those areas that improve efficiency and meet performance expectations.

Bar charts may be a programming tool of choice, regardless of whether the ASTM standards of facility functionality are used, because the format lends itself to simple graphic clarity and communication. For example, a bar chart can show circulation and traffic flow within a cafeteria according to movement within, into, and out of the functional area. Some of the movement will be critical to operations (food, cleaning supplies), other movement will be incidental (periodic inspections, occasional visitors), and other movement will be dictated by schedule and capacity (e.g., at lunch hour). By analyzing the anticipated circulation of commodities, people, and other traffic, the programmatic bar charts can convey potential efficiencies for adjacent spaces.

Diagrammatic matrix. Construction of a diagrammatic matrix for a facility requires that the various spaces be listed, each on a perpendicular axis, so that a symbol can be placed at the intersection of spaces showing whether the spaces (1) must be located adjacent to one another; (2) would benefit from being located adjacent to one another; (3) are neutral in terms of being together or separate; (4) should be separated, if possible, from the other space being examined; or (5) must be separated because of incompatibility in use, safety concerns, or other reasons. A challenge for the programmer (whether working for the owner to define the facility or for a design-builder to respond to an owner's statement of need) is to name-related spaces at an appropriate level of detail. For example, spaces can be named at the departmental level or at the employee activity level for an office building (or at levels in between), which will require multiple matrices to fully define the facility. A second challenge is to develop sound reasons for aligning some of the spaces together and holding others apart in anticipation of tough questions about later design decisions. Specific spaces may have both a positive and negative relationship, such as a childcare facility at a textile loom (the workers want to be within relative proximity of their children, but the children must be separated from the dust, noise, or dangerous machinery). See Fig. 3.4.

A programming activity that employs a diagrammatic matrix for a business facility rushes headlong into identification of organizational operations, functional components, or work activity areas. The owner (and its predesign consultants) must decide whether the definition of spaces is more appropriately done by the owner-consultant team or the owner-design/builder team (given the type of design-build delivery used). Additionally, the diagrammatic matrix may also accommodate related information, such as the relative size of each of the facility operations and spaces, mandatory equipment that must be housed in specific spaces, critical dimensions of various spaces (e.g., exit hall widths or manufacturing ceiling heights), and critical services or environmental requirements (e.g., oxygen in hospital rooms, humidity control in historic archival storage).

Attaching an overall estimated size to each space permits the reader of the matrix to interpret relative sizes of like spaces through the symbols that depict beneficial, neutral, or separation characteristics. Density or diffusion of spaces will be shown by the diagram; with careful coding, interior or exterior areas, estimated energy requirements, human occupancy characteristics (children, physically impaired, etc.), and openness or enclosure can be shown. Simpler or smaller projects do not require the rigor of the diagrammatic matrix, and can be programmed through basic user requirements and needs statement techniques. Larger and more complex projects will benefit from the methodical problem analysis accomplished through the development of a diagrammatic matrix.

Narrative space descriptions. Programming language, written clearly and without professional jargon, can be understood by owners, consultants, designers, constructors, and others who contribute to the creation of facilities. Narratives describing functional or space needs and characteristics are a universal communication tool for facility programming. Departmental or space descriptions can be written by the potential user/occupant of the planned space. Provided with a template, the programmer can enlist the owner or user to record functional goals, services, and activities within the space, need for flow or accessibility, dedicated use or flexibility, energy and environmental needs, safety and security, and importance of the space to the overall mission of the department or organization. By describing the goals, processes, and problems of the space in explicit terms, the narrative program document will contain implied adjacency requirements for other spaces within the facility.

Bubble and zoning sketches. Using circles or bubbles to represent spaces, the programmer can show the relative size of spaces and the importance of connections between the various facility areas. A positioning guideline has grown

Since we may have many different reasons for relating spaces to one another in the building, our reasons code could be quite long.

Example Reasons Code. An example reasons code for a small office building is presented below. The reasons for required spatial adjacency (or the need for spatial separation) are numbered. These numbers would appear on the matrix where spaces intersect to explain why those relationships between the spaces are required. A relative importance decision code is also included together with a completed matrix to illustrate the application of the reasons code.

Relative Importance Code

◆ Mandatory adjacency

◇ Desirable adjacency

◇ Neutral

◈ Negative (separation)

Reasons Code

1. Regular office conferences
2. Paperwork flow. Convenient client access to manager's office from reception.
3. Intermittent use of files by manager.
4. Regular interaction with clients.
5. Zone offices away from support areas.
6. Frequent use of calculators/computers.
7. Some use of copy machine.
8. Frequent use of offices for client meetings.
9. Frequent use of copy machine and access to clerical supplies.
10. Convenient client access to conference room.
11. The receptionist should greet approaching clients.
12. Intermittent retrieval of files during conference meetings.
13. Intermittent need to copy items during conference meetings.
14. Delivery of supplies once a month.
15. Client use of restrooms.
16. Visual image conflict between client parking and exterior service area (trash, etc.).
17. Consider service drive to mechanical area as extension of staff parking.
18. May be related to simplification of paving but not related functionally.

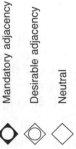

A BRANCH BANK BUILDING

Customer Service	Manager's office
	Loan/Account Specialists (3)
	Tellers (4)
Clerical	File room and vault
	Clerical supply and workroom
Support	Conference room
	Restrooms
Exterior	Mechanical-janitor-general storage
	Client parking
	Staff parking

Figure 3.4 Programming diagrammatic matrix.

up around bubble diagrams that has established rules for graphic representation, including (1) bubbled spaces should not touch each other, (2) lines showing the importance of adjacencies between bubbles should not cross each other or into bubbled areas, and (3) lines showing connectivity between bubbles should be weighted (usually by width) to show the importance of the adjacency.

Spaces that have been identified during the creation of a diagrammatic matrix or for narrative space descriptions are used to label the bubbled areas. Initially, the bubble diagram may appear to be an unrelated maze, but successive iterations can reveal spaces with the most connections. The spaces with the most connections can form the nodal points or nuclei leading eventually to a conceptual space plan. Once the clear options for a workable spatial arrangement emerge through a bubble diagramming exercise, the diagram can be used to explore circulation among the spaces, how the spaces should be walled off or open to other spaces, connectivity of spaces to the exterior of the facility, and the importance of space-based activity zones to the organization (e.g., the primacy of the trading floor to a stock exchange or the manufacturing line to a sailboat company).

Once the bubble diagram has been stabilized, a facility programmer may carve out zones within the bubble sketch, separating groups of uses by operational activities, departmental titles, private versus public areas, primary and support service areas, and other clustering issues. A design-builder, working with team-generated bubble diagrams in response to an owner's narrative space description, may organize its facility program around the inputs or outputs of the various spaces, or by codes and regulations that dictate types of construction, fire separations, and exit travel distances. A series of zoning diagrams, each dedicated to showing the aspects of the facility, will allow the program to convey design problems for solution by the owner's consultant or the design-builder.

Image, character, and broad design goals.   Architects and engineers are often too solution-oriented and may resist the effort required to define design problems in verbal or programmatic terms. However, a true focus on customer satisfaction will lead designers and design-builders back to original *narrative* statements of project and image goals. For example, legislators appropriating money for a new county courthouse building may ask for a structure that is solid, timeless, respected, secure, institutional, and barrier-free. The county department of public works may have further image goals relating to low-maintenance materials for exterior and interior surfaces, and an exterior look that fits within the context of the town's mid-nineteenth-century Victorian style.

Certainly there are gradations of foreground and background structures depending on the owner-user requirements and budget. A bridge connecting Annapolis (the historic capital city of Maryland and home of the Naval Academy) across the Severn River to suburban communities received enormous attention by state highway planners and local citizens groups because of its highly visual location. The state Department of Transportation (DOT), with input from the public, determined that the location warranted a structure that was above average in appearance but not necessarily detailed with expensive stone arches or cast-iron decorative castings as were found on the predecessor bridge. By contrast, research-development (R&D) buildings that are placed within a well-landscaped industrial park may have an image that it is clean and work-conducive, but few would find the structures visually attractive or suitable for a public community area.

A statement of facility image goals concentrates on the visual environment, and does not include concerns about facility function or serviceability. Words that convey values held by owners or users of facilities can range from pretentious to unpretentious, from grand to petite, from traditional to modern, and from welcoming to private. Expressive words such as "institutional," "cozy," "bold," "friendly," "strong," "formal," "informal," "forward-looking," "stable," "accessible," "practical," "attention-getting," "humane," "outrageous," "health-giving," and "fun" can capture the essence of image in the program.

Not every project is image-driven. But for those government buildings, resorts, corporate headquarters, retail stores, and other structures that depend on image to help their facility serve and succeed, the symbolism of architecture and engineered facilities is of critical importance to the programming document. The owner's image goals, conveyed in a verbal format, obligate the design-builder to respond with a facility that possesses particular visual characteristics.

3.10   The Decision to Proceed[2]

The renewed focus on owners/users of facilities (and their organizational needs and requirements) by the multiple disciplines within the design and construction industry will speed the adoption of integrated services delivery. Interest in more efficient delivery will have other consequences, as well; corporations and organizations wishing to relocate will factor in the time that it takes the local county or city to process zoning and building permits, and will consider the host community's willingness to allow adaptation of the facility to other uses as markets change.

Those who make decisions about acquiring facilities assets are realizing that modern but inflexible buildings, the latest PCs (personal computers) and local area networks, and the newest workstation furniture are not the answer in an information-based economy. Successful organizations are aware that people, and their ability to acquire knowledge and apply what they have learned, is the timeless asset and one that must be supported through infrastructure. A sociotechnical infrastructure, consisting of buildings, civil infrastructure (roads and bridges, water and wastewater facilities, environmental projects, etc.) and process/power facilities are required to sustain the capital-based economy. The ability of a project delivery process to complete the needed facilities efficiently, while encouraging innovation in procedures and products, will work hand-in-glove with the knowledge-information-technologically based age. Design-build offers maximum flexibility and an extensive range of possibilities for creators of that built environment.

---

[2]See Martha O'Mara's book, *Strategy and Place* (Free Press, New York, 1999).

Where and When to Use Design-Build Delivery

4.1 The Owner's Wants, Needs, and Constraints

4.1.1 Needs create a market

In today's market, the design-build project delivery method can respond to a variety of owner needs related to facilities and other physical improvements. As a rule, a design-builder can or will undertake any risk for which it has both the financial capacity to accept and the skill, ability, capacity, and authority to manage. The design-build contract method, which typically combines architectural and engineering design with construction, can also include site procurement, construction and permanent financing, physical plant maintenance, and, if necessary, facility operation. The flexibility of the design-build methodology is evidenced by the evolving multiplicity of contract examples. These new contracts combine design and construction in various combinations and often add related products and services, including furniture, fixtures, and equipment; long-term maintenance contracts and building system guarantees; project financing; and facility staffing and operations. However, these combinations did not occur without motivation from the marketplace, that is, from owners' needs and constraints. Practitioners of design-build have recognized that the design and construction industry is driven by owner needs and not by any limitations on the supply of design or construction services.

4.1.2 Needs must be defined

A design-build offer is an answer to a question. Unlike traditional design-bid-build procurement, which asks the simple and direct question "What will my building cost?," design-build proposals respond to the more complex and difficult request "Show me the best facility with the greatest value that meets my needs and budget." Thus we see that needs define responses, and responses define the design-build contract. In order to utilize the design-build contract method, owners must clearly characterize their needs and provide considerable technical detail as to their resources, including program, budget, site conditions, schedule, staff, and the facility's functions.

Traditionally in a divided design and construction procurement process, the owner will sit down with the designers and, over the course of several meetings, will define its facility needs in response to the designers' questions and as reactions to their preliminary or conceptual designs. This procedure is slow and ineffective. It very likely will not discover needs that the owner has but is technically qualified to describe only in the broadest terms. Knowledgeable owners that build often will sometimes employ facility program specialists skilled in the owners' building types to develop a detailed program of facility requirements in advance of any meaningful design efforts. Architects and engineers recognize that a good program is the basis of any successful design. In a competitive design-build procurement process, a detailed program is an absolute necessity. With some methods of design-build selection, such as direct selection or negotiated selection, the program can be developed by the design-build team in consultation with the owner after selection, but before design begins or firm prices are determined. However, if the owner/design-builder team does not include the skills of experienced program specialists to define the owner's needs, an incomplete or unsatisfying project may result.

4.1.3 Constraints must be recognized and risks assigned

Every owner, from the young couple building a house to the corporation building a complex industrial manufacturing facility, is faced with a series of similar limitations. Restraints on resources mean budgets, schedules, and site constraints. Staff capabilities, availability, or the lack of appropriate experiences and skills may restrict the manpower the owner can dedicate to the coordination of design and construction for a new facility. Local zoning ordinances and land-use regulations impose qualifications on a new improvement's use, size, height, and setbacks from property lines. They will likely impose parking and loading requirements and mandate formal design reviews, to name only a few. Public agencies must deal with the constraints and requirements of public contract law. Careful cataloging and prudent analyses of these constraints and limitations should precede a decision to employ any of several available project delivery methods.

Most of the risks associated with these restraints can be transferred to the design-builder if the risks are clearly defined in the request for proposals, and the respondent has the financial resources and management skills to control them. For instance, if one wanted to control the initial capital cost of a new facility, the design-builders could be asked to respond to a stipulated-sum request for proposals where the cost is fixed and the size and quality of the facility design is adjusted by the proposers to fit the owner's budget. Most design-build organizations—either associations of designers and builders or integrated design-build firms—that work in the commercial, institutional, or industrial construction market are modern management organizations with multiple skills and sufficient resources for the size and type of project that they pursue. The addition of unusual or complex requirements outside normal design and construction definitions are a challenge to these firms, and they often seek projects suited to their particular experiences, skills, and resources, projects that might discourage their competition. The decision of which risks to transfer to the design-builder should be made by owners familiar with the marketplace, or with the advice of those that are. Likewise, owners should assess their own ability to accept risks in order to arrive at the most cost-effective balance for their organization.

4.2 Conditions Conducive to Design-Build

4.2.1 Project size

In competitive design-build selection programs, the size of the project must be sufficiently large (and potentially profitable) to warrant the proposer's risk to obtain the contract. In other words, is the risk and effort to propose proportional to the reward? Is the cost to prepare a comprehensive design-build proposal commensurate with the opportunity to profit from the contract to complete the design and construct the facility? While size alone, namely, the cost value of the contract, is only one factor in a potential design-builder's evaluation of the risk to pursue the contract, it is one of the first measures that will be examined in their analysis. If the project is too small, it probably will not justify the expense to prepare a design-build proposal. This factor can be offset by the owner's offer of an honorarium to the unsuccessful proposers to partially offset the cost to participate in the selection process. Likewise, the owner's limited submittal requirements can mitigate the cost to prepare the design-builder's proposal.

Ironically, the firms and individuals responsible for some of the smallest construction projects in any community, speculatively built single-family houses, are in fact design-builders. House builders will routinely pick the site and design the house; or more likely, select a stock house plan. Small builders will make decisions on siting, materials, colors, fixtures, and appliances; chose subcontractors; obtain construction financing; assist in the marketing of the house; and in some instances, provide their own labor to build the house. The house builder performs all the functions and assumes similar types of risk that the commercial or industrial design-builder does, except at a much smaller scale. This might suggest that no project is too small for design-build procurement.

However, there is a considerable gap between the skills necessary to design and build repetitive single-family houses and those required for a unique "one of a kind" commercial building. In spite of their modest size, small nonresidential buildings (excluding farm buildings) require some knowledge of planning and zoning, building codes, commercial construction, and sophisticated building components such as central multizone heating, cooling, and ventilating systems and fire protection systems. Therefore, small nonresidential buildings may not be good candidates for design-build. This is particularly true if the contractors that normally build these projects are hesitant to assume design risks, or lack the management skills to coordinate design and cost estimating in a concentrated bidding period.

If a project is too large, it may exceed the capabilities of the design-builders (or potential design-builders) in the region. If the location of the design-build team is not an issue with the owner, then contract scope, in and of itself, will not likely be a problem. Through joint ventures and subcontracting arrangements, design-builders can devise contracting organizations that can adequately accept the risk of very large contracts. Typically, the very largest construction projects have always been the purview of the design-builder. These projects include power-generating facilities, paper plants, nuclear power plants, and complex chemical processing plants and refineries. The risk of separating the responsibility for design from that of construction is too great for the owners of these large facilities.

4.2.2 Project complexity

Complexity of program, design, or construction need not be a cause for an owner to avoid the design-build project delivery. For reasons stated above, complexity may be the raison d'être for an owner to select it as an effective project delivery method. However, if complexity causes confusion in the owner's program, it will cause proposers to have differing understandings of the owner's needs and expectations, and lead to unsatisfactory results.

Building types that often have complex or loosely defined programs are state-of-the-art research facilities, manufacturing plants with new production technology, and other high-technology buildings where the facility's needs are constantly evolving. These types of facilities may also require new and untried building design solutions that are clearly experimental. Up-front pricing of these types of facilities is not practical. However, design-build contracting methods can still be effectively employed in these situations if contract clauses for compensation recognize the need to keep the design decisions (and their corresponding budget and schedule impacts) fluid as long as possible. Unit prices and guaranteed maximum pricing in design-build contracts are two methods that permit competition among proposers, while allowing the program and the designs to evolve right up until the time of construction of the specific building systems affected.

4.2.3 Caveats

Design-build competitive selection and contract methods work best when these caveats are observed:

- The project's financing is secure, and that fact is communicated to the proposers.

- The owner employs qualifications-based selection procedures leading to a reasonable number of proposers in the final proposal stage.
- The owner's needs and expectations are clearly stated, and the proposers are likely to have consistently similar understandings of those needs.
- The single point of responsibility for design and construction is maintained.
- The contract terms make reasonable assignments of risks between owner and design-builder.
- The owner's organization is able to make decisions in a timely manner.
- A working environment of trust and mutual respect can be established among the owner's organization and the design-build team.

Conversely, design-build project delivery *will not* likely be effective if the following conditions exist:

- The owner requests complete design-build proposals and the project is speculative and may not be built.
- The number of design-builders proposing are so many as to discourage real competition, or cause the "A teams" to drop out.
- The criteria for selection are not clear or their relative values to the owner are not stated.
- The project's program is unclear or ambiguous.
- The selection panel is not well informed about the project's requirements, or is not truly capable of acting independent of outside influence.
- The owner's program limits innovation and flexibility in facility design, system selection, or materials.
- The owner insists on absolute design control over all aspects of a facility, and is not satisfied by general design definitions and performance guarantees from the design-builder.

4.3 Applicable Market Sectors

If it can be designed and can be built, it can be design-built. If the project must be designed by professionals and has to be constructed by experienced builders, it is a valid candidate for the design-build project delivery method. Subject to the criteria listed in Sec. 4.2 concerning size, complexity, and the owner's ability to act on the proposals, almost anything in the built environment can utilize the design-build methodology to advantage. Rarely does building type alone define the design-build market. When evaluating the applicability of design-build contracts to a project, the following factors weigh heavily in favor of utilizing the process:

- *Program requirements that can be defined by performance criteria.* If the owner's needs can be expressed in objective performance terms that are readily understood by the industry, and the results can be accurately measured by unbiased testing procedures, the project delivery method allows the design-builder broad opportunity to design cost-effective solutions. Building types that fall readily into this category are power-producing facilities, water and wastewater treatment plants, food processing plants, and bulk material handling facilities such as coal handling, wood chip processing, and solid-waste handling and disposal facilities.

■ *Program requirements that are largely prescribed by industry or regulatory standards.* Typically, transportation facilities (roads, bridges, culverts, highway excavation, grading, and drainage and retaining structures) are described by detailed, technical, federal, state, and municipal standards, and because of those standards, civil infrastructure projects are excellent candidates for design-build procurement procedures. Other types of facilities that fall into this category are correctional facilities, primary and secondary schools, and community hospitals. Design solutions for similar building types will vary considerably, but the program elements will all meet common standards for size, function, proximity, performance (of building systems), and the strength and durability of materials.

■ *Constructability challenges.* Some projects will have considerable challenges to their construction that must be taken into account in their design. Difficult construction circumstances can be present in any type of project. They may include radically accelerated construction schedules, remote or unusual project sites, extreme weather conditions likely during the construction period, severe labor or material shortages, and the need to keep the owner's critical functions in operation during construction. Designs that effectively mitigate these factors can be devised only with the meaningful participation of the builder. In these extreme situations, "meaningful participation" means that the builder controls the design and guarantees the cost, that is, serves the owner as the design-builder. Projects that are often subject to these challenges include airports, teaching hospitals, arctic construction, military facilities in remote areas, community facilities in areas of rapid population growth, computer and telecommunication facilities, and high-tech and other manufacturing facilities in rapidly changing markets.

■ *Up-front price guarantees required.* It is not unusual for the institutions financing the design and construction of a project (where the improvement and/or its revenue stream is the primary security securing the loan) to require cost guarantees from the owner before significant funds are released. Lacking this requirement, a financial institution may offer a significantly lower interest rate if the design-to-budget risk is removed. This requires the owner to expend other funds to finance the design and bidding phase of a project or to seek early cost guarantees. In order to offer early cost guarantees, the design-builder must be able to control the design, subject to the owner's program and specifications, after the cost guarantees are accepted. An effective design-build procurement program will balance the need for the owner to have wide latitude in design choices, with the requirement of the design-builder to control design details after its design proposal has been selected.

4.4 Predictors of Success

4.4.1 General

As with any method of project delivery, success is typically measured in terms of the results observed after project completion, when compared to the owner's expectations at project inception. Elements common to all forms of contract for design and construction that can generally predict a successful outcome are (1) a definite and consistent statement of the owner's needs and a budget proportional to the owner's program defined at project initiation; (2) precise contract language that clearly assigns manageable risks among the principal project participants, owner, designer, and builder; and (3) the establishment of trust and mutual respect among the project principals. Beyond these broad definitions, one must look to definitive research comparing various project delivery methodologies available to facility owners for predicators of success.

4.4.2 Research studies available

If success is defined in terms of time, cost, and quality, one of several available studies can be used to measure the relative levels of success achieved by three distinct project delivery methods. One such study is described in the *Journal of Construction Engineering and Management* [vol. 124, no. 6 (1998), pp. 435–444]. Specifically, it is a report entitled "Comparison of U.S. Project Delivery Systems" by Mark Konchar and Victor Sanvido, researchers in architectural engineering at Pennsylvania State University. The study, principally sponsored by the Construction Industry Institute (CII), empirically compared cost, schedule, and quality performance of the three primary U.S. project delivery methods. It utilized project-specific data collected from 351 building projects. The results of this study are summarized in Fig. 4.1 and below.

A similar and contemporary study was conducted in the United Kingdom by the University of Reading Design and Build Forum.

4.4.3 Quality as a measure of success

In the Konchar-Sanvido study, quality performance was measured in seven specific areas (Table 4.1). The facility owner was asked to rank the actual performance of the facility versus expected performance. A high score of 10 indicates that the listed system had exceeded the quality expectations of the client. A score of five shows that the owner's expectations were being met, whereas a system scoring zero was not meeting the owner's expectations.

It is interesting to note that the design-build method significantly outperformed design-bid-build in every quality measure criterion, and that design-

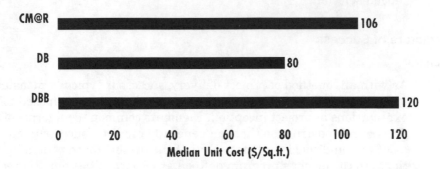

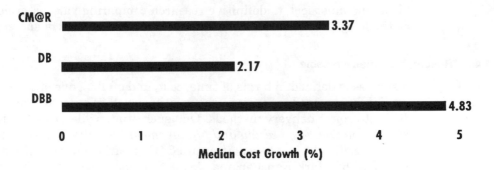

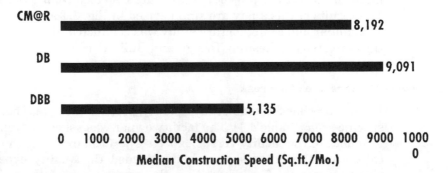

Figure 4.1 Construction Industry Institute study findings. (*From V. Sanvido and M. Konchar, Selecting Project Delivery Systems, Project Delivery Institute, State College, PA, 1999.*)

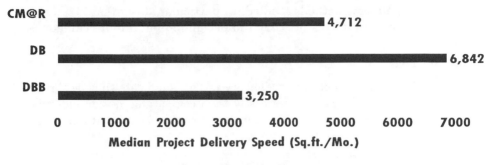

Median Project Delivery Speed By Delivery System

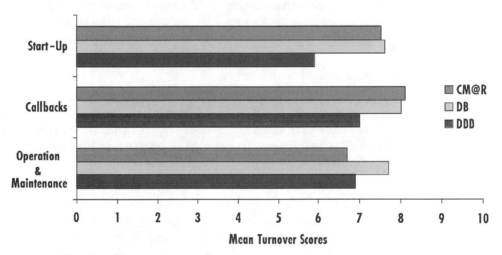

Turnover Process Quality

Figure 4.1 (*Continued*).

TABLE 4.1   Mean Quality Scores by Project Delivery Method

Quality measure	CMR	D/B	D/B/B	MSE
Start-up	7.43	7.50	5.96	0.19
Callbacks	8.07	7.94	7.04	0.19
Operation and maintenance cost	6.69	7.67	6.88	0.19
Envelope, roof, structure, and foundation	5.36	5.71	4.95	0.19
Interior space and layout	6.28	6.15	5.19	0.19
Environment	5.34	5.24	4.86	0.19
Process equipment and layout	5.63	5.61	5.07	0.19

Note: D/B = design-build; D/B/B = design-bid-build; CMR = construction management at risk; MSE = maximum standard error.

build and construction management at risk were comparable in many categories (Table 4.2). It is clear from this study that design-build projects achieved quality results equal to, if not better than, those of other project delivery methods studied. In particular, design-build offered significantly better quality results than did design-bid-build in all categories.

4.4.4   Unit cost

The effects of project delivery methods indicate that design-build projects are at least 6.1 percent less costly than comparable design-bid-build projects, and 4.5 percent less costly than construction management at risk projects on the average.

4.4.5   Construction speed

When all other variables were held constant, the effects of project delivery methods indicate that design-build projects are at least 12 percent faster than design-bid-build projects and 7 percent faster than construction management at risk (CMR) projects on average in terms of construction speed.

4.4.6   Project delivery speed

Comparing similar projects (i.e., size, unit cost, predesign, facility type, and team communications), design-build projects were found to be at least 33.5 percent faster than design-bid-build projects and 23.5 percent faster than construction management at risk projects, on average, in terms of project delivery when design and construction durations were included.

4.4.7   Cost growth

When comparing the degree of cost growth over the term of projects' design and construction phases, the study found that, on the average, design-build projects' cost grew at least 5.2 percent less than did design-bid-build projects and 12.6 percent less than did construction management at risk projects.

4.4.8   Schedule growth

In a similar fashion, the study compared the relative growth rates of the projects' design and construction schedules. It found that design-build projects

TABLE 4.2   **Percentage of Average Difference between Project Delivery Methods by Measure**

Measure	D/B vs. CMR	CMR vs. D/B/B	D/B vs. D/B/B	Variation
Unit cost	4.5% less	1.5% less	6.0% less	99%
Construction speed	7% faster	6% faster	12% faster	89%
Project delivery speed	23% faster	13% faster	33% faster	87%
Cost growth	12.6% less	7.8% more	5.2% less	24%
Schedule growth	2.2% less	9.2% less	11.4% less	24%

Note: D/B = design-build; D/B/B = design-bid-build; CMR = construction management at risk.

had at least 11.4 percent less schedule growth than did comparable design-bid-build projects and 2.2 percent less than did construction management at risk projects.

4.4.9 Study's summary and conclusion

It is the conclusion of the authors of this study that projects administered using design-build as a project delivery method can achieve significantly improved cost and schedule advantages. Likewise, design-build projects produce equal and sometimes more desirable quality and performance results than do CMR and design-bid-build projects. In their opinion, the study offers proof that design-build, on average, is the optimum project delivery method.

Roles of Individuals in a Design-Build Project

5.1 Owner's Team

5.1.1 Owner creates the project

The owner's primary function is to create the project and provide the wherewithal to pay for it. Beyond those two basic responsibilities, the owner may take a relatively passive role or it may elect to actively participate in the management of the project. It is in this latter mode that the owner, through the mechanics of the design-build project delivery method, can have a considerable effect on the success of the undertaking. The owner, whether an individual or an organization, whether public or private, profit or nonprofit, will impart to the project the objectives of the parent organization. This imprinting comes with the definition of the project as it is created. A store must "sell"; a jail will "imprison"; and a house must provide shelter, security, and privacy. These objectives come with the project automatically. It is in the project's details and in the variations to these basic facility functions that the owner further defines the project and adds value to the end product. The owner's management of the design-build process starts with the owner's team.

5.1.2 Owner's representative

As an employee of or a consultant to the owner's organization, the owner's representative is empowered to implement the project and is responsible for ensuring that the project meets the organization's objectives. The representative will select the project delivery method, elect the procurement options available within the project delivery method, and chose the type of contract to recommend to the owner's governing authority (board of directors, chief executive officer, president, partners, etc.). It is the representatives' task to communicate the project objectives (function, scope, schedule, budget, and character) to their own staff and consultants, and ultimately to the design and construction industry. Conversely, representatives are also responsible for keeping the organization's decisionmakers informed of the progress of the project and of the significant choices or options available along the way. Finally, owner's representatives must also ensure that the methodology chosen and the procurement and contract options selected are meeting the project objectives. If budgets are missed, schedules exceeded, or function compromised, the owner's representative must make or recommend corrections to the initial procedures. These responsibilities are common to all project delivery types. The design-build project delivery method will demand more of the owner's representative.

Design-build can facilitate the active involvement of the owner in the creative process to a greater extent than can other methods. It is the responsibility of the owner's representative to take advantage of the flexibility inherent in the design-build method to charge the design and construction industry to create a product with the greatest worth to the organization. A common procedure owner's representatives use to motivate design-build teams is proposal competition, not competition for price but competition for best value, a proposal evaluation method described later in this publication. Some representatives involve the organization's operational staff or users in direct communications with the competing design-build teams through design *charettes* or proprietary conferences in the proposal preparation phase, and later include facility users in the proposal selection process. In all instances, it is the owner's representatives' duty to leverage the project's competitive position in the design and construction marketplace to the organization's greatest advantage.

5.1.3   Design criteria professional

As a representative of the owner and an intermediary between the owner and the design and construction industry, the design criteria professional is expected to address the interests of both groups. This role requires the individual to be both professional and technically qualified to translate the owner's needs into objective, quantifiable terms that can be readily understood by the industry. Likewise, the design criteria professional must appreciate the design and construction marketplace and should attempt to present or offer the proposed project to the industry in a manner likely to attract exceptional offers, namely, best-value proposals. The owner will expect the professional to enforce the rules of the selection process. The proposers will expect the representative to require the owner to evaluate the proposals and select the winning proposal according to the published selection criteria in the RFP.

In a design-build contract, *design criteria* is the collective term used to identify those documents that define the scope, quality, and function of the proposed facility. It can be expanded to outline project cost limitations and schedule requirements. In a competitive selection process, the design criteria professional will be expected to develop the terms and conditions of the competition and to assist the owner to administer the process. In this latter role, the professional must be impartial and fair-minded in all issues effecting both the proposers and the owner. The position is similar to that of the professional advisor in an architectural design competition.

The work products of the design criteria professional in a competitive design-build selection process will include

- Recommendations for design-build solicitation (including the procurement budget, schedule, and jury composition)
- Request for design-build qualifications (RFQs)
- Request for design-build proposals (RFPs)
- Instructions to design-build proposers (including the weighted selection criteria)
- General conditions of the design-build contract
- Program of facility requirements
- Program narrative
- Functional descriptions of the users' operations
- Tabulation of programmed spaces and usable floor areas
- Environmental conditions of programmed spaces by type (finishes, light, sound, utilities, etc.)
- Performance specifications
- Technical evaluation report to the jury
- Final report (or recommendations) of the jury

In a direct or negotiated selection process, the same information will be created by either the design criteria professional or the owner's representative. However, it will likely be communicated to the prospective design-builder in an informal manner at conferences, by letter or transmitted documents.

5.1.4   Design professional accomplishing "preliminary design"

In all project delivery methods, an important function of the owner is to communicate its needs to the designers and builders in an effective manner. In a competitive, design-build selection process, the customary document used to affect this communication is the *program of facility requirements,* or simply *program.* It will be similar to the program the owner transmits to its architect or engineer in the traditional design-bid-build project delivery method, but perhaps with more technical detail. However, sometimes the narrative and tabular text of the program are inadequate to fully communicate the owner's requirements. In these instances, an owner will often resort to graphic images to communicate its functional or aesthetic intent. These graphic images or drawings are referred to as *bridging designs.*

When an owner deems it appropriate to communicate to design-build proposers with a bridging design, the owner will typically engage a design professional to prepare a preliminary design of the proposed facility. More extensive than a master plan, but less detailed than construction documents, these preliminary designs may represent as much as 35 percent of the entire design task. The owner may insist that the design-builder follow the bridging design exactly, may allow some minor deviations, or may simply use the designs to communicate the owner's intent to the proposers. In the latter instance, the design-build proposers would be free to offer original designs as long as they met the implied intent of the bridging design.

The role of the design professional preparing the preliminary design is to assist the owner to communicate its intent to the proposers and to evaluate the technical proposals submitted in response to the RFP. During the proposal stage, the design professional must be independent of the proposing design-builders. After a contract has been awarded to a design-builder, the design professional may continue to assist the owner to review the design-builder's design development and construction documents. Alternatively, the owner could assign the design professional's contract to the design-builder for the purposes of completing the design and construction documents. This latter option is more common in the private sector than in public work. In either case, the owner would remain accountable to the design-builder for any errors or omissions in the bridging design. An exception to this liability would be created if the bridging documents were issued exclusively to communicate the owner's intent and the design-builder were to submit an original and complete design. To further reinforce the single point of responsibility for design, the owner's preliminary design must be specifically omitted from the contract documents attached to the owner/design-builder agreement.

5.1.5 Construction manager

Depending on the owner's level of technical expertise, a degree of prudent oversight is expected of the owner during the design and construction phase of a design-build project. If technical capability is not available from the owner's staff, then a construction management team or CM can be engaged to look after the owner's interests. This role is made much more significant in a design-build project because the normal safeguards of an architect and an engineer acting as the owner's exclusive agents during the design and construction phases are absent. Although architects and engineers in a design-builder's organization have professional responsibilities to the facility users and to the public, vis-à-vis health, safety, and convenience issues, the owner is still obligated to take some responsibility for quality assurance through a construction manager. Tasks that a construction manager, acting as the owner's agent, could perform in a design-build project would include

- Advising the owner on the advantages and disadvantages of various procurement procedures and design-build contract forms and provisions
- Developing the owner's preliminary project budget and schedule
- Coordinating the contract's postaward submittal requirements
- Developing the requirements for the design-builder's critical path method (CPM) schedule and reviewing their proposal and periodic submittals
- Reviewing the proposers' on-site construction management plan
- Reviewing postaward design submittals for conformance with the owner's design criteria and the accepted design-build proposal
- Reviewing the design-builder's quality control plan
- Reviewing the periodic inspections and reports of the project's design professionals
- Coordinating the activities of independent special inspectors and reviewing their findings
- Coordinating postaward scope changes and change orders
- Coordinating the activities of the owner's separate contractors
- Performing general inspections, and recording existing conditions and the design-builder's progress
- Certifying the design-builder's periodic payment requests
- Coordinating the owner's separate commissioning activities
- Performing final inspections
- Coordinating postoccupancy performance testing and warranty claims

5.2   Design-Builder's Team

5.2.1   Design-builder as team leader

The design-builder is the legal entity in privity with the owner and obligated by contract to design and construct the subject facility to the requirements of the program, performance specifications, and other restrictions. The design-builder may be an individual, a partnership, a joint venture, a single firm, or a corporation. The only absolute requirements to be a design-builder are to have the financial and management capabilities to accept the risks and undertake and, if necessary, to assign and/or subcontract the design-build tasks and responsibilities. Rarely, and only in the instances of very simple structures, is the design-builder an individual. In most instances, the design-builder is a developer, design professional, contractor, or some combination thereof. Because of the risks involved, an entity not familiar and experienced with facility management, real estate development, design, or construction would be reluctant to enter into a design-build contract at a competitive price. In practice, the design-builder is already a player in the industry fulfilling one of the key roles on the design-build team and functions as the team leader for the other member firms and individuals.

As an entity considering the role of design-builder, the design-builder must first access the risks and rewards associated with the prospective design-build contract. If the situation is competitive or the project tentative, it must also estimate the investment to develop a proposal and the chances of being successful. In light of these considerations, the design-builder will determine whether it wishes to assemble a team and offer a proposal (or to offer its qualifications in a two-phase selection process). The design-builders refer to this evaluation process as the *go/no-go decision*.

During the proposal preparation phase, the design-builder will organize its team, focus the team's creative efforts, and set objectives, parameters, budgets, and schedules. In order to be successful at this stage, the design-builder must offer more value to the owner than its competitors. If the proposal is to be negotiated with the owner rather than offered competitively, the proposal must meet or exceed the owner's expectations. In either case, the design-builder will want to clearly understand the owner's objectives, values, and methods of evaluating the proposal. An effective design-builder will be in direct communication with the owner's representative and will use the opportunity to increase its team's understanding of the problem to be solved by the design-build proposal.

On the proposal side, the design-builder must provide leadership that encourages creative suggestions from all members of the design and construction team, regardless of the source. Successful design-build programs are those that integrate, at the conceptual design stage, the entire industry from design architect through design-build subcontractors, trade subcontractors, suppliers, vendors, building systems manufacturers, and craftspeople. To provide effective leadership for such a diverse team, the design-builder must have the professional and personal respect of the team members. Moreover, it should be able to recognize and appreciate each team member's individual capabilities and contributions to the overall team effort.

After the award of the contract, the design-builder will facilitate a broadened partnering process among the project team. That team, expanded to encompass additional industry members now associated with the project, must now include the owner's team. Although the design-builder's remaining task is to simply implement the project described in the proposal, a constant vigilance and numerous "midcourse corrections" are required to maintain quality, schedule, and profitability.

5.2.2 Design professionals as design-build team members

The design professionals, architects, or engineers depending on the type of project, are, obviously, half of the design-build team. They may be individual employees or principals in the design-builder's organization, or they may be consultant firms subcontracted to the design-builder. Most projects require more design skills and capabilities than can be provided by a single individual, and the design professionals required by the project are referred to collectively as the *architect of record,* or *engineer of record,* depending on the nature of the project. The architect-engineer (A/E) of record will be the contractual lead of the design team and, if the capability is not available internally, will subcontract with design architects, engineers, interior designers, specification writers, and other specialty consultants. Within the parameters set by the design-builder, and under its leadership, the A/E will lead the design team to produce the design portion of the design-build proposal. This task, if it is to be successful, must be conducted in close coordination with the construction members of the design-build team and mindful of the owner's program and project objectives. After contract award, the A/E will continue to lead the professional design team. Their responsibility will be to develop construction documents that implement the design proposal and meet or exceed the requirements of the owner's program.

Beyond the A/E's contractual obligations to the design-builder, and indirectly to the owner, the architects and engineers have additional professional responsibilities associated with their work, regardless of the project delivery method. As professionals, they are expected to put the interests and well-being of those they serve above their own interests. As designers, their first responsibility by law is to protect the health, safety, and convenience of those who will use the facilities they design. Also, as professionals, they are expected to reasonably serve or protect the interests of all parties to the design-build contract beyond the commercial obligations of the designers' contract. To illustrate, if the designers became aware of new products, materials, equipment, or processes that noticeably improve the quality or productivity of the facility, but which the owner apparently has no knowledge of, it would be the designers' professional obligation to apprise the owner of the options available, regardless of the financial consequences to the design-builder or to the designer. Conversely, there are limits of liability afforded to design professionals that provide a "commonly accepted standard of care" that are not present in a strictly commercial contract. In this manner, the services of the entire design-build team rise to a professional level, rather than to the usual level of mutual self-interests implied in a typical construction contract.

5.2.3 Constructor as a member of the design-build team

The constructor or builder completes the other half of the design-build team. If not the design-builder or the entity in direct contract with the owner, the constructor will function as a separate subcontractor to the design-builder and will be contractually responsible for construction activities and costs alone. In practice, the constructor must function as a participating team member during all phases of the project.

In addition to providing a single point of responsibility, the design-build project delivery method allows (or requires) the design-builder to vertically integrate the design and construction industry for the benefit of the project. This means that the creative efforts of all participants, from the lead A/E designer to the trade subcontractors and the individual craftspeople, must be focused on the project at every phase, from proposal preparation to facility commissioning and start-up. To the extent that each team member can contribute to a successful proposal and project, the design-builder must solicit that contribution and provide a methodology to evaluate its worth and, if acceptable, integrate it into the project. Because of their traditional relationships with subcontractors, suppliers, vendors, and manufacturers, much of that responsibility falls to the constructor. Except for manufacturers' representatives and salespeople, most construction industry participants have little contact with designers, particularly during the creative design phases. It is the responsibility of the constructor to find ways to involve industry members at the points they can add the greatest value to the project. The early involvement of lower-tier trade subcontractors and building system manufacturers creates a sense of ownership and pride in the project that manifests itself in higher-quality work and fewer conflicts in the construction phase of the project. The constructor has the ability to involve the industry members "at risk"; that is, if a subcontractor's contribution, whether creative or financial, helps the design-builder to win the contract, the subcontractor will be rewarded with a contract or purchase order. Likewise, if the subcontractor promised to meet a specific performance criterion or quality level, the subcontractor will be held contractually accountable. In a similar fashion, the contributions of vendors, suppliers, and building system manufacturers are also recognized. This function as construction industry integrator at the creative design stage is new to most builders; however, it is an attribute that will distinguish the real design-builders from building contractors in the future.

In the traditional design-bid-build project delivery method, the constructor commences construction with a complete and detailed set of construction documents. A design-build constructor will likely start construction with only partial plans and specifications and must anticipate the requirements of the following construction phases. Further, the nature of a design-build contract based primarily on performance specifications is one that requires a continuous "value engineering" evaluation throughout the design phase and well into the construction phase. This changes the role of the constructor significantly. Instead of only attending to the means and methods of construction and accepting the design as a given and irrefutable fact, the constructor may consider other options. It may investigate and implement any design change that meets the owner's design criteria and does not compromise or diminish the value of the design-builder's initial proposal. This flexibility allows the constructor to engineer its way out of design and construction conflicts, and perhaps improve the product and increase its own profitability.

Most design-build contracts are based, in part, on specific design criteria and performance specifications. For this reason, the constructor's involvement in the construction of the project may be significantly expanded in the warranty phase of the project. Until the required levels of facility performance are met and verified (e.g., environmentally, acoustically, and operationally), the construction contract is not complete. When a performance level cannot be achieved by the completed facility, the constructor and the designer must cooperatively assist the design-builder to determine the cause and suggest appropriate remediation. This is a much greater warranty to the owner than the typical callback to repair faulty work found in traditional construction contracts.

5.2.4 Design-build subcontractors

Subcontractors that represent building systems with significant engineering content, such as heating, ventilating, and air conditioning (HVAC), structural steel fabrication and erection; or building curtainwalls, have an opportunity to act as specialty design-build contractors within the larger context of the general design-build contract. Their role toward the design-builder is similar to those of other subcontractors, except that they are in a much better position to contribute to the creative aspects of the design-build proposal because of their role as specialty designers. Internally, the principals of a design-build subcontracting firm must act in a fashion similar to that of the general design-build contractor; that is, they must integrate the various activities of design, manufacture, and installation into a mutually supportive organization whose members attempt to add value to their company's products, regardless of their individual functions within the organization—representing a true team effort.

Design-build subcontractors may also accept responsibility for the performance of entire building systems. For example, a curtainwall contractor may accept the contractual responsibility for the building's exterior skin, even if it has to subcontract other portions of the work such as masonry veneer, storefronts and entrances, flashing, and waterproofing. This gives the general design-build contractor a single seamless source of responsibility for a major building system covered by a single set of design criteria and performance requirements, thereby reducing the general's risk. Because of this heightened level of risk assumption, it is likely something that only larger subcontracting firms with significant capital resources can afford to do.

5.2.5 Trade subcontractors

Trade subcontractors and their crafts employees can make meaningful contributions to the design-build process by bringing their practical field experience and suggestions to the attention of the designers during the design development stage. The design-build project delivery method is the one procedure that facilitates and encourages one end of the process, design, to learn from the other end of the process, field construction. Not all design-builders have learned how to overcome the obstacles of timing and work culture differences to use this procedure successfully. However, the hallmark of the successful design-builder of the future may well be its ability to accomplish this integration for the betterment of the project.

5.3   Independents in the Design-Build Process

5.3.1   Quality assurance professionals

There is a common misconception in the design and construction industry that the role of independent materials testing laboratories in a construction project is to assure the owner that the construction meets the specified quality. This is not their purpose. The reason for independent testing is to confirm to the design professional that the assumptions made during design can be confirmed on site, such as soil bearing, soil moisture content, and water table. Another purpose is to confirm to architects and/or engineers that the materials they specified for construction are being used, such as concrete mixtures, concrete slump, and structural steel strength (mill tests). In the traditional design-bid-build project delivery method, the designer is an agent of the owner, so the testing program is specified by the owner (in addition to code-mandated tests) and paid for by the owner. Typically, test reports are addressed to the owner in care of the designer and information copies sent to the builder.

However, in the design-build project delivery method, the design-builder is both the designer and the builder. In this instance the design-builder should determine the extent of the materials testing program (in addition to code-mandated tests) and pay for them. The owner's quality assurance is in the performance testing of the finished whole project, not in testing the materials used in the construction of project components. However, as part of the owner's obligation to provide prudent oversight of the design and construction efforts, the owner should require copies of the material testing reports during construction. The owner, of course, may exercise its option to conduct additional material testing and special inspections, for which the owner should assume responsibility for payment.

Postoccupancy testing to determine the facility's performance capabilities is an option of the owner. Normally, testing of a completed facility is done only (1) in response to problems perceived by the owner or the occupant (e.g., excessive room-to-room noise transmissions) or (2) when the tests are simple and easy to conduct (e.g., lightmeter tests of the occupied spaces). When performance testing is conducted by the owner to verify the facility's compliance with the performance specifications, the design-builder should be given an opportunity to observe these tests. If the building system fails to meet the specified performance, the design-builder should have the right to conduct its own tests. The testing agency should be independent and well qualified to conduct the necessary tests, and it must remain objective and unbiased in its professional evaluation and reporting of the test results.

5.3.2   Building officials

From the perspective of building code officials, projects employing design-builders by direct selection or sole-source negotiated would not vary significantly from the design-bid-build projects in their building permit documentation. However, design-build projects selected by competitive procurement procedures would require the officials to review several preliminary facility designs for code compliance, with only one proceeding to an actual permit application. Code interpretations rendered by building code officials during the proposal preparation phase become the basis of the proposal price offered to the owner by the design-builder. This aspect of competitive selection of design-build proposals puts a burden on building officials to review multiple designs for a single site, and to carefully consider code interpretations requested by the proposers. Design-builders would also be likely to request partial building permits on their fast-tracked projects, not unlike any other project delivery methods that feature accelerated completion schedules.

5.3.3 Users

The needs of the user were conceivably the raison d'être for the project's initiation. The users of the facility can be intimately involved in the programming and design criteria phases of the project. Likewise, a prudent owner will devise an appropriate forum for the users to review the design proposals and offer comments and opinions to the owner's selection panel. Multiple design-build proposals submitted in response to a competitive RFP, with their ample graphic exhibits and eager proponents, are well suited for user and community review prior to selection. Because each proposal would be presented to the owner, and subsequently to the users, with firm fixed prices, the owner should have no concern that a design solution with an unacceptable cost would be selected by the users and/or community members. Proposals that do not meet budgetary requirements would be deemed nonresponsive by the owner and would not be shown to the users or to the community.

The needs of the user's special interests groups must be collected, analyzed, and, if appropriate, included in the RFP by the design criteria professional. These special interests may include provisions for the disabled, security, maintenance, energy conservation, indoor air quality, and global environmental concerns. User participation at the programming stage is critical to a successful design-build project because there is limited opportunity for the designers to meet informally with the users during the proposal preparation phase.

6

The Owner's Responsibilities in Design-Build

6.1 Introduction

As discussed in Chap. 1, project delivery systems establish an organizational framework for a construction project. This framework not only defines how the parties on the project relate to each other, but also creates the structure for determining what roles and responsibilities they will have as the work is executed. Because each project delivery system has unique features, the specific role of a party under one system can look quite different from the role assumed by that party under another system.

The owner is the most important party on any construction project, as it creates the need for the project. The owner decides every key aspect of the project—including what type of project will be built, where it will be built, when it will be built, and who will build it. As a result, the owner, unlike any other party, has the ability to control the destiny of the project and everyone working on it, from conception to completion. The owner also has the corresponding duty to manage how the project will be developed. This, of course, includes the responsibility of selecting a project delivery system.

Owners who choose design-build frequently do so to reduce the complexity of developing the project and to make their lives easier. Under any other delivery system, including at-risk construction management, the owner procures, manages, and coordinates the A/E and contractor teams. This role includes the obligation to resolve disputes between the design and construction teams and to take on the risk, vis-à-vis the construction team, of a defective design. Through design-build, the owner eliminates one of the key trouble spots on the project by shifting to the design-builder the responsibility for managing and coordinating the design and construction.

Although the use of design-build does simplify the owner's role on the project, it does not mean that the owner will be passive until the project is completed.

In fact, many of the responsibilities that an owner has under other delivery systems remain responsibilities under the design-build approach. This chapter reviews some of the owner's most important duties on a design-build project. (See Fig. 6.2 at end of chapter for a checklist of owner's responsibilities.)

6.2   Management Structure and Project Team

One of the first things that an owner must do on a design-build project is establish its project management structure and assemble an internal and external project team. Part of this process is to ensure that the owner's "personality" is reflected in the management structure and that all critical players within the owner's organization are involved to the extent necessary. Critical questions to be asked to assess the owner's personality include:

- Is management collectively on board and fully supportive of the project?
- Will the owner's management structure empower the project team or require that the members simply act as conduits for information?
- Can the decisionmakers in the organization make decisions expeditiously?
- Are the decisionmakers driven by the same goals as the project team?
- Do the decisionmakers have to be involved in most of the design and construction meetings for things to be accomplished?
- Do the financial controls in the owner's organization necessitate multilevels of approval before changes can be made to preapproved budgets?
- Will there be strong views from uninformed sources who have power to influence the project?
- Does management frequently use outside consultants, and, if so, is it willing to follow the advice of these consultants?
- Does the owner need hands-on involvement to be comfortable with the design and construction process?

The answers to each of these questions will be critical in several respects. First, they will determine the type of procurement and contracting system that will work most effectively, how the owner will require the design-builder to execute the project (including the frequency of design reviews), and what changes may need to be made within the owner's organization to facilitate design-build. Second, they may establish whether the owner needs to use an outside party—such as a developer, construction manager, or program manager—to give the project its greatest opportunity for success.

The benefit of design-build is that an owner with experienced in-house project management personnel can run the project without hiring an outside consultant. The owner's project team should be led by someone who is familiar with the internal operations of the organization, trusted by the organization's hierarchy, and capable of obtaining full and complete support for the development of the project. This individual should candidly assess whether the owner has the in-house depth to effectively manage all aspects of the process—including whether the in-house resources have sufficient design-build experience.

As a result of corporate downsizing and the limited experience that many owners have with design-build, owners seldom have sufficient internal capabilities to fully handle their responsibilities under a design-build contract. As a result, these owners retain consultants to perform specialized services necessary to the process. This process requires the owner to first determine what services are needed and then how to procure the consultants to provide them. The owner's consulting team typically consists of

- *Technical advisors.* The owner often needs technical assistance in (1) developing its programmatic requirements for the project and the request for proposals (RFPs), (2) evaluating the RFPs, (3) selecting the design-builder, and (4) monitoring the design-builder's performance, including a review of the design-builder's design submissions and quality control programs during construction.

- *Legal and procurement advisors.* The legal and procurement issues associated with design-build are unique and should be subject to a specific review by capable professionals. The owner's in-house lawyers and procurement experts are often not familiar enough with the process to fully represent the owner's interests.

- *Permitting experts.* Because the permitting process can be complex, it often behooves the owner to use external consultants who understand the relevant requirements.

The role of the owner's consultant is defined more fully in Chap. 5.

6.3   Owner's Program

Central to the success of a design-build program is a well-conceived owner's program. This program should address, among other things, (1) what the owner intends to do with the facility, (2) a proposed site location, (3) special needs of the owner that may impact the facility, and (4) any areas of the project that will have design limitations or specific design requirements. The owner's program in a competitive design-build selection process is generally defined before any meaningful direct discussions with the design-builder take place. If, however, the owner uses direct design-build and engages the design-builder under a preliminary agreement, the design-builder often serves as the owner's consultant to help develop its program. The DBIA *Standard Form of Preliminary Agreement Between Owner and Design-Builder* provides for this arrangement by stating the following:

> If Owner's Project Criteria have not been developed prior to the execution of this Agreement, Design-Builder will assist Owner in developing Owner's Project

Criteria, with such service deemed to be an Additional Service pursuant to Section 2.7 hereof. If Owner has developed Owner's Project Criteria prior to executing this Agreement, Design-Builder shall review and prepare a written evaluation of such criteria, including recommendations to Owner for different and innovative approaches to the design and construction of the Project. The parties shall meet to discuss Design-Builder's written evaluation of Owner's Project Criteria and agree upon what revisions, if any, should be made to such criteria.[1]

Regardless of how or when the owner develops its program, it should remember that the design should be defined in a specific, but not overly rigid, manner. The owner should state its needs and constraints in terms of performance specifications, rather than by providing solutions to those needs and constraints. Once the owner specifies the solution in the form of a design, it may defeat many of the benefits of the design-build process, particularly the ingenuity of the design-builder in initiating a cost-effective design. The owner who defines the solution also runs the risk of losing the single point of responsibility benefit, particularly if what the owner specified has a flaw.[2]

6.4 Procurement and Contracting Methodology

Once the owner decides to use design-build, it must decide how the design-builder will be selected. If the owner is in the private sector, it can use any type of procurement process, ranging from direct selection on a sole source basis to a structured competition. Most private-sector owners will forgo a rigid competitive process in favor of one that is based largely on qualifications, with pricing information, if any, limited to the design-builder's proposed fee and general conditions expenses. Even owners who use a qualifications-based process will have the responsibility to decide several important issues relative to selection, including:

- How will the shortlist of prospective design-builders be developed?
- What criteria will be used to make the selection?
- What information will be required from the shortlisted firms in their proposals? Will any substantial design submittals be required?
- What process will be used to obtain proposals? Will there be oral presentations and, if so, how will they be conducted?
- Who will be on the management team which selects the design-builder?

The answers to these questions will be a function of (1) the owner's general procurement philosophy, which may dictate the formality of the selection

---

[1]Section 2.2.2, DBIA *Standard Form of Preliminary Agreement Between Owner and Design-Builder.*

[2]This process where an owner develops design solutions to a fairly advanced level (30 to 45 percent) and lets the design-builder complete the design is often called *bridging* or *draw-build*. Although this may be appropriate on some projects, the owner should not overspecify areas of the design where the design-builder can reasonably have some flexibility.

process; (2) the amount of time the owner has to select the design-builder; and (3) whether the owner has a clear understanding of what it wants or whether it needs the selection process to clarify its overall program.

Public-sector owners are more constrained than private-sector owners in what procurement methods can be used since they are bound to the procurement laws which govern them. While there are myriad variations of procurement laws around the country, many public owners are following the federal government two-step process enacted in 1997 for design-build selection.[3] This process first requires the owner to issue a request for qualifications (RFQs) to solicit expressions of interest from design-builders. After reviewing the responses to the RFQ, the owner develops a shortlist of three to five qualified firms who are then sent an RFP. The shortlisted firms submit formal proposals, basing their submittals on the factors identified in the RFP.

The responsibility of the public owner is to develop a procurement plan that not only satisfies the requirements of applicable procurement laws but also enables it to obtain the design-builder that will meet its needs. To accomplish this, the public owner must answer many of the same questions that were posed earlier relative to private sector owners. Additionally, the public owner must

- Conduct a procurement review to determine what laws and regulations will affect development of the project on a design-build basis and its selection of a design builder.

- Formulate selection factors and weights to be assigned to them for the RFP.

- Ensure that steps are taken to avoid any protests—which can have a major effect on the ability of the owner to proceed in a timely manner.

To competently accomplish these responsibilities, the owner's project management team should include a person who is intimately familiar with design-build as well as with the procurement process governing selection.

In addition to deciding on a selection process, the owner must establish at the outset of the procurement process the payment methodology to be used on the project. The available systems—lump sum, cost plus, cost plus with a guaranteed maximum price and unit prices—are discussed more fully in Chap. 13. The owner should ensure that it is choosing a system that meets its project goals for a reasonable price. For example, selecting a lump-sum approach when the project is not well defined may result in the owner paying more than if a cost-plus arrangement is used.

Finally, the owner needs to establish its contracting philosophy for the project. It can begin this process by deciding which risks it will assume and which ones it will shift to the design-builder. At one time, many public and private owners followed the philosophy that all risk should be shifted to the

---

[3]Federal Acquisition Reform Act of 1996, Title XLI, Section 4105.

contractor, on the theory that contractor acceptance of risk was the price of doing business and that the profit potentials justified the contractor assuming considerable risk.[4] This philosophy manifested itself in one-sided contracts that sometimes made the contractor the de facto insurer of project risk. As discussed in Chap. 14, history has shown that by shifting all risk unreasonably to contractors, owners paid significantly more for the constructed project through increased bid prices, change-order disputes, and litigation costs.[5]

Owners should not use design-build as an opportunity to abandon sound risk allocation philosophies and attempt to shift any and all conceivable risk to the design-build team. Integrating design and construction does not improve the design-builder's chance to control many of the risks that are inherent to the site and the project. As a result, the design-builder will have no choice but to place contingencies in its price to compensate it for these unforeseeable and uncontrollable risks. More importantly, many competent design-builders will simply refuse to submit proposals on projects where they are required to bear substantial and unreasonable risks. A smart owner will carefully and thoughtfully analyze project risk, and then decide who is in the best position to absorb the responsibility for this risk.

6.5 Services of the Design-Builder

The design-builder's scope of work will be defined by the owner's program and the risks that the owner requires the design-builder to assume in executing the program. However, because design-build is so flexible, some owners are interested in having the design-builder provide more than simply design and construction services. These owners will evaluate their overall project needs, and their own abilities, and have the design-builder take on

- Development-type services, such as site acquisition, environmental permitting, and governmental approvals
- Project financing
- Operations and maintenance
- Move-in logistics

The addition of these responsibilities broadens the contract from *design-build* into systems that are often referred to as *design-build plus, build-operate-transfer, build-operate-lease,* and *build-operate-own-transfer.* Each of these variations on design-build results in the owner outsourcing more

---

[4]See R. J. Smith, "Risk Identification & Allocation—Saving Money by Improving Contracts and Contracting Practices," *International Construction Law Review,* 40 (1955).

[5]See, e.g., R. E. Levitt, R. D. Logcher, and D. B. Ashley, "Allocating Risk and Incentive in Construction," *Journal of the Construction Division* (ASCE), **106**(3):297–305 (Sept. 1980); *Preventing and Resolving Construction Disputes,* Center for Public Resources, Inc. (1991).

of its ownership role to the design-builder, sometimes to the point where the owner is actually receiving a product, with very little control in its development.

6.6   Site Investigation and Information

Unless the owner is using a "design-build-plus" process, most owners agree to retain most of the responsibilities that relate to the physical site conditions. The DBIA *Standard Form of General Conditions of Contract Between Owner and Design-Builder*[6] addresses this responsibility by requiring that the owner furnish the following information and services:

1. Surveys describing the property, boundaries, topography, and reference points for use during construction, including existing service and utility lines
2. Geotechnical studies describing subsurface conditions, and other surveys describing other latent or concealed physical conditions at the site
3. Temporary and permanent easements, zoning, and other requirements and encumbrances affecting land use, or as necessary to permit the proper design and construction of the project and enable the design-builder to perform the work
4. A legal description of the site
5. To the extent available, as-built and record drawings of any existing structures at the site
6. To the extent available, environmental studies, reports, and impact statements describing the environmental conditions, including hazardous conditions, in existence at the site

The DBIA not only requires that this information be provided by the owner but also states that the design-builder is entitled to rely on such information in performing its work.[7]

Taking responsibility for site characteristics under design-build is consistent with the owner's responsibilities under other delivery systems.[8] The owner is in the best position to obtain information relative to site characteristics, since it can require this as a condition to closing on the purchase of the property. Moreover, the owner should be responsible for securing and executing all necessary agreements with adjacent land or property owners that are necessary to enable the design-builder to perform its work.[9]

---

[6]Section 3.2.1, DBIA Document 435 (1998).

[7]All the standard form design-build contracts, including Document A191, AGC 410, and EJCDC 1910-40, consistently place these responsibilities on the owner.

[8]See, for example, Article 4 of EJCDC 1910-8, Standard General Conditions of the Construction Contract (1996 edition), for a typical provision under a design-bid-build delivery system.

[9]Section 3.2.2, DBIA Document 435 (1998).

6.7 Owner's Representative

The owner must designate an individual to serve as its representative in its dealings with the design-builder. As noted in Sec. 6.2, this individual should have a strong understanding of the overall project goals of the owner and be empowered to administer the contractual obligations of the owner (Fig. 6.1). The DBIA *Standard General Conditions of Contract between Owner and Design-Builder* defines this responsibility as follows[10]:

> Owner's Representative shall be responsible for providing Owner-supplied information and approvals in a timely manner to permit Design-Builder to fulfill its obligations under the Contract Documents. Owner's Representative shall also provide Design-Builder with prompt notice if it observes any failure on the part of Design-Builder to fulfill its contractual obligations, including any errors, omissions or defects in the performance of the Work.

If the owner needs to place limitations on the authority of its representative, this should be clearly identified in documentation provided to the design-builder during the execution of the work.

6.8 Permits and Approvals

A design-builder generally assumes more permit responsibility than a contractor operating under another type of delivery system. The owner of a design-build project, however, typically remains responsible for obtaining many permits and approvals, including air quality and environmental permits. The current standard form contract language requires that the owner provide reasonable cooperation with respect to the design-builder's obligations to secure such permits,[11] although the AGC form contract imposes specific

QUALIFICATIONS OF OWNER'S REPRESENTATIVE	
1.	Understands Owner's project goals
2.	Understands internal structure of Owner's organization
3.	Empowered to administer Owner's contractual obligations
4.	Trusted by Owner
5.	Understands the design-build process

Figure 6.1

---

[10]Section 3.4.1.

[11]Paragraph 2.3, Part 2 of AIA Document A191; Section 3.5.2, DBIA Document 435.

responsibility on the owner to obtain the building permit and requires the design-builder to merely assist in this endeavor.[12]

DBIA and EJCDC specifically make the design-builder responsible for obtaining permits except as stated elsewhere in the contract documents.[13] The DBIA clause states as follows[14]:

> Owner shall obtain and pay for all necessary permits, approvals, licenses, government charges and inspection fees set forth in the Owner's Permit List attached as an exhibit to the Agreement.

It is vital for the owner to have a clear understanding of its permit obligations and to ensure that it has a plan in place for obtaining such permits and approvals. An owner's inability to meet its commitments in this area can result in major delays in the project.

6.9   Owner Deliverables

The owner is typically responsible for providing some specific deliverables to the project site that are integral to the completion of the design-builder's work. In the contract for construction of a facility, these deliverables may include furnishings, information technology systems, special landscaping, or other discrete items. Typical deliverables on a power project include fuel, potable water, telephone service to the site, or interconnections into the power grid. It is incumbent on both parties to identify specifically what will be provided by the owner and when the design-builder should expect to receive these items at the site. The dates on which the owner is to furnish these deliverables should be set forth in the project schedule.

If services will be performed by contractors or suppliers who have direct contracts with the owner, the delivery approach becomes less like design-build and more like a multiple prime contracting process. This will require the owner to integrate the work of these separate contractors and suppliers with the work performed by the design-builder. Because scheduling and coordination of multiple prime contractors can be a challenge, the owner must have effective in-house or outside resources to accomplish these tasks. The owner should also ensure that its contracts with the various vendors address, on an integrated and consistent basis, important commercial terms, such as liquidated damages, indemnification, and the duty of each separate contractor to collaborate with the other separate contractors.

6.10   Review and Approval of the Design

Owners on most design-build contracts expect to have the right to review the design and construction for conformance with the contract requirements. In

---

[12]Subparagraph 3.3.5, AGC 410.

[13]Subparagraph 6.07.A, EJCDC 1910-40.

[14]Section 3.5.1, DBIA Document 435.

performing this work, however, it is critical that the owner not interfere with the design-builder's progress or needlessly increase the design-builder's cost of performing the work. This potential interference may be a particular problem in the design phase of the work, since owners who are accustomed to contracting directly with the A/E, and thus paying the A/E on a fee basis, often feel that they have the unfettered right to be involved in monitoring the design.

Given that design for the project will be developed in stages and be performed after the contract is executed, it is important for the parties to understand how the owner and design-builder will work together and communicate with each other during development of the design. The owner's interest in having access to and evaluating the design must be balanced with the design-builder's interest in achieving its quality, cost, and schedule goals. This is particularly true if the design-builder is working on a lump-sum or guaranteed maximum price (GMP) basis and was selected on the basis of price after competitively developing a proposal based on a detailed RFP and owner's program. On the other hand, if the parties are contracting on a cost-plus basis, with the intent of converting to a lump sum or a GMP, one would reasonably expect the owner to have more involvement in the design process.

The DBIA *Design-Build Contracting Guide* recommends that this issue be addressed as follows[15]:

> The design process is an area where owners may choose to play an active role to ensure that their goals are being achieved. This is particularly true when price is being established during the design process (as opposed to when price has already been established on a competitive basis). However, the owner's participation must be balanced with the design-builder's needs to avoid impact to the project schedule or the project budget. To balance these competing interests, it may be appropriate for the parties to agree that the owner's review will be limited to matters affecting technical accuracy, conformance with previous design submission and compatibility with the design concept.
>
> Considering all of the factors associated with design review, DBIA believes that the contract should contain language that formalizes how the design iterations will be presented. DBIA suggests that a specific design review workshop be used for each major submission to allow interactive dialogue that promotes clearer and more comprehensive understandings as to what has taken place relative to the design.

Consistent with this philosophy, DBIA's standard form contracts call for the parties to specifically agree on what design submissions will be required for the owner review and the turnaround times for these submissions.[16] This principle has application in all industry sectors using design-build. For example, on a typical engineer-procure-construct process or power generation project, the owner and design-builder should meet and confer about what the owner will have the right to review. A typical list might include the following:

---

[15]DBIA *Design-Build Contracting Guide,* DBIA Document 510 (1997), Article 8.

[16]DBIA Document 435, Section 2.4.1.

- Electrical one-line diagrams
- Piping and instrumentation diagrams (P&IDs)
- Process flow diagrams (PFDs)
- Equipment location plan
- Underground piping plans
- Electrical protection setpoints
- Control system functional logic diagrams
- Cable and raceway schedules
- Connection report/loop diagrams

Although this list may be representative of what a design-builder in the process industry is willing to provide to the owner, it is incumbent on the parties to any project to discuss and identify what is an acceptable design deliverable and what is "off limits."

The parties should also reach specific agreement on what action the owner has the right to take after it has reviewed the design—including whether the owner needs to signify its approval of the design—and how quickly it needs to respond. Although industry contracts address the design review and approval process similarly, some subtle differences should be noted:

- *AIA contract.* Part 2 of AIA Document A191 states that the design-builder will submit "Construction Documents for review and approval by the Owner."[17] No time period is established for this review and approval process. Moreover, since the term *construction documents* is a defined term, it does not appear that the AIA format allows for interim submissions of designs for review.

- *EJCDC contract.* EJCDC 1910-40 provides for a preliminary design submission that will be reviewed and approved by the owner, after which the design-builder will prepare and submit final drawings and specifications for owner review and approval.[18]

- *AGC contract.* AGC 410 and 415 require three phases of submissions: schematic design documents, design development documents, and construction documents.[19] Each set of documents is subject to the owner's "written approval," with the owner having the general obligation to "review and timely approve schedules, estimates, Schematic Design Documents, Design Development Documents and Construction Documents furnished during the Design Phase."[20]

---

[17]Subparagraph 3.2.3, Part 2 of AIA Document A191.

[18]Subparagraphs 6.01.B and 6.01.C, EJCDC 1910-40.

[19]Subparagraphs 3.1.4 to 3.1.6, AGC 410.

[20]Subparagraph 4.2.1, AGC 410.

■ *DBIA contract.* DBIA Document 435 establishes a more flexible process, requiring that the parties agree on what interim submissions the owner will be reviewing and the time periods for such review.[21] The owner is required to approve both the interim submissions and the final construction documents.[22]

Each sponsoring organization requires owner approval to ensure that there is no confusion between the owner and design-builder as to the design that is being proffered.

Finally, unless the parties intend otherwise, the owner's review and approval should not relieve the design-builder from performing its work and having single-point responsibility for the design and construction. A typical clause stating this proposition is contained in the DBIA *Standard Form of General Conditions of Contract Between Owner and Design-Builder,* which reads as follows[23]:

> Owner's review and approval of interim design submissions and the Construction Documents is for the purpose of mutually establishing a conformed set of Contract Documents compatible with the requirements of the Work. Neither Owner's review nor approval of any interim design submissions and Construction Documents shall be deemed to transfer any design liability from Design-Builder to Owner.

Despite this general proposition, owners should be mindful of the axiom "with control comes responsibility." The more the owner is involved in reviewing the design, the greater will be its accountability in the event a problem arises. Consequently, it is prudent for the owner to have a good reason for seeking review and approval rights and to have processes in place that enable the owner to competently and timely perform these functions.

6.11   Inspection and Approval of the Construction

Under the traditional delivery system, the owner is responsible for inspecting the contractor's work for general conformance with the contract documents. This responsibility, frequently performed by the owner's A/E, is performed on a periodic basis throughout the construction of the project. Those owners who prefer to have a greater assurance of contract compliance will employ a full-time "clerk of the works" at the site to monitor the work. "Acceptance of the work" occurs when the owner deems the work to be substantially complete, except for punchlist items that do not impact the owner's ability to use the project for its intended purpose.

---

[21]Section 2.4.1, DBIA Document 435.

[22]Section 2.4.2, DBIA Document 435.

[23]Section 2.4.3, DBIA Document 435.

The inspection and acceptance process for construction under design-build is substantially the same as under the design-bid-build approach. On most design-build projects, the owner will

- Retain the right to inspect the work as it progresses, including work which takes place off site

- Review, approve, and monitor the design-builder's quality assurance/quality control (QA/QC) program

- Determine when the project is substantially and finally complete

The manner in which the owner will perform these inspection and approval duties will depend on the type of design-build project being constructed. For example, on industrial projects—including power generation, pharmaceutical and process facilities—acceptance is tied into a specific performance test of systems, subsystems, and the project as a whole. This requires that the parties agree on specific acceptance criteria and testing protocol. On hazardous-waste remediation contracts, acceptance may be conditioned on approval by a government agency. Regardless of the industry sector, it is critical for the contract to clearly identify the full extent of the owner's obligations relative to inspection and whether the owner or design-builder will take the lead in declaring the project successfully completed.

6.12 Changes and Claims

Changes and claims will occur on design-build projects, although the frequency is generally far less than on traditionally delivered projects. The owner is responsible for developing a system for processing change orders and claims in a fair and expeditious manner. Part of the system will be established by the contract and part in a project procedure manual.

Contractually, the owner should have the right to direct the design-builder to make changes, regardless of whether the parties are able to agree on the commercial terms associated with the change. This unilateral right, often referred to as a *construction change directive,* is established in the AIA, DBIA, and EJCDC changes clauses.[24] In exchange for this right, the owner will be obligated to provide the design-builder with appropriate adjustments to the contract price and time. The contract should further

---

[24]See Paragraph 8.3, Part 2 of AIA Document A191; Section 9.2, DBIA Document 535; Subparagraph 1.01.A.50, EJCDC 1910-40.

identify the actions to be taken if the parties are ultimately unable to agree within a reasonable period of time.[25]

The contract terms should not be a substitute for good contract administration practices. Experience shows that the longer changes and claims go unresolved, the more likely the project will be prone to major disputes and problems. Owners should have procedures to establish (1) who has the authority to commit to a change, (2) the time periods within which changes should be resolved, and (3) the manner in which time extensions will be analyzed. The owner should also establish a structure and chain of command for the prompt resolution of disputed changes. This is particularly true for changes that impact the schedule, since the design-builder may be forced to incur even more costs in attempting to accelerate and overcome the delay if a time extension is not promptly recognized.

6.13 Insurance

Obtaining and maintaining appropriate project insurance is an obligation for both the design-builder and owner. Both parties should have appropriate commercial general liability coverage, builder's risk, and pollution coverages.

Creative owners are implementing insurance programs that attempt to provide seamless coverage with minimal duplication. These programs, often called *owner-controlled insurance programs, project wrapups,* and *consolidated insurance programs,* are written for specific projects and intended to provide sufficient insurance to cover the needs of the entire project team. The primary benefits of these programs are that (1) the limits of insurance are dedicated to the owner's specific project, rather than to several projects; (2) there is a coordinated loss control program for the project which eliminates gaps in coverage; (3) there can be a clear definition of when specific insurance is triggered so that a project team member knows what to do when a claim must be submitted; and (4) consolidating insurance coverages typically enables the

---

[25]As discussed more fully in Chap. 14, DBIA has an innovative way to deal with unresolved claims. The DBIA *Standard Form of General Conditions of Contract Between Owner and Design-Builder* establishes the following process:

If Owner and Design-Builder disagree on whether Design-Builder is entitled to be paid for any services required by Owner, or if there are any other disagreements over the scope of Work or proposed changes to the Work, Owner and Design-Builder shall resolve the disagreement pursuant to Article 10 hereof. As part of the negotiation process, Design-Builder shall furnish Owner with a good faith estimate of the costs to perform the disputed services in accordance with Owner's interpretations. If the parties are unable to agree and Owner expects Design-Builder to perform the services in accordance with Owner's interpretations, Design-Builder shall proceed to perform the disputed services, conditioned upon Owner issuing a written order to Design-Builder (i) directing Design-Builder to proceed and (ii) specifying Owner's interpretation of the services that are to be performed. If this occurs, Design-Builder shall be entitled to submit in its Applications for Payment an amount equal to fifty percent (50%) of its reasonable estimated direct cost to perform the services, and Owner agrees to pay such amounts, with the express understanding that (i) such payment by Owner does not prejudice Owner's right to argue that it has no responsibility to pay for such services and (ii) receipt of such payment by Design-Builder does not prejudice Design-Builder's right to seek full payment of the disputed services if Owner's order is deemed to be a change to the Work.

owner to obtain a better quote for the desired coverage and, again, to avoid the gaps and duplication associated with multiple policies.

6.14 Dispute Resolution Program

As on any construction project, the owner of a design-build project plays a major role in establishing the manner in which disputes will be resolved. Although the contractual disputes process will be part of contract negotiations, the owner's draft contract will create the setting for what, if any, alternative dispute resolution (ADR) processes (such as dispute review boards, project neutrals, and standing mediators) will be considered and whether arbitration or litigation will be the process for formally and finally resolving disputes that cannot be resolved through ADR. The owner will also be the key in deciding whether to use a formal partnering program on the project.[26]

6.15 Payment and Financial Guarantees

A fundamental duty of any construction owner is to meet its financial obligations on the project. This is particularly true when the project will be developed on a fast-track basis, since the owner may be relying heavily on estimates until final pricing information is obtained from trade contractors. Unless the design-builder is providing some form of financing on the project, it is reasonable for the design-builder to require the owner to furnish evidence that it has sufficient funding to meet its commitments. This right should apply not only prior to the parties' execution of the contract, but also during the course of the work—particularly if there is evidence that the project is in trouble. To protect their interests, some design-builders also require the owner to demonstrate, through a device such as a letter of credit or an escrow account, its specific ability to finance changes that exceed a certain price.

Depending on who it is, the owner may also be required to post some form of security that it can meet its payment obligations. Security is often required when the owner is (1) a limited purpose corporation with limited assets, (2) a relatively new company with little credit history, or (3) experiencing financial difficulty. The form of the security can vary, but is most frequently one of the following:

- A letter of credit for some portion of the contract price
- A third-party guarantee—often from the corporate parent
- A substantial advance payment, coupled with limited retention and prompt payment rights

Because there are many creative ways for an owner to provide the design-builder with reasonable assurances that it will be paid, the parties should not

---

[26]See Chap. 16 for a more comprehensive discussion about dispute resolution and ADR.

let this issue be a stumbling block to entering into a contract. They should, however, spend time at the outset of the contract to understand each other's constraints relative to this issue and reach an appropriate accommodation.

6.16 Obligation to Cooperate

The nature of the design-build approach by necessity creates a close and open working environment among the parties. Because the use of the process is often driven by the need for expedited project completion, the owner and design-builder are generally motivated to work closely together and attempt to meet each other's legitimate needs.

Although there is a practical need to work together, owners should remember that they will be charged, as a matter of law, with providing reasonable cooperation to the design-builder to achieve the goals of the contract. This implied duty of cooperation has been a long-standing principle of both public- and private-sector construction law[27] and is also frequently integrated into the design-build contract. A typical clause reads as follows[28]:

> Owner shall, throughout the performance of the work, cooperate with design-builder and perform its responsibilities, obligations, and services in a timely manner to facilitate design-builder's timely and efficient performance of the work and so as not to delay or interfere with design-builder's performance of its obligations under the contract documents.

As a result, even if the owner is not charged with a specific duty or role under the contract, it may nevertheless have numerous other duties (see checklist in Fig. 6.2) imposed on it by virtue of the duty to cooperate. These duties may include (1) providing reasonable assistance to the design-builder for permit acquisition, (2) furnishing the design-builder with access to information that could assist it in the performance of the work, and (3) performing expedited reviews of submittals when needed to meet the reasonable needs of the design-builder.

---

[27]See, e.g., R. J. Bednar et al., *Construction Contracting,* George Washington Univ. National Law Center (1991), pp. 544–551.

[28]DBIA Document 435, Section 3.1.1.

OWNER'S RESPONSIBILITIES: A CHECKLIST
1.	Identify project goals
2.	Decide upon a project delivery system which will give Owner best chance to achieve project goals
3.	Determine whether Owner needs assistance of a consultant
4.	Select a representative & determine any limitations on the representative's authority
5.	Decide upon a procurement process for Design-Builder
6.	Determine the appropriate payment methodology (e.g., cost plus, lump sum) for Design-Builder's contract
7.	Identify and allocate project risk
8.	Obtain project financing
9.	Furnish site information
10.	Secure agreements with adjacent property owners
11.	Identify Owner-furnished deliverables (e.g., permits, furnishings) and develop process for timely obtaining such deliverables
12.	Review design submissions of Design-Builder
13.	Inspect construction for compliance with contract
14.	Develop efficient change order process
15.	Obtain appropriate project insurance

Figure 6.2

Design-Build Entity Organization and Management of Projects

7.1 Organizing the Design-Build Entity

Assembling a highly effective project team from among the many disciplines and vocations within the design and construction industry is at the heart of design-build delivery. The focus on common goals and the single point of overall project control are basic ingredients that make up design-build services.

Many individuals with different professions and skills participate in integrated services projects. Each member of a design-build team has basic personal motivations and needs that must be at least partially satisfied if the project is going to progress with harmony and efficiency. Needs that motivate individual team members range from physical and biological (shelter, food, rest) and security needs (safety, removal of dangerous threats), to societal (acceptance, friendship, sense of belonging to a larger organization), and self-esteem (pride, self-confidence, status, recognition, ego), to self-actualization (continuing to seek further development of one's potential).

Societal needs are often satisfied by working with like-minded individuals in an environment where friendly relationships and job satisfaction are found. At work, the ability to provide input into problem solving, to build business relationships and to develop trust in coworkers can translate into project-based productivity and teamwork. Self-esteem is satisfied when each person on a project team is given responsibility and the opportunity to meaningfully contribute to the success of the endeavor. Recognition through incentives and public acknowledgment rewards the individual and encourages continued behavior that produces high-quality services, decisions, and work. Self-actualization is the result of achieving meaningful goals, producing high-quality work, and exhibiting initiative in areas that will benefit the project and the individual. The project team's capability in working through disputes and change requests with an interdependent problem-solving focus can provide each individual with a sense of professional satisfaction and job fulfillment.

An organizational plan that shows the functional professions and trade skills necessary to integrate the design and construction process is an excellent starting point in organizing the design-build team. A systems-based structure will show functional systems and subsystems necessary to implement project goals by assigning key responsibilities to firms and people. Some projects may be handled entirely within the human resources capabilities of an integrated design-build company, whereas other projects may require outside team members. The size and complexity of the design-build project are two factors that strongly influence the organizational structure. Additional influences include demand for cost control or pressure of an early completion schedule. Principals of the prime design-builder must determine how their business entity will be structured, by examining their markets, capacity for risk, availability of human and financial resources, and the degree of business integration (internally or through alliances) with which they will be comfortable.

7.2 Multidisciplinary Culture

The design and construction industry has been described as the most fragmented and disparate of all the major industries in North America. University education, industry training, and business operations have perpetuated the separation of facility design from facility construction for more than a century.

Yet the process of design and construction relies on the contributions of multiple disciplines. Successful design-build depends on a team that will work harmoniously and efficiently toward shared goals. Careful thought and sustained effort is required to achieve the right balance of ideas and business chemistry among the team. If any one of the teammates is unwilling to personally adopt the team's goals, or is so versed in the traditional adversarial way of prosecuting the work that team-based solutions are difficult, the potential exists for the process of design-build to be compromised. When key teammates are unable to break out of their single disciplinary thinking, characterized by protectiveness or unwillingness to recognize the contributions of other disciplines, the process can become bogged down and careers jeopardized.

Design-build is a collaborative process with an underlying single-source contract acting as the glue to hold all the parties together in the interest of shared goals. Design-build, implemented with best practices, is partnering with teeth. The collaborative process is characterized by

- *Focus on shared goals.* Each design-build teammate understands and shares the mutual goals for the project. Goals and objectives are generally set by the owner, then implementation strategies are discussed and agreed on by the team. The sense of common vision contrasts with traditional design-bid-build, wherein the designer's goal is to create design documents and the constructor's goal (under its contract) is to do the "work" shown on and in the A/E's contract documents.

- *Commitment over time.* A true design-build relationship will outlast a single project and can be applied to other projects and markets over time. The benefits of the relationships developed through a design-build partnership aid solving problems quickly as they arise and can help the team gain a competitive advantage.

- *Flexibility and trust.* The design-build collaboration can respond to changing conditions, whether they are customer-driven or as a result of shifts in the economy, better than traditional systems. Trust among the teammates enables specialized knowledge and resources to be combined with the other's special talents. There is no suspicion about the goals and expectations about one's teammates, but an understanding that leads to acceptance and good will. Difficulties on the project can be settled at the lowest possible level, before those problems become an administrative burden or a question of who's at fault.

7.3   Teaming Agreements

Successful design-build delivery is dependent on the assembly of the right team: one that will bring expertise, assets, value, experience, and standing to the venture. Team members, beginning with the key roles of designer and constructor, must ascertain whether they have compatible business philosophies, cultural values, understanding of the design-build process, and complementary capabilities. The agreement to work as a team to satisfy an owner's design-build requirements should be established in advance of the decision to pursue a project. The format for establishing the relationship among the team members may be an informal letter or a formal covenant. Provisions of the agreement set forth the principles that will guide team communications, allocate responsibilities, and serve as a basis for building on the trust and confidence that the teammates have in each other. The teaming agreement does not substitute for the contract that will be signed by the parties if they are successful in landing the project. However, the contract will likely incorporate the teaming agreement and all the organizational provisions that will assist the team in executing the owner's project.

7.3.1   Questions to answer before teaming on a design-build project

The formation of design-build teams usually begins far in advance of the issuance of a request for qualifications (RFQs) or a request for proposals (RFPs). Teams often coalesce on the basis of interest in a specific design-build market opportunity or past working relationships on traditional projects. The ad hoc design-build entity should explore whether the parties share common goals for quality, creativity, and innovation. Potential teammates will also want to know whether the other parties have the necessary resources (funds, workforce, time) to ensure a successful project. At least one of the primary team members (designer or constructor) should have significant design-build experience in the type and size of project that is being offered.

Potential design-build teams may want to consider the following issues:

With whom should we team?
- Should relationships be established prior to or at the RFQ stage?
- Are levels of experience appropriate for the projects in question?
- Do all or one of the parties have a previous relationship with the owner?
- Given the team's attributes, can the team win the project?
- Have the team members approached the alliance disclosing all information related to the project(s), including conflict of interest issues?
- Can the team afford the resources necessary to compete in the process (time, money, personnel, good will,...)?

What type of business establishment should be formed for pursuing the work?
- What form of legal entity should be created?
  Joint venture
  Limited-liability corporation
  A/E-led–constructor subcontractor
  Constructor-led–designer subcontractor
  Developer-led
  Other
- What is the business approach for pursuing the project?
  Who leads the marketing effort?
  Who leads the proposal preparation?
  Development of a schedule for team meetings
  Setting up of communication channels
  Key responsibilities/personnel

What are the important roles and responsibilities given the demands of the project?
- Owner (primary role)
- Owner representative (support role; optional)
- Design professional as teaming partner (primary role)
- Constructor as teaming partner (primary role)
- Specialty and subcontractors (support role, but may serve in primary role on some projects)
- Subconsultants (support role)
- Construction manager (optional)
- Other potential team members and roles

How does one conduct a teaming risk assessment? (The assessment helps in assigning risks and responsibilities to the parties best able to assume the risks—during the proposal phase and after contract award.)

- What are the proposal phase risks?
  Adequacy of project financing
  Design-build entity selection process

Whether there is a realistic project budget

Who leads the project marketing effort for the team?

Review of the site and soils report

Whether the agreement is exclusive to the parties

What happens if a team member exits before submission?

What are the performance requirements and whether they can be met by the team?

What are the legal requirements and constraints?

Whether an honorarium is offered to unsuccessful proposers; if so, how much is the honorarium

Who are the political and community stakeholders?

What are the state licensure and procurement statutes?

Whether the project objectives (technical performance, aesthetics, functional efficiency, safety, etc.) are clear

Other proposal phase risks

- What are the design-build contract risks?

Design reviews by the owner and its agents

Differences between program or design criteria and completed design

Project site safety

Permits and approvals

Environmental impact review

Coordination of the work

Quality control and quality assurance

Differing subsurface conditions

Design defects

Construction defects

Strikes or labor disputes

Weather conditions

Fire, flood, earthquake, and other catastrophes

Utilities maintenance or relocation

Hazardous-waste containment or removal

Third-party litigation

Warranty for facility performance

Bonding and insurance coverages

Liquidated damages

Other contract risks

How does the team allocate the costs for preparing the submission?

- Who pays for the marketing costs?
- Who pays for the conceptual design costs?
- Who pays for the estimating costs?
- Who pays for the proposal coordination and management costs?
- Whether the venture is split 50/50 between the designer and the constructor
- In a prime/sub relationship, is the arrangement is for services only (not at risk) or is the subconsultant or subcontractor participating in the risk?

- If the team is successful, will the revenues from the contract be sufficient to cover the cost of pursuing the project for both parties?
- If the team is unsuccessful, is there an arrangement to share in the costs of preparing the submission?
- Who pays for owner delays in the proposal preparation process, or in delays in the award decision process?
- Other proposal cost issues

Does the team understand the owner's needs through RFP analysis, research, and communication?

- Has the team analyzed the requirements in the RFP to uncover both the qualitative and the quantitative issues to be solved?
- Have the teammates set goals for winning the project and assigned responsibilities accordingly?
- Should the team conduct its own soil borings or other site research?
- Has each RFQ or RFP element been "brainstormed" and ideas captured narratively and graphically?
- The team must attend all the owner's preproposal meetings.

What steps will the team take in preparing and presenting its design-build proposal?

- Analyze program and design criteria
- Develop conceptual estimate
- Develop conceptual design
- Write proposal narratives
- Create project staffing plan and organizational chart
- Write project approach, with phasing, cost control, quality assurance, and other features intact
- Complete concept design submission, with code analysis, life safety checks, drawings, renderings, and engineering calculations, as required
- Write outline specifications
- Practice continuous value analysis
- Make an effective oral presentation
- Submit a complete and responsive package

What considerations are important if the team is successful (receives the award)?

- How the proposal development costs are divided or shared
- Whether the design fee and contingency have been determined
- The commercial terms of the contract for the project (cost-plus fee, guaranteed maximum price, lump-sum fixed price, unit price, etc.)
- Whether there is a shared savings provision between the owner and the design-builder

- How insurance responsibilities and long-term liabilities will be allocated
- Other terms

What considerations are important if the team is not successful (does not receive the award)?

- How costs of pursuing the project will be shared and how the honorarium (if any) will be shared
- Whether there is protection of the design submission or other innovative ideas in the proposal
- Whether proprietary data and information are returned to the proposer
- Other considerations

Should the team continue beyond the subject project(s) or terminate?

- Does the strategic alliance have continuing value?
- Should termination be exercised to allow pursuit of other work with other partners?
- Are there clear provisions for termination of the teaming alliance?

7.3.2 Unique aspects and special issues in teaming

Proposal development costs. On design-build projects where services are procured via a request for proposals and the owner stipulates that a concept design must be submitted as part of the response package, preparation of the proposal can be difficult and time-consuming. Cost considerations include marketing, conceptual design, preliminary schedules, estimates, coordinating specialty subcontractor and manufacturer involvement, and managing the team during the proposal process. Rarely does the owner reimburse the team for these items, and it is important for the teammates to decide in advance how to pay for direct proposal development costs. Design firms are usually less prepared to devote significant resources to design without some expectation of recovering a portion of their actual costs; however, constructors are also incurring costs as estimating, scheduling, and coordination of specialty contractor input must keep pace with the conceptual design effort.

Because the makeup of each design-build team is different, and because every design-build project has differing characteristics, the participants must discuss the alternative ways for allocating the costs of proposal development. On some projects the parties may agree to share development costs equally. Teammates may decide to help defray the costs of the party that has expended a significant effort to land the project (often this party is the A/E firm that is creating design concepts). Other alternatives include sharing the proposal development costs at a fixed percentage based on the level of effort and financial capability of each teammate; or having each team member bear its own costs, but agreeing to reimburse specific team members if they are successful in winning the project. For projects with an honorarium or stipend, the parties may include language stating that the lion's share of these funds will go to the party that made the most significant contribution to the proposal; or it can be a share on a fixed basis as defined in the teaming agreement.

Noncompete provisions.   To maintain the highest level of focus and trust, design-build teams can be well served by including a noncompete provision in their teaming agreement. The team members should think through whether the arrangement is an exclusive one, or whether the individual parties to the agreement can pursue the project (or other projects) with different teams. Serious questions may arise if teammates are working with other teams, such as divided loyalties, professional ethics, and business jealousies. Unless these concerns are aired and resolved, and the consent of all team members obtained, it is more appropriate for the teaming agreement to contain a provision that prohibits joining other teams or submitting an individual proposal on the project. The agreement should also contain language that precludes team members from abandoning the initial team and joining another team (where intellectual property from the first team may be disclosed to a competitor). The parties may want to add a provision to the teaming agreement that requires the team member who joins another team (or merely abandons work on the proposal effort, jeopardizing the team's ability to pursue the project), to compensate the initial team for financial harm.

Proprietary information and innovative concepts.   Preparation of a winning proposal will require the teammates to develop innovative concepts or to share proprietary business information with other members of the team. Disclosure of this confidential business information to another team or outside party would severely hamper the team's ability to succeed on the project, and may have consequences beyond the project itself. The information may include creative ideas or solutions that can be used for business advantage (or copyrighted or patented by one or more of the team members), including new products or applications; improvements on means and methods of design and construction; data or software; unique programming, estimating, or scheduling techniques; or facility/process design solutions that are innovative or confidential.

To protect these ideas and information, the teaming agreement should include language that requires all team members to maintain confidentiality unless express written permission is obtained. The teammates will agree that the protected information may be presented to the owner, and if the proposal is successful, the information and ideas may be used during the execution of the contract. The provision should also exempt information from the nondisclosure requirement that is in the public domain at the time of the proposal. For unsuccessful proposers, information should be returned to the rightful parties within a defined period of time (such as 30 days) so that the technical information may be further developed by its originators or applied on a future project.

Termination of the teaming agreement.   The team members will want to establish the duration of the agreement. In many cases, the teaming agreement or its provisions will be rolled over into the contract if the team is awarded the project. The agreement may also be terminated in the event that the team is unsuccessful; however, the teammates may want to keep the strategic alliance in place for pursuit of similar work in the future. It may be advisable to place anniversary dates (or sunset dates) within an ongoing teaming alliance to reunite the parties to determine whether the teaming agreement should be extended.

7.4   Team Leadership: The Design-Build Project Manager

The role of the project manager in the collaborative process of design-build is complex. The individual must understand the requirements of design management as well as construction management, leading the design and construction team in meeting the project objectives of quality, cost, schedule, and performance (e.g., safety, environmental protection). General skill areas recognized by the Project Management Institute include

- Integration
- Scope
- Time
- Quality
- Cost
- Risk
- Procurement
- Communications
- Human resources

For fully integrated projects, the design-build project manager serves as the single point of responsibility and coordination for planning and executing all the project activities. The project manager is the extension of the design-build entity's senior management to represent the team (integrated firm or consortia of companies) to the client. Serving in this capacity, the project manager is the point of client interface and trust for the overall project delivery. Within the project team, the manager coordinates and delegates to get the project accomplished.

A top-flight design-build project manager will provide leadership, vision, and direction rather than "follow the landscape." It is the mission of this key position to create results, rather than to give reasons of why goals cannot be achieved. Top-notch project managers will anticipate and avoid problems, but if confronted with problems, will have a system in place for working toward a rapid solution. With "time to market" becoming an important issue for facility owners, the project management role also requires promoting and maintaining a sense of urgency.

The primary responsibilities of a project manager range from the project's inception to the completion of the task, incorporating a variety of resources to reach the following objectives:

- *Understand and/or define the scope and objectives of the project.* The design-build project manager must understand and embrace the owner's goals for the facility; often balancing the competing goals of increasing design quality versus maintaining cost control. The design-build system broadens the scope of the contract and therefore extends the responsibilities of the project manager. Success is not achieved by managing a successful design phase alone (or a zoning permit phase, a successful construction phase, or even a commissioning phase), but is achieved only when the owner's hotel is open for business or the municipality's bridge is carrying traffic.

- *Develop and communicate a strategy for managing the project.* A strategy for executing the project involves organizing a scheme to accomplish the project objectives. The project manager must divide the project into major phases and identify important milestones. Methods of doing the work must be established because they will show both management and the team a workable approach for accomplishing the goal. In addition to methods, the project manager should determine the types of skills, materials, and equipment that will be necessary to accomplish the kinds of work required by the project.

- *Establish and mobilize the project team.* Other than the selection of the project manager, attracting good team members to the project organization is the human resources key to design-build project success. Experience with similar kinds of projects is necessary, along with problem-solving ability, initiative, and energy. Mobilize the team by zeroing in on why their skills are critical for project success, engage them in the planning process, and emphasize ongoing communication and involvement to address problems and monitor progress.

- *Create and implement project operating procedures.* Each project has unique characteristics, and it is the project manager's role to understand and communicate procedures on the basis of the contract, owner expectations, permitting agency requirements, and design or construction strategy. The procedures will contain guidelines for project reporting, management control, quality assurance, and work problem or dispute resolution. Operating procedures can be placed in a manual or document that contains the business and administrative functions of the project team.

- *Plan and control budget, schedule, and quality.* A detailed project plan consists of direct costs, contingencies, and indirect costs. It details tasks and work items and places them in an expanded schedule with sequencing and milestones. Cost and time management control techniques work in tandem with technical and performance objectives to produce project quality. The design-build project manager uses these tools to develop an integrated control system starting with organization, and continuing with phased design, construction, and commissioning.

- *Coordinate, communicate, and execute the work.* The project manager's duties begin with coordinating the work and directing the project's resources. The coordination requires regular communication with the owner, team members, specialty and subcontractors, inspection agencies, firm principals, and others. Poor communication is the primary source of project problems. Moving the project forward with a sense of urgency by supervising, checking, delegating, and doing are the marks of a well-rounded manager.

- *Administer changes, direct personnel, and solve problems.* Day-to-day operations will require that the project manager respond to changes by adjusting costs and schedules to accommodate the modifications. The manager will also assign personnel to tasks and activities to maintain productivity. It is the project manager who focuses on issues and problems while maintaining organizational momentum (with administrative reporting and communication) who understands the balance of skills and temperament necessary to carry out the work.

- *Maintain safety.* Insurance advisors, owners, and government safety officials have been advising design and construction industry practitioners for years about the necessity for project safety. Enlightened constructors and design-build firms understand that safety, like investing in zero-coupon bonds, takes an initial investment but will pay back later in the project. The project manager should insist on a safe work site, not only because of financial reasons but out of empathy for the workers.

- *Exhibit maturity, steadfastness, and a sense of humor.* To motivate the project team, the design-build project manager must appeal to the collective interest (success of the project means success of the team) and to the individual interests (in making money, gaining recognition, etc.) of the team members. Having a good sense of timing, maintaining one's convictions despite outside pressures, and exhibiting a sense of humor about life's vicissitudes are traits of a mature project manager.

- *Satisfy the client.* The development of an excellent working relationship with the owner should be a goal of the design-build project manager. This arrangement is similar to the role of a traditional design professional, that is, having an ethical responsibility to represent the project fairly and accurately while still being under a paid contractual relationship with the client. The closer the relationship between the design-build team leader and the owner's representative, the better the communications and the easier the resolution of differences. If each party believes that the other party has its best interest at heart, the project concludes with a feeling of "win-win."

7.5   Responding to the Owner's Requirements

Success of the design-build entity is measured by the contributions made by the team toward achieving or surpassing the owner's objectives. Depending on how the project is procured (purchased) by the owner, the task of defining those project objectives may be the responsibility of the design-builder, a third-party consultant, or direct employees of the owner.

Defining the objectives and the scope of a project are not simple tasks; yet there needs to be a brief and clear statement summarizing the overall objective. Fundamentally, there is a need or problem on the part of the owner or user due to some lack of facility or structure. The problem can be delineated, described, or specified through a facility program or engineering study and report.

Regardless of whether it is the design-builder's responsibility to define the project, the owner will expect various submittals or deliverables in response to its solicitation. At minimum, the contract between the owner and its integrated services provider will enumerate the size and nature of the work to be performed. Design-build solicitations may specifically set forth the deliverables to be provided to the owner prior to award of the contract. Deliverables can include the concept drawings, specifications, commentary, models, and supporting data prepared by an offeror in response to an RFP.

When there are two or more design-build teams competing for a contract, the proposers will want to demonstrate that they possess the capacity and experience to successfully create a capital facility that meets the expectations of the owner. Design-build teams may consider some or all of the following steps when responding to an owner's solicitation for services:

Step 1. Understand the objectives, scope, and program of the owner's project. The level of quality for the project is established at the outset by the owner's objectives and requirements. The design-build team must analyze the objectives to determine how the owner has outlined the proportion of scope contrasted with the degree of complexity. Projects may have limited scope and low complexity (e.g., small storage building), or they may have expansive scope with incredible complexity (e.g., a nuclear power plant). A few projects have limited scope and high complexity, or expansive scope with low complexity, but most fall somewhere in between. Together with the budget and schedule (whether given or estimated by the team), the design-build entity will ascertain whether the program is reasonable. The teammates should ask "Can we provide a complete solution (a fully functioning facility) within the owner's budget, time allowance, objectives and expectations?"

Step 2. Confirm the owner's capacity to execute the project. Verifying the owner's ability to finance the project is a sound business practice. For private projects, usually a combination of current cash-on-hand and debt financing is in place to pay for design and construction costs. Public projects, due to sunshine laws, are more easily verified; however, it is often important to have project funds appropriated early, rather than waiting for budget cycles. Design-builders should check out references with service providers who have done work with the owner. References can reveal how representatives of the owner may communicate, react to problems, and facilitate project execution. In brief, the team must determine if the owner is a potential client with whom they would want to do business.

Step 3. Explore the project site. The design-build team should visit the project site to verify access points, terrain, site cover, and environmentally sensitive areas. If soil borings are not provided by the owner, site investigation is probably necessary for excavation planning and foundation design. Discussions with the local government will provide zoning information, as well as instructions on required permits (zoning, grading, environmental protection, building, occupancy, etc.). Local governmental agencies may also have applications for professional and business licenses and valuable data concerning the labor pool (availability, skills, average earnings, etc.).

Step 4. Plan to prepare deliverables in accordance with the owner's guidelines. Most owners have specific expectations about the proposal deliverables. To avoid wasting the owner's (and the team's) time, create checklists of required submittals, criteria, and timelines. For busy owners, less is probably more; however, all forms and supporting documents (conceptual design; outline specifications) should be complete without adding unnecessary volume. The process of design-build includes designer-of-record services; hence, understanding and clarifications are critical to meeting the owner's needs. Therefore, attend preproposal conferences and ask questions (in writing if necessary) to clear up any doubts about project objectives or scope.

Step 5. Work with the design build team toward a common vision. Given the owner's expectations and priorities, obtain consensus from the team about the needed solutions. By analyzing the criteria, the design-build entity can demonstrate its unique capacity to satisfy the owner's concerns. Owner selection criteria on most design-build projects will include some combination of aesthetic image (or engineering solution), value (in terms of level of quality to cost), time, efficiency, access, and additional amenities that are project-specific.

Step 6. Organize the appropriate team and select the best team leader. Experienced and long-standing design-build teams may or may not have the specialized consulting and construction experience necessary for performing work on the owner's new project. Additional teammates should be added to the entity who possess experience with the facility type or systems/assemblies that are expressly suited to the final product. Technical strengths are essential to the success of the design-build team, but the team leader should be chosen on the ability to manage relationships and motivate personnel.

Step 7. Dissect the project criteria and assign tasks for preparing the deliverables. A full understanding of the program and evaluation criteria should focus the organized team on what is required to attain the award of the contract. A checklist of required submittals along with technical work behind the scenes necessary to create the submittals will help distribute individual tasks among the teammates. The checklist will include but not be limited to

- Chart or matrix showing evaluation criteria and their weights
- Documentation that shows the design-build team's financial capacity
- Listing of the team's project experiences and references
- Projections of preliminary schedule and milestones for the project
- Budgets and estimates, identifying the cost evolution
- Drawings and specifications that are needed both prior to and after award
- Other deliverables such as a model or special engineering (e.g., structural, mechanical, or electrical) calculations

Step 8. Analyze the project program. A thorough evaluation of the project program by the design-build team will uncover the owner's facility needs and design intent. Elements of the project scope can then be quantified and qualified. For building projects, the program will include an enumeration of functional areas, plus required adjacencies or proximities of spaces. The program will also state the environmental expectations for the project, both within the structure(s) and on the facility site. Using all items listed in the program, the design-build team should create a running tally of the inclusions and exclusions, keeping in mind that the owner has chosen design-build delivery to arrive at a fully functioning facility.

Step 9. Establish a project schedule. Speed of delivery is a priority for many project owners, especially those in the private sector. During the response-to-solicitation phase, the team should lay out the expected sequencing and expected durations to ferret out any special problems or opportunities. Critical long-lead items can be itemized, and a buyout sequence for complex elements or equipment can be added to the preliminary project schedule. The design-build team can now determine, on the basis of the scheduling process, whether the expectations for delivery time are conservative or aggressive.

Step 10. Determine the budget and elemental estimates. A general understanding of the owner's project budget is usually apparent to the design-build team. Knowledge of similar project's and quick comparison estimates will verify whether the overall budget is realistic. More reliable budget estimates should be developed by the design-build team using facility systems; by working from an ASTM UniFormat or similar breakdown. The team can determine the scope and quality of each system, and then trade off costs among the systems to bring the project within cost expectations. Design-build also encompasses design fees for various systems; and the process may be inclusive of finance, warranty or life cycle expenses, all of which have to be added to preliminary estimates.

Step 11. Develop conceptual design-engineering solution. With the project criteria listing and project program analysis in hand, the team is prepared to develop the conceptual design in accordance with the budget and elemental estimates. Regular reviews of the design should be conducted by the team for the purposes of evolving the project estimate and gaining consensus on the proposed solution. As the concept design matures, team members must confirm that the scheme meets the scope, quality, cost, and schedule goals for the project. Input by major subcontractors is essential during the process, because those key teammates often determine the success of the entire entity. The team leader (relationships manager) should remind all involved about the bases for the award: What combination of best design, lowest cost, quickest delivery, most efficiency or other significant goal is the owner seeking?

Step 12. Assemble the proposal and submit deliverables to the owner. The design-build team should now be in the position of assembling the project proposal documents according to step 4, which recommends having a checklist of required deliverables. The proposed solution should be tempered by an exhibit listing what is not included. This list of exclusions helps manage expectations (for both the owner and the design-build team members) and it is a good check for completeness. In the final response package, include an item or two that helps the team stand out. For example, provide an unsolicited alternate that helps provide an additional value to the project at no additional cost to the owner; show precedent-breaking innovation in design, process, or materials; use the owner's logo within the proposal package, and include the owner's logo on signage or in a design element within the project. In addition to a quality, bound proposal submission, the team may have an opportunity for an interview with the owner. When meeting with the customer, stress the team's common vision and have thoughtful answers for the team's solutions to the owner's priorities. At the conclusion of the solicitation process, most owners will gravitate to the "best value" solution for their project.

7.6   Executing the Design-Build Contract

Design and construction of a concrete office building in San Diego requires different technical knowledge and skills than, say, a steel bridge across a wide estuary in Norfolk, Virginia. But the costing, scheduling, management, and control tools and techniques are remarkably similar. In design-build project management, business, organizational, and employee systems enable the team to execute the contract.

After establishing an organizational strategy at the project level, the design-build project leader can work from a general systems breakdown of the contract to refine a set of milestones and tasks. (The major design-build project phases are illustrated in flowchart format in Fig. 7.1.) Listing of the tasks and activities, plus the dependencies between the activities, permits the development of a model of project execution. Although this book does not provide a detailed description of CPM (critical path method) network development, the benefits gained by answering how each activity fits within the logic of an overall project chain are considerable. The CPM diagram is a graphic representation of the project wherein activities are arranged so that constraints are taken into account, time is plotted, and spare time (float) is identified. Constraints may include resources, workforce, safety, technology, site or physical form, and other issues surrounding the project.

Planning of the project through definition of a series of tasks and relationships helps the team better understand its roles and challenges. By discussing tasks in terms of their complexity, effort, and duration, the team members can debate alternative approaches and can single out bottlenecks or areas of flexibility.

A fundamental competency of a design-build project manager is the ability to solve resource allocation issues. The project plan is often an idealized schedule or baseline that does not include resource constraints that inevitably occur over the life of the project. In practice, limited resources must not be overcommitted. For example, an experienced concrete slip-forming crew cannot work on two building sites in different parts of the state at the same time. And a single tunnel-boring machine cannot start work on separate faces of neighboring mountains on the same day. The baseline project structure will have to be adjusted by moving tasks to other time periods, adding more resources, delaying tasks to other time periods, delaying tasks that are concurrent, substituting one important task for another in the same time period, or using some other resource.

Project planning marks the start of many design-build contracts. (Note, however, that many full-service design-build teams begin their project involvement at project inception, often providing feasibility and programming services for owners.) After planning, which can be accomplished by the owner and its consultants or the design-builder, typical design and engineering activities undertaken by the design-build entity during the proposal stage and after award of the contract include

- Conceptual design and design development with responsibilities and delivery dates
- Performance and detailed specifications
- Design packages for permitting
- Long-lead equipment and materials lists
- Design packages for specialty and subcontractors

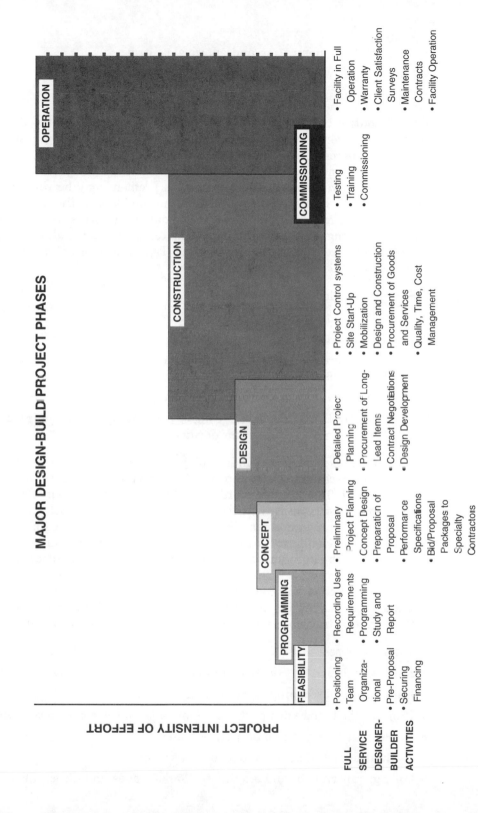

Figure 7.1 Project tools.

- Submittal administration and approval control
- Construction documentation; field design assistance
- Commissioning, testing, and systems start-up

Project control techniques allow the team to keep on top of the design-build contract. Systems must be in place that will allow ongoing cost analyses, schedule progress reports and forecasting, cash flow planning, and change-order management. Issues and problems show up on projects like unannounced relatives arriving for a visit. The problem may be resolved quickly and happily, or it may stay around for the remainder of the project. Regular project control monitoring, using both formal and informal indicators, should aid in conveying project health signs. The team can make adjustments to the project to ensure stability and continuity. Typical design-build project control activities include

- Schedule modeling and updating
- Budget and cost control systems development
- Creation and use of cost versus schedule forecasting system
- Regular progress reporting including "earned value" to determine overruns and underruns
- Using a system to monitor and track changes
- Having a system of document control (contracts and subcontracts, project graphics, submittals, and other documenting data)

If design represents the words in a play, then construction is the spoken language and the cast, the sets, and the choreography, all successfully staged on opening night. The director coordinates the disparate company, coaching the bit players to perform at the best of their abilities and smoothing the ruffled feathers of the prima donnas. The days are long and the tasks numerous, but there is no production unless the words are brought to life, unless the design is fulfilled in three dimensions. Completed construction embodies successful staging of the owner's goals, the designer's interpretations, and the project manager's direction and coordination.

Construction activities in design-build range from critics' reviews to curtain calls; from constructability reviews during design to commissioning, including

- Value analyses
- Review of design for constructability
- Definition and scooping of work packages
- Project scheduling with particular emphasis on the construction phase
- Project start-up and site preparation and layout
- Evaluation, negotiation, and award of subcontracts
- Hiring, supervision, and administration of trades and direct labor

- Acquisition and timely delivery of construction materials and equipment
- Planning, coordination, and monitoring of construction
- Quality assurance/quality control of in-place work
- Site safety and security
- Change negotiation and administration
- Measurement of progress and authorization of payments
- System and equipment testing and start-up
- Facility completion, acceptance, and turnover

Executing a design-build contract differs from traditional design-bid-build because the detailed project scope is not clearly defined when price and schedule are fixed. Impacts of any design errors and omissions will be absorbed by the design-builder, and in-place quality and work quality also remain the responsibility of the entity. By combining design and construction under a single contract, the project itself becomes the focus of a temporary utilization of a set of resources to achieve a specified purpose. A team concept is the most effective means of realizing multiple project objectives. Flexibility in adjusting to the client's goals, contract terms and conditions, financial approach, project geographic location, schedule, and environmental issues also contribute to project success.

Nevertheless, some projects fail, not as the result of the project delivery system, but because of the misapplication of concepts, poor use of skills and techniques, and nonproductive relationships. Among the many pitfalls of design-build contract execution, the major causes for failure include

1. *Inability to grasp the designer-builder culture.* Owners have gravitated to design-build because it helps focus the parties on a combined set of objectives and extricates the owner from its traditional role as arbiter between designer and constructor in the event of any disagreements. Some traditional hard-bid general contractors have experienced difficulty in transitioning to team-based relationships, where the constructor (regardless of whether they are in the lead) must also be a team player, willing to lead, follow, or pull oars together as the situation merits. In fairness, many of those same challenges exist for independent design consultants trying to overcome cultural biases and preconceived notions.

2. *Shoehorning inappropriate procurement and purchasing methods with specific design-build variations.* Those who are unfamiliar with the differences between project delivery and procurement, or who are unable to match a design-build variation with a workable purchasing scheme, tend to chip away at the potential for project success. For example, the insistence on low bid procurement on a design criteria design-build RFP is questionable. The design-build entity will, in this instance, be regarded as a vendor or commodities supplier, and will have little incentive to provide innovative solutions.

3. *Having owners or upper management who are not fully supportive of the delivery system.* Support of the process and decisions made within the process must be provided by both the owner and the business entities that make up

the design-build team. Lack of commitment and sense of mission will undermine the project team leader.

4. *Inadequately defined tasks and responsibilities.* A well-thought-out organizational format plus delegation of sufficient authority lay the groundwork for efficient execution of the work. However, the project strategy should also foster relationships among key team members and the client to aid in timely decisionmaking and problem solving.

5. *Potential misuse of management techniques.* Technical systems, data, and reporting are essential to the professional management of complex projects, but overreliance on technical detail without the balance of professional judgment, business oversight or macromanagement, and maintenance of key relationships can contribute to project failure.

6. *Lack of communication up, down, and sideways.* Balancing reporting relationships is a difficult task within a design-build project. Communication with line managers, interaction with general management and the client, and resolution of issues and problems on behalf of the entire project team demands concise information flow. More design and construction managers lose their positions because of communications problems than for any other reason.

7. *Incapacity to adjust to changing demands.* The fluidity of project issues requires continual adjustments to accommodate problems in resource availability and productivity. Development of alternative courses of action and having the strength to implement tough decisions in the face of serious problems can avert project failure.

8. *Insufficient focus on project goals and completion.* Every project must have a long-term perspective. The goal of the project is not limited to the design process or the construction process, but is concentrated on producing a completed facility ready to serve the owner or user. Planning for completion by stressing overall performance, cost, and schedule is healthy for the project; obsession with management tools rather than the application and use of information supported by those tools could have adverse consequences for all concerned.

7.7   Key Design-Build Project Management Issues

7.7.1   Predesign services

The bundle of predesign services for design-build projects can encompass a variety of preproject planning activities that normally occur subsequent to the owner's business planning effort. The Construction Industry Institute has published guidelines for major planning processes, which if performed early in the project's life, positively affect outcomes during project execution (and facility operation) at the lowest possible investment.[1] Loosely summarized, the preproject and predesign services for design-build delivery would consist of three parts:

Organize for front-end planning and feasibility determination. One of the most interesting characteristics of integrated services delivery is the

---

[1]A. N. Tortora, *Project Success and Pre-Project Planning Effort,* Univ. Texas, Austin, 1993.

opportunity for either a preproject planning team or an integrated design-build services provider to develop project objectives. The predesign team should consist of skilled and experienced members who can provide inputs based on business objectives, technical applications, project management best practices, and facility operations.

Develop project approach alternatives. The team should consider existing and emerging technologies to find opportunities for innovation. Some approaches will use available resources more efficiently than others, or will be more appropriate for the geographic region or site. Given the project objectives, the team can evaluate combinations of alternatives that fulfill the owner's objectives. Alternative approaches can be ranked according to their sustainability, value, or advantage to discrete business objectives.

Create design criteria and a performance definition package. This third step in the series of predesign services can, as mentioned earlier, be accomplished by the owner's team, an independent third-party multidisciplinary team, or a design-build team. At this stage, the team develops the design criteria and/or conceptual design to a point that clearly takes into account the commercial and technical risks. In conjunction with the design criteria or concept design, a management plan is crafted documenting the project execution approach, including the methods of performing detailed design, procurement, validation, construction, start-up, and other issues. Prior to project design, the team also establishes project control guidelines, such as milestone CPM schedules, procurement schedules, safety guidelines, and a control plan that can track and compare planning, scheduling, changes, and management information systems.

Additional preproject or predesign services that have not been mentioned but are available from multidisciplinary A/E or integrated services firms include financial feasibility studies, zoning analysis, site or soil explorations, public hearings, environmental assessments and permitting, user requirements surveys, programming, utilization studies, and facility and/or product life-cycle studies.

7.7.2 Planning and estimating

A comprehensive design-build project management model consists of a number of systems that are put into place to facilitate the execution of the contract. The planning system consists of all the plans, strategies, goals, and schedules. Planning allocates scarce resources. The information management system deals with communication and retrieval of information, which provides intelligence for decisionmaking. The project control system gathers and disseminates data and information on cost schedules and technical performance.[2]

Calculating planned costs consists of a summation of individual work items that is based on assigning resources to individual tasks. Earlier in this

---

[2]H. N. Ahuja, S. P. Dozzi, and S. M. Abourik, *Project Management, Techniques in Planning and Controlling Construction Projects,* John Wiley, New York, 1994.

chapter, a brief overview of CPM scheduling discussed the sequencing of activities. To determine the cost of the project plan from the schedule, the following steps are recommended:

Determine a baseline for efficient project duration. Project planning involves placing inputs through filters to separate what is physically and economically feasible from that which is not. From what the team deems to be a feasible project template, more detailed tasks, resource assignments, and activity durations are formatted.

Calculate direct costs for the resources identified in the baseline. An entire chapter in this book is devoted to cost estimating for design-build projects; however, it is useful to keep in mind that an estimate is an approximation of the cost of the project based on the information available at a point in time. The budget evolves from a feasibility estimate and transitions through design criteria, concept design, and design development stages. Estimating methods tend to track the evolution of the project budget, starting with facility type and cost indices and progressing to elemental and parametric systems and ultimately to detailed component, task, and productivity costing.

Factor in management and capital costs. After the direct costs of material, labor, and equipment have been satisfactorily assembled, the project team will account for all indirect costs (note that owner estimates early in evolution of the project are inclusive of these costs). Classic indirect costs for construction encompass overhead; payroll burden; management and supervision, bonds, taxes, and insurances; smaller tools; site utilities and rentals; and other construction costs. Added to direct and indirect costs are costs of design (some consider these expenses a direct cost within a design-build firm), owner allowances, contingencies, and design-builder profit. Concerns over overall project economics have helped drive the increase in design-build delivery. The design-build organization, then, may also need to consider the cost of financing, permits and licenses, site acquisition, and public outreach during the early phases. At start-up, significant costs may also be associated with commissioning, including systems operation, training, equipment debugging or adjustment, and inventory stocking.

One of America's largest design-build firms has devised the following system for estimating costs for commercial projects such as a hotel building:

First, identify major cost elements:
- Land costs
- Off-site costs
- A/E costs
- Financing and interest costs
- Legal costs
- Tap and development fees

- Data and communication costs
- Fixtures, furniture, and equipment costs
- Taxes
- Other subgroup costs

Second, quantify the project:

- Name the major project elements.
- Associate volumes for all of the spaces.
- Quantify the spaces, rooms, beds, modules, etc.
- Enumerate site activities and quantities.
- Cite major building systems.
- Identify HVAC and electrical needs and solutions.
- Determine exterior skin (wall and roof systems).
- Zero in on the level of finishes (primarily interiors).
- Review level of quality for each area and trade off if necessary.

Third, apply estimating methods to the facility systems breakdowns:

- Employ a parameter estimating format.
- Communicate initial costs to the team.
- Design to the owner's program and team's budget.
- Meet all requirements of project objectives and performance goals.
- Track the evolution of design, budget, and any changes initiated by owner communications and discussions.
- At the appropriate point, convert the estimate from a systems basis to a materials basis (generally, this requires using CSI 16-division to communicate with trades and suppliers).
- Transition the costed work breakdown structure to a cost control mechanism.

7.7.3 Planning and time management

The attractiveness of planning a project using systems-based organizational structures such as UniFormat becomes apparent with design-build. Tasks and activities are part of larger tasks and activities; components are part of major facility elements. Starting with macrosystems, one looks from a mile-high view at a project, and then cascades downward through levels of strata until reaching a density of data necessary for project control.

There are a number of excellent computer software programs available for producing CPM networks and schedules. Most can produce bar-chart reports from the database for use with owner representatives, lenders, and others who are not familiar with CPM. By working with those closest to the execution of an activity (with individuals such as the design project manager, the foundation foreman, key subcontractors), the initial schedule is likely to come closer to realistic project duration. The team can provide input and build consensus for the components of the schedule by collectively answering the following questions:

- What activities must be completed before this activity can begin?
- What tasks must be carried out at the same time as this activity?
- What activities are dependent on the completion of these tasks?

In lieu of (or in addition to) a project plan based on facility systems, a case can be made for constructing a plan around innovative methods of construction. For example, if a multifloor office building is being constructed out of a combination of slip-formed and posttensioned modules, the plan could be organized on a floor-by-floor basis.

Owners are usually far more interested in having the project reach various milestones than in the performance of individual activities and tasks. Yet on-time completion is an ongoing challenge for the design-build team. Usually time or schedule management problems can be traced to ill-defined project objectives, inadequate quality management during design or construction execution, or unrealistic estimating (wherein the figuring of how many labor-hours that it takes to perform a task is one of the most volatile variables in design and construction). The goal of "completion on time" is more readily attainable through the integration of cost, quality, schedule, management, risk control, and collaboration, which are characteristics of a well-conceived and competently delivered design-build project.

7.7.4 Cost management

Among the critical skills required for design-build project management, perhaps the most important are (1) planning and scheduling and (2) estimating and costing. Without these two components, meaningful project control is impossible.

Design-build project management is somewhat different from traditional design-bid-build with respect to controlling schedules and costs. Detailed project scope is not fully defined when price and schedule are fixed (usually the act of signing a design-build contract with the owner occurs while the project is still in the performance requirements or conceptual design stage). At this point in the project development, subcontractors and suppliers do not have final quantities and specifications to provide firm pricing.

It is also likely that the permitting process is only in its early stages, and final design may have to be modified to accommodate jurisdictional requirements. Regardless, the design-builder must begin procuring long-lead items early to meet the schedule. Design is often scheduled to meet a "work package" plan that may be dictated by procurement and construction needs.

The basic techniques for building a cost management system is to create a work breakdown structure (WBS) that breaks down components of the work in ways that are familiar to the team (some prefer systems-based; others use

CSI's MasterFormat; others want both). The work breakdown structure is the basis for the ongoing project cost breakdown system. A technique for marrying the owner's and design-build entity's need for functional element costing and the subcontractors' or suppliers' needs for materials and trades-oriented costing is to earmark those work items that may be assigned to the various functional elements. With this coding method, cost breakdowns can be separated or merged as necessary to satisfy the individual tracking needs of the owner and the project team members.

On many design-build contracts, the project budget is the "as sold" or "as conducted" price broken down in the work breakdown structure. The structure is coded to identify all significant scope items and work activities in a system that incorporates expected versus actual costs, hours, duration, time, quantities, and other variables. The smallest unit for which a budget may be established (such as for a single task or activity) are cost centers at or close to the "front line" of the project. Cost centers, implying responsibility for expenses, tend to motivate personnel by committing them to measurable project objectives. Summaries of cost centers can be aggrandized at increasing levels to show performance by income statements (revenues and expenditures) on various portions of the contract. Naturally, at the highest level, the income statements show entire project performance, or total firm performance, for a given period of time.

A good cost budgeting system is built on the framework of cost coding. Having a single and inviolable coding system not only reduces confusion but also facilitates its adaptation to computer processing. Properly constituted, a cost code can signify, in a single digit or letter, the difference between a labor, material, equipment, or other job expense item. Using an alphanumeric system, any type of cost can be stipulated, such as direct, indirect, financing, commissioning, and maintenance.

Another issue in cost management is the forecasting and tracking of cash inflows and outflows. Every design-build contractor relies on progress payments (or alternatively, a turn-key lump-sum payment) as inflow, and experiences cash outflows to designers, subcontractors, suppliers, insurance and surety professionals, and others. The design-builder would like to have positive net cash flow throughout the project, but such in-the-black financial standing is often difficult to attain early in the project. Owners, on the other hand, have different motivations toward cash flow. Owners usually want to defer paying the integrated services provider as long as it does not constrict the performance of the design-builder. Understanding the owner's psychology and taking a cold-hearted view toward the time value of money will help the design-build team anticipate any necessary borrowing expenses during the term of the contract.

7.7.5 Design management

Reverence for good design is a philosophy of quality design-builders. From quality design springs facility functionality and longevity. The role of design management in design-build lays the groundwork for reliable facility operations and customer satisfaction.

The definition of competent design management is the ability to provide concrete direction to the design team without foiling their creativity. Excellent design can result despite the severity of this limitation; indeed, an expanded recognition of successful design would include design to cost, innovative use of inexpensive materials, new design systems, and other forms of design innovation, while still embracing aesthetic attainment as a high art form.

The role of the design manager on a design-build project is fulfilled in three distinct areas: (1) the design manager develops graphic concepts that respond to the owner's goals and limitations; (2) once the concept has been accepted as the preferred option, the design manager leads design development, generation of work packages, and preparation of construction documents; and (3) during construction (which almost always overlaps the second role), the design manager provides solutions to project design issues (by tracking problems or changes, issuing clarification sketches or drawings, and maintaining and communicating up-to-date record documents.

During the creation of a design-build proposal, the design manager may lead a design-build charette, which consists of an intense activity to complete an architectural design effort within a tight schedule (often within 24 to 48 h). The term *charette* originated from the use of a cart that students used to rush drawings to the Ecole de Beaux Arts in nineteenth-century Paris. A design-build charette relies on a multidisciplinary team using their imagination to develop a realistic and creative scheme within a crucible of time and funding. The design manager (who may also serve as the overall project manager) is the likely designee for presenting a summation of the charette results to the owner.

As the project moves into design development and construction, design management can employ a number of good practice tools to assist the process:

- *Computed-assisted design* (CAD) *coordination.* Place all team members on the same CAD, scheduling, and other project software systems.

- *Design schedule.* Specify detailed schedules for each discipline to control the procurement of construction-ready documents. This technique is especially important for work activities that begin early because of the fast-track process.

- *Action list.* Have an electronic format for listing all design issues to be tracked and resolved. The list should be sorted by discipline, trade, facility system, date, and whether the item is resolved or unresolved.

- *Design review comments.* A method for reviewing and responding to comments made during design review sessions should be created. All comments requiring clarification and/or costing should be followed up immediately to avoid any impact to the schedule for either design or construction completion.

- *Design binder.* A compendium for facility systems information is recommended; it should include CSI specification division 1-16 data, owner requests, permit requirements, meeting agendas and minutes, design schedules, action lists, concept sketches, and clarification sketches. Some of the newer Web-based programs are useful in creating fully accessible project binder information.

7.7.6 Subcontract management

An interesting evolution in design and construction has been the growing recognition of the enormous contributions that major subcontractors provide to the overall success of projects. When specialty and subcontractor services are procured on a basis of best value or qualifications, the prime [whether a general contractor (GC), construction manager (CM), or design-builder] is communicating that the relationship is based on more than simply a commodities basis. Furthermore, the specialty or subcontractor may be furnishing design services (or early design input) that are critical to the overall success of the project. The net result, from an organizational standpoint, is the creation of a design-build entity that has a relatively flat (team-based) management hierarchy rather than a traditional vertical hierarchy that places designers and prime/general contractors at the top and subcontractors/suppliers at the bottom. (See, e.g., the organizational chart in Fig. 7.2.)

A team working together on a design-build project should have common aspirations and goals. The team members recognize that their tasks and activities must be executed in a timely manner in order for other team members to complete their tasks and activities. This interdependence makes it mandatory for the teammates to be able to identify and solve problems together. It is nearly always to the prime design-builder's advantage to use proven performers and to avoid people or firms that have a "hard bid" mentality. Problem-solving skills, initiative, energy, and the ability to integrate are the valued attributes.

The role of mechanical, electrical, and plumbing (MEP) subcontractors on building projects is significant, often accounting for 20 to 40 percent of the contract costs. When engaged on the team at the outset of the project, these team members can help the design-build entity organize parts of the project and manage a portion of the risks. Subcontract work packages that are critical to schedule include MEP, fire protection, sitework, foundation, structural frame, elevators and escalators, and building envelope (exterior walls and roof). A knowledgeable design-build project manager will scout for those subcontractors who are willing to jump in early, provide feedback regularly, who stress quality design and work, and who are thinking about repeat work and long-term relationships.

Development of subcontract work packages (also called "bid" or "proposal" packages) helps the design-build project manager define issues for major subcontractors. The subcontract packages should

- Define the scope and limits of the work.
- Include performance and quality criteria established for the project.

SAMPLE DESIGN-BUILD PROJECT ORGANIZATION
FOR A CIVIL INFRASTRUCTURE PROJECT

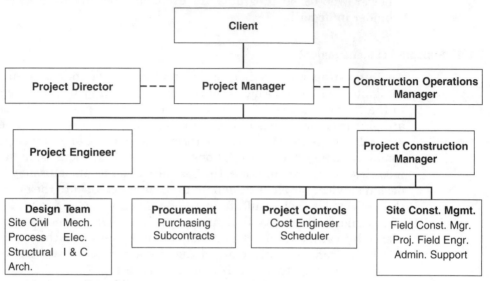

Figure 7.2 Design-build organizational chart.

- Clarify budget control and design responsibilities.
- Define and require acceptance of the project schedule.
- Address bonding, sales tax, and other indirects.
- Leave the door open to innovations and value engineering ideas.

A design-build entity realizes benefits from subcontracting portions of the project. Design and construction work is done by people who specialize in specific systems, equipment, or materials, often improving the efficiency of the project and saving money.

The prime design-builder does not have to tie up additional capital for work that is assigned to a subcontractor. However, the design-builder must determine whether the subcontractor will work as a design-build subcontractor, taking design responsibility as well as construction responsibility as a key participant in the team; or whether the subcontractor is in a traditional construction-only role, relying on the plans and specifications of the design-builder. The ability to vertically integrate the design-build process with subcontractors and manufacturers will distinguish future design-build teams.

7.7.7 Changes and trend management

The management of changes is a significant part of every design and construction project. During the project's early phases, which may include feasibility, programming, and criteria development, wholesale changes may be made in material selection (e.g., steel vs. concrete; precast vs. aluminum cladding) or in approach (e.g., tunnel boring vs. cut-and-cover; cast-in-place vs. prestressed or posttensioned) that cause substantial variation in cost of delivery time.

In later stages of the project, changes may be necessary because of an unforeseen condition or a design omission, or the change may be discretionary on the part of an owner adding or deleting any time from the original contract. How changes are administered depends on the terms of the legal agreement: in lump-sum contracts, the change usually results in a request for additional time and money to complete the work; in guaranteed maximum price (GMP) or cost-plus fee contracts, the change may be processed routinely as long as the parties are in general agreement and the project is projected to remain within the cap or budget.

Change orders emanate from a number of sources. Owner scope changes can be instituted as adjustments are made to facilities' performance or capacity needs. An owner may want to adjust quality standards higher or lower depending on budget tradeoffs or expected returns on investment. In other instances, owners alter scope because of equipment preferences of the operations staff, or make changes in the project as a result of local concerns or community influence.

It is during the design development phase that the design-build project manager must guard vigilantly against scope creep. Changes in site conditions, alterations in quantities of work beyond that which was originally scoped, issuance of new industry standards, and designer enhancements that are added after the contract cost is fixed can run roughshod over the original budget. Other changes may be necessary because of building code requirements, permitting, right-of-way or easements, new sales tax rules, weather problems, or other "acts of God." Financial issues may also cause pressures on the project budget, such as interest rates, currency exchange, and outstanding or poor economic conditions.

To manage a change on a design-build project, some teams have instituted a "trend program" as a process to identify, evaluate, manage, and resolve changes that occur during the design, procurement, and construction of the project. The two trend baselines consist of the project budget and the project schedule. A trend program usually covers a longer period of the project, from early design and procurement continuing through preoperational testing and start-up. Trend programs are geared to cover all changes with emphasis on timeliness and immediate knowledge of trends rather than on scientific accuracy.

Implementing a trend program requires the development of a trend register, which may consist of a form that lists all trends, monitors their status, and summarizes their evaluation (see example in Fig. 7.3). Each trend issue also should have its own notice form for recording the trend definition, data related to the trend, team evaluation, and disposition of the item. The responsibility for maintaining the trend program should not necessarily reside with the project manager; however, the person should be knowledgeable and experienced; respected by peers; familiar with scope, cost, and schedule aspects of the project; and able to interact with or lead the design process.

Trends that are typically flagged as the responsibility of the project team are

- Design development changes
- Missed items that were in the scope but left out of the proposal
- Work quantity adjustments
- Price escalation of materials, labor, or equipment
- Labor and equipment productivity

Notice of Trend

Project No._____   Project Name:_____

Initiator Name:_____   Discipline / Group:_____   Location:_____
Trend Title: _____   Phone: _____
Description/Scope:_____   Effect on Other Disciplines _____
_____   _____
_____   _____
_____   _____
Reason:_____   Effect on Cost _____
_____   _____

Cause: Engineering Development ☐   Regulatory/Tech. Code ☐   _____
        Client Preference ☐   Construction Site Requirement ☐
                                     Supplier / Supplier Requirement ☐   Effect on Schedule _____
                                                                        _____
        Other_____   _____

Send to: Trend #_____   Client Issue #_____
Trend Engineer_____ Date Rec'd:_____

	Staff Hours		Estimated Cost Impact ($000)
Group	To Evaluate	To Execute	Direct: _____
Engr./Design	_____	_____	Engr./Indirect:_____
Procure.	_____	_____	OH &P _____
Const.	_____	_____	Total _____
Est./Sched.	_____	_____	
PM	_____	_____	Estimated Sched. Impact
Other	_____	_____	_____ wks (+,-)
Total	_____	_____	Critical Path _____ (Y,N)

Disposition   Project   Client   Reference & Date

Approved: Scope Change   ☐   ☐   _____
        : Other Trend   ☐   ☐   _____
Disapproved: ☐   ☐   _____

Recommended Action: _____
_____
_____
_____

Comments:_____
_____
_____

Trend Engineer, date   Project Controls Specialist, date   Project Manager, date   Client, date (as required)

_____   _____   _____   _____

Source: Chuck Williams, Senior Vice President and COO, HDR, Inc.

Figure 7.3 Trend forms.

TREND REGISTER

Project No. _____   Name: _____   Prepared By:

Trend No.	Client Issue No.	Date Ident.	Key Word/Description	Orig. By	Cost Impact	Schedule Impact (wks)	Scope Change PCN # To Client	Status Process	Status Decision DB Team	Status Decision Client	Next Step

SCHEDULE IMPACT
C = Critical (Weeks)
N = Noncritical (Weeks)

PROCESS STATUS
R1   Resolved w/ D/B Team & Client
R2   Resolved w/ D/B Team; Unresolved w/ Client
R3   Resolved w/ D/B Team; Client not applicable
UR   Unresolved w/ D/B Team and Client
D   Unresolved; Under Development/Evaluation

Source: Chuck Williams, Senior Vice President and COO, HDR, Inc.

Figure 7.3 *(Continued)*

- Transportation and storage costs

Trends that are reviewed with the owner are

- Client-driven scope changes or delays
- Changes in site conditions
- Permitting process requirements
- Force majeure events
- Industry standards changes or owner-specified items
- Financial, tax, or legal issues

The benefits of a trend program are considerable. The system provides a formal tool for identifying and managing changes around which the design-build team can build synergy. The trend program encourages a proactive approach with the owner to resolve any changes (and mitigate any disputes before they become larger issues). The resulting data incorporate a log of changes

Owner / Client : _____ Prepared By: _____
Project No: _____ Date: _____
Project Title: _____ Page: _____ _____

SUMMARY TREND REPORT- Period Ending: _____

Trends
to Date
(2a & 2b)

($000)

500
400
300
200
100
0
-100
-200
-300
-400

Months Since Date of Current Trend Base

1. Current Trend Base (date and basis reference)

2. Trends to Date Approved by Project Team:
   a) Submitted to Client:
      - Client Approved
      - Pending w/ Client
   b) Other Approved Trends:
3. TOTAL (1+2) Trend Base Plus Approved Trends to Date:

4. Other Submitted Pending & Unresolved Trends

5. CLIENT CONTRACT AMOUNT
Original: $_____

Changes: $_____
Current Total: $

All ($000)   Sched. (wks)

_____

_____

_____

_____

_____

6. Notes / Comments:

Source: Chuck Williams, Senior Vice President and COO, HDR, Inc.

Figure 7.3 (*Continued*)

history, both internally to the team and externally with the owner, and provides a cross-check for cost and schedule forecasting. Finally, the trend program improves risk management for the project team and the client, with the structure allowing maximum time for open discussions, corrective action, or negotiation over changes.

7.7.8 Project safety management

Design-build entities assume prime responsibility for both design and construction, and are clearly responsible for project safety. This risk assumption stance differs from traditional design-bid-build, in which architects and engineers have maintained that they do not control the building site (and therefore do not have responsibility for site safety). The general contractor, as the employer of the trades and director of the site means and methods, accepts responsibility for the physical safety of persons and property in the traditional building process.

When the design-build entity is an integrated firm or a construction-led team, the legal and business relationships fit traditional expectations and underwriting tests. However, when the design-build entity is designer-led or fee-based and developer-led, coverage under general liability and builder's risk policies may contain additional limits or exclusions based on the organization's commitment to project safety.

Insurance advisors recommend that the insureds tell employees what safety measures are required and then hold the employees responsible for measurable results. Design-build firms should maintain a safety and health procedures manual that describes methods and identifies safety goals for industry-accepted safety practices and regulatory requirements. A project safety action plan reduces the potential for injuries and unwanted losses by calling out potential hazards or high-risk operations.[3] Project managers and job superintendents can devise strategies for eliminating or controlling hazards which should be documented and communicated to the design-build team.

Although the number of construction workers constitute only about 5 percent of the workforce, almost 25 percent of all occupational fatalities occur in construction, primarily because of constantly changing site conditions, untrained workers (or, at least, lack of training in safety procedures), and the sheer number of hazardous factors. The leading causes of injury or death in construction are falls, being struck by falling objects or moving equipment, electrocution, and excavation cave-ins. Crane-related accidents can be very serious on large projects, but small projects have nearly the same incidence rate of lost time because of the previously mentioned hazards plus burns, explosives (chemicals, dust), and lack of ventilation in confined spaces.

The knowledgeable design-build project manager communicates with all others associated with the project, including the job-site workforce, about the goals of the project. In addition to the general messages of "do it right" and "do it well," the project manager must add "do it safely." Strategies to improve safety include

- Developing project safety goals and objectives
- Communicating and implementing the safety action plan
- Conducting employee safety orientation and training on a regular basis
- Recording injury, illness, and accident occurrences

---

[3]W. J. Palmer, J. Maloney, and J. L. Heffron, *Construction Insurance, Bonding and Risk Management*, McGraw-Hill, New York, 1996.

- Imposing subcontractor safety requirements that match the prime's
- Holding work area safety evaluations—by the crews themselves and by senior office management or an outside consultant
- Instituting employee screening (especially for any type of substance abuse that would impair performance)
- Installing incentive programs—rewarding workers for safety awareness and performance

Another key safety component on design-build projects is the influence of design on the performance of structures. Constructability reviews of design can improve site means and methods by uncovering difficult details or resolving material size issues. Design challenges (and errors) may occur in the areas of deep excavations (e.g., problems with soil stability), construction loads may exceed design loads, there may be strength and stability problems with materials and assemblies that are not complete, and alterations and demolition can destabilize existing structures. Involvement of construction professionals in the design process and design professionals in the construction process through design-build portends safer projects by anticipating and mitigating problems and hazards before they occur.

7.7.9 Human resources

Management of people constitutes a fundamental role for the design-build project manager. Each project has phases for which the personnel mix must be organized according to project needs. Generally, these phases may be grouped into a project life-cycle progression of feasibility, design criteria, conceptual design, design development and early construction, final design and construction, and commissioning or project turnover. These design-build phases have distinct organizational, management, and decision challenges and involve different professional and functional groupings of employees. Some employees will begin the project cycle in the home office, move to a remote site during construction, and be absorbed back into the home office after substantial completion.

Although there are many approaches to overall project organization, it is good practice to create a job-specific organizational chart showing key supervisory positions and their interrelationships. The organizational plan will allocate personnel between office versus site activities, and it will likely be backed up with a description of responsibilities for each position.

The major roles that must be assigned and accounted for include design management, procurement, finance and administration, supervision of the work, changes to the prime contract, subcontract administration, budget and schedule control, progress reporting, quality control, safety, and document processing (e.g., submittals, invoices, and approvals for payment). Naturally, on smaller or less complex projects, many of the roles can be combined or handled by fewer supervisory personnel.

The field construction management organization is a subset of the overall project organization. The design-build project manager must determine the size of the field workforce, coordinate the acquisition of workforce services throughout the project phases, and structure the project workforce into management sections and work crews. The creation and control of a field workforce is an ongoing management problem for a design-builder, who must react to changing technical demands of the project and meet the operating needs for each step in the construction process.

Confronted with the high turnover of today's employees, and the employees' perceived lack of stability and long-term work prospects with a particular firm, it is important to make the attainment of goals of the design-build entity and the worker mutually positive and rewarding. If members of the team, including site labor, achieve specific goals in schedule completion, quality, and lack of lost time due to accidents or other identified goals, enlightened firms are finding ways to offer incentives for rewarding workers. Progressive firms are also providing training, allowing workers to become multiskilled and to gain new technical proficiency, which enables the company to offer more long-term employment opportunities through external transfers and promotions.

Workers' motivation to carry out their own specific duties is influenced by the work environment, peer pressure from other trades, perceived management attitudes, and the disposition of the worker toward physical labor and skills application. Workers are usually aware of the effectiveness of management's site organization and direction of the work, by observing or participating in rework, discussions regarding coordination of the drawings and details, or lack of tools or materials. Because most site problems result from inadequate planning, or materials or equipment at the site, not from worker attitudes, the design-build project manager may find it more profitable to spend time on organizational efficiency rather than be preoccupied with individual worker behavior.

Nevertheless, work is a personal and intimate pursuit, and it is incumbent on the project manager and superintendent to gauge the attitudes of individual workers. Indicators that a worker is well motivated and responsive to the work environment include positive physical behavior patterns such as prompt arrival on site, prompt response to direction, efficient use and care of tools, constant production of good-quality work, and willingness to give assistance to others. Obvious negative behavior patterns include sullenness; slowness to respond to requests; lax behavior in the care of tools, materials, or approach to work; and unwillingness to initiate new actions or offer suggestions.[4]

7.7.10 Materials and equipment

The editorial trade press has sometimes commented on the inefficiency of the construction industry, especially in regard to its slow adaptation of manufac-

---

[4]Anderson and Woodhead.

turing concepts for materials management. Unorganized purchasing procedures and shipping delays can have negative cost impacts on projects. Early delivery of materials rarely results in accelerating a project because the progress of individual activities controls start times. On the other side of the coin, however, delayed material delivery can stymie the progress of the entire project.

Through pioneering studies by such organizations as the Lean Construction Institute, the design and construction industry is beginning to recognize the enormous potential for increasing productivity through better material management. When indirect work (such as waiting for materials or handling materials) exceeds direct work (installation of materials), on-site productivity is in need of serious attention.

Projects with good materials management systems in place generally have the correct materials arriving at the right place at the right time. But a series of steps are necessary to ensure materials availability:

1. Front-end planning that assigns responsibility among the project team and designates site access, anticipates material shake-out, and schedules workflow will provide organizational dividends.

2. Establishment of a material control system can provide procedures whereby estimates based on performance criteria can evolve into bills of material suitable for ordering. At this stage, the timing of material and equipment flow to the site is also planned—some products will be scheduled to arrive "just in time"; others will be purchased in bulk and inventoried either at the supplier's warehouse or in the design-builder's (or owner's) storage facilities.

3. The materials management effort requires the procurement of commodities, supplies, equipment, and engineered products. To optimize purchases, it is necessary to optimize tradeoffs among categories of costs: the basic original purchase costs, shopping costs, holding costs (storage over time), and shortage costs (not having the stock when the customer wants or needs the items).[5] Despite the claims of some Web-based purchasing sites, the process is far more complex than just filling the shopping cart with the cheapest buys.

4. Another concern about materials management involves site handling of products and equipment. On average, workers spend up to one-quarter of their time unloading, storing, sorting, loading, moving, or hoisting materials and equipment on site. Project managers who devise more efficient solutions for on-site materials handling will help in meeting or exceeding financial performance and schedule goals of the firm.

Expediting is that group of actions undertaken to assure timely delivery of material and equipment, and it is a process that is routinely applied on fast-

---

[5]D. S. Barrie and B. C. Paulson, *Professional Construction Management,* McGraw-Hill, New York, 1978.

track projects. The design-build project manager, using project control reports, will monitor the progress of materials and equipment procurement to check submittal status and manufacturer or supplier actions. The most critical phases of the process are fabrication and delivery, since the greatest amount of time can be lost (or gained) during these phases.

The expediting function serves to keep deliveries on track, and it stimulates solutions when problems are encountered. Corrective actions that are used in expediting the design-build project include

- Reinforcing, with the fabricator or supplier, the criticality of the delivery dates

- Improving the delivery status by bringing the order to the attention of more senior company officials

- Adjusting the schedule slightly to accommodate different delivery times by adjusting other activities to bring the overall project in on time

- Changing to a faster means of shipment

- Asking the fabricator or supplier to work overtime or additional shifts

- Dividing the work among multiple vendors

- Using backup vendors, fabricators, and suppliers

- Changing to an equal or better quality material if it can be delivered on or ahead of schedule

7.8   Project Finance

A design-builder will review a project to determine if it is financially feasible, although in a manner somewhat different from that of the owner. Design-build projects can make significant demands on the contractor's cash on hand, from the initial proposal development costs to covering the gap between project outlays and progress payments (which are often paid 30 to 45 days after the work is in place and are reduced by retainage).

To determine the project's financial viability, a study of the projected inflows and outflows of funds to determine the degree of (typically) negative cash flow will allow the design-builder to plan for use of borrowing until the flows become positive. The design-build entity should add interest on its investment in the cost of the project (i.e., financing of the construction) if they are to realistically determine the best funding scheme. Restructuring the financial aspects of the project so as to permit certain costs to be paid earlier, or to allow earlier funding for capital expenditures such as major equipment purchases, can help the design-builder accrue income faster than it accrues costs.

Most projects are still financed with straightforward owner's equity and commercial bank debt; the growing popularity, however, of combined design and construction contracting has encouraged some practitioners to add finance to the equation (design-build-finance). At first, design-build consortia were providing a significant share of the equity on large design-build-finance projects, but as developers and investors saw the success of the projects in the 1980s and 1990s, the more highly leveraged design-build firms were able to reduce their stake.

A definition of true project finance (as opposed to project financial management as discussed earlier in this section) is an arrangement of debt, equity, and credit enhancement for the design and construction or refinancing of a facility in which lenders base credit appraisals on the projected revenues from the operation of the facility rather than on the general assets or the corporate credit of the developer or owner of the facility.[6] The assets of the facility, including revenue-producing contracts and cash flow, are the collateral for the debt.

Design-build finance opens the door to a new owner and project base wherein the clients, although they may not be cash-rich or well capitalized, have excellent potential for return on investment. Tight design-build contracts and robust financial returns ease lending institutions and institutional investors toward limited-recourse or nonrecourse financing, provided that the risk underwriters and financial engineers are convinced that completion is guaranteed and are 99 percent certain that the project will pay back all funds that have been advanced.

Interestingly, the difficulty in expanding the range of the design-build-finance market is not simply the availability of funding; it is also the inability of financiers and developers to locate projects that meet the normal tests of loan underwriters. A more likely testing ground for small and midsized design-build-finance projects may be through state and municipal public-private partnership ventures.

7.9 Quality Assurance/Quality Control

Quality is not a single absolute, but rather is a variable that is placed higher or lower on an ascending scale based on a mix of project characteristics, such as performance and price, that are established by the owner. Facility managers and plant engineers sometimes use the phrase "fitness for use" to describe the level of quality of a project or system. An element or item that is fit for use will address the user's needs with a relatively reliable product that can be designed, fabricated, and installed without incurring excessive costs. Quality may also be achieved by systems, materials, and work quality that meet or exceed project performance requirements (and, to a lesser degree, on design-build projects, by conformance with prescriptive specifications).

Quality assurance is generally regarded as an overall process for applying accepted procedures and industry standards to verify whether a piece of equipment or an entire structure conforms with or exceeds expected performance.

---

[6]A. Wehner, "Project Management—Financial and Legal Issues," paper presented at Arizona State University Alliance for Construction Excellence Conference, April 2000.

Quality control is a subset of quality assurance wherein steps are taken to establish baseline standards and to measure outcomes that vary from those standards. The design-build project manager (or third-party quality control organization) can initiate actions to correct greater-than-normal variations, which may be caused by fabrication difficulties, lack of worker training, design or field errors or omissions, and improper means and methods.

The design-build delivery method has inherent advantages over noninte-grated approaches in the quest to achieve a quality project. The balance of design and construction disciplines and their focus on common performance goals helps point the way toward improved facility procurement. By contrast, where design firms are unfettered in setting design quality, the A/E may tend toward conservative overdesign, sometimes because the designer is unaware of true construction costs. Where the designer also inspects the work of the constructor, the constructor may have legitimate concerns about fairness issues related to the quality of the design as compared to the quality of the conformance with the design.

There are a number of quality "hooks" that can be built into design-build contracts. The services of a third-party inspection and testing agency can be paid for by the owner or by the design-build entity. The implementation of a quality assurance plan will include measures for quality attainment, and the design-build contract will reinforce the plan with incentive-based (or punitive-based) quality requirements clauses.

Another form of quality verification is available to owners employing design-build delivery—the A/E of record is part of the design-build entity along with the constructor, subcontractors, manufacturers, and other key players. Together, they are responsible for both the quality of design and the quality of conformance. Once the work has been put in place and inspected or tested by the team, the A/E can certify to the owner that it meets the performance requirements set forth in the owner's scope and design criteria. While period-ic third-party inspections may be included to reassure owners and lenders, cer-tification by the design-build team is recognition that the internal workforce is responsible for creating quality—not external fee-based consultants.

7.10   Measuring Performance/Best Practices Issues

Project performance, in an overall sense, may be characterized as the desired outputs being the result of a combination of project inputs, such as expertise and skills, money, materials, and time. Among the primary measures of per-formance for design-build projects are

- *Quality.* Have prime contract responsibilities been achieved? Does the proj-ect meet or exceed industry standards? Are rework or callbacks required for systems or equipment? Were there innovations in the systems or the execu-tion of the project?

- *Schedule.* Have contract milestones been met? Is the facility open on time or ahead of time?

- *Cost.* Is the project in accordance with the projected budget? Are there any shared savings that can be split between the owner and the design-builder?

- *Safety and security.* Were lost-time accident rates low? Were safety measures embraced from top to bottom? Did the project maintain security and protection for the public?

- *Customer satisfaction.* Were the owner's expectations met with professionalism and business ethics? What was the ease of start-up and reliability of the systems? Will the firm or team be considered for the next design-build project?

Measuring performance of the design-build project through a management-led system, such as total quality management (TQM), involves the entire team with metrics for design and design reviews, estimating and cost control, labor costs, productivity, materials and equipment costs, and operations and overhead costs. The system should help identify wasteful efforts at every stage, improving its processes with a constant eye on satisfying the customer. Meaningful improvement comes when each member of the design-build team embraces the quality-oriented customer-focused concept and understands that the goal is not just some lofty, static number, but continuous improvement. The reward, naturally, in this competitive marketplace, is repeat business from the customer.

Constructability is a term that came into use during the 1980s as a process to bring greater efficiency to the design and construction process. According to the Construction Industry Institute, "Constructability is the optimum use of construction knowledge and experience in planning, design, procurement and field operations to achieve overall objectives." The design-build delivery method, which integrates construction expertise into the design process and design sensitivity into construction, has the best potential for moving innovation into practice.

Application of constructability begins at the earliest stages of the project, often during the stages of user requirements analysis and economic studies. The procedures engage the major stakeholders immediately, and sustain their input and involvement throughout the design and construction process. A newer term now coming into vogue for facilities delivery, operation, and maintenance is *sustainability*. Not only will constructability concepts such as procurement and design or construct strategies be considered during the project's early gestation period, but so will the owner's long-term operating objectives, maintainability, reliability, environmental impact or enhancement, and other broad horizon issues.

Successful management of the design-build project depends on the application of "best practices" as developed by the best-in-class industry practitioners. The following practices are loosely based on the key elements enumerated in *Executing the Design-Build Project*, Document 303 of the Design-Build Institute of America.[7]

---

[7]DBIA, *Executing the Design-Build Project*.

- *Engage all stakeholders at the start of the process.* At a minimum, elicit meaningful input from the designer, constructor, and owner at the project outset so that objectives and potential solutions can be considered before designs, costs, and schedules harden.

- *Verify programmatic and technical requirements.* Have a multidirectional dialog with the project performance team, the owner or user, and project consultants such as facility planners or programmers who have taken user needs and translated the data into performance requirements and design criteria.

- *Develop a comprehensive design and build schedule and budget.* The team can think through issues of overlapping activities, phasing, sequencing of construction, and systems costs as the project moves from negotiation to notice to proceed.

- *Establish clear lines of responsibility, authority, and problem resolution procedure.* The project personnel organizational chart plus a step-by-step procedure for resolving problems expeditiously are sound management tools in a team-based environment.

- *Involve both design and construction professionals in design development and content and format of construction documents.* The sum of the disciplinary inputs can be greater that its parts, and the collaborative process allows the team to play to its combined strengths in design and construction.

- *Focus on quality and value as success measures for single-source responsibility.* The design-builder cannot shift responsibility for any shortcomings or defects in the completed product to any other party, and is therefore motivated to emphasize quality throughout the design and construction process.

- *Establish a permitting and inspection or testing process for the project.* Because design-build work is often done on an accelerated schedule using work packages, the permitting process (e.g., environmental, zoning, building) affects scheduling and must be diligently pursued and monitored. Similarly, the testing and inspection schedule can be programmed with the work schedule by engaging testing agencies, code officials, and others in ongoing communications about audits, reviews, and approvals.

- *Create, in addition to design-build entity project control documentation, a periodic project process report.* A summary monthly progress report, containing a narrative summary of the work status, a schedule showing progress to date, a financial overview of total contract amount expended plus any relevant changes, and progress photographs can be provided to the owner, project team, lenders, and others either in hard copy or on a Website.

- *Execute and complete the project with professionalism and dignity.* Work through field construction issues as a team and maintain a flexible "can do" attitude. Provide an updated project operating manual to the owner or client at closeout.

- *Use postconstruction reviews to reinforce commitment to the project and the client.* Schedule meetings with the owner or customers after the facility has been in use, and continue to demonstrate sincere team commitment to performance.

Design-Build Process Variations

8.1 Introduction

A systematic acquisition planning process (discussed in Chap. 3) can determine what form of overall project delivery is appropriate for the project. Whenever an owner decides to employ the design-build project delivery method, an important next step is to determine which variation of design-build is most appropriate for meeting the owner's and the project's needs. Like Baskin-Robbins ice cream (which comes in many different flavors and textures and may be composed of varying amounts of cream, milk, or yogurt), design-build is not just one thing; it is many things. "Design-build," says architect and design criteria professional Edward Wundram, "is an entire range of possibilities."

Current writings about design-build have focused almost entirely on the structural variations of design-build. Structural variations are characterized by the roles of the parties within the design-build entity, including joint-venture arrangements, designer-led, contractor-led, integrated firm, and developer-led (see Fig. 8.1). While structural variations are important, another way to analyze and categorize design-build is to consider the operational variations of design-build. A simplified breakdown of operational variations would include direct design-build, design criteria design-build, and preliminary design design-build. Explanations of how both *structural* and *operational* design-build variations may affect an owner's project are included in this chapter.

8.2 Structural Variations of Design-Build

Each of the five structural forms of design-build is identified by the entity that contracts directly with the owner for combined design and construction services. The five structural variations include

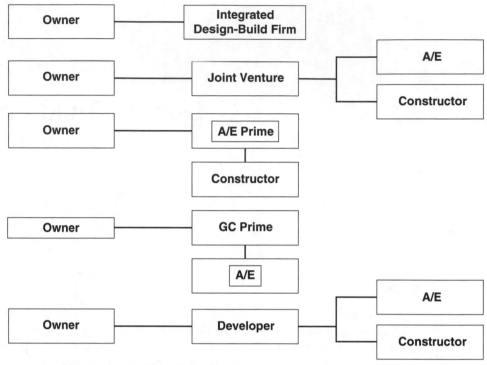

Figure 8.1 Structural variations of design-build.

1. Owner and joint-venture design-builder
2. Owner and constructor-led design-builder
3. Owner and designer-led design-builder
4. Owner and integrated firm design-builder
5. Owner and developer-led design-builder

As a caveat, although private owners can exercise a preference for one type of structural variation over another, it is unlikely that public owners would discriminate against some forms of design-build entities simply because of their structural arrangement. Public owners must not show favoritism or bias toward multiple types of structural variations because they run the risk of violating anticompetition statutes enforced by the Department of Justice. On the other hand, public owners can, in their procurements, ask for specific thresholds of design-build experience from either the design-build teammates or the design-build entity itself. A public owner will also require the design-build proposers to show that they have the required professional and business licenses in place, and will ascertain whether the entity is bonded and insured to the limits prescribed in the jurisdiction's design and construction procurement regulations.

8.2.1 Structural variation 1: joint venture

A *joint venture* is a contractual collaboration between two or more parties for the purpose of carrying out the design and construction services associated with a design-build project. The joint venture may be project-specific, or it may be organized to bind the parties for a specified period of time. Longer-term joint-venture arrangements are often employed by A/E firms and construction firms to pursue work within a specific market segment. Joint ventures may be formed among corporations, partnerships, or sole proprietorships.

Joint-venture agreements can be complex, with risks and responsibilities carefully assigned to parties within the venture. The joint-venture agreement delegates responsibility for services to the individual parties; for example, on a highway project, the constructor may have responsibility for schedule management and control, and the design engineer for quality of materials.

A joint venture can exist "only on paper"; that is, the joint venture does not own equipment or have employees but is merely a pass-through entity, and all profits and losses are passed through to the joint-venture members. The joint venture itself does not perform services; instead, it serves as a vehicle for subcontracting those responsibilities to the member parties.

Alternatively, the joint venture can be created as a business and management structure wherein specific aspects of design or construction management are performed by individuals directly employed by the joint venture, rather than by employees of the two or more coventurers. Usually the parties to a joint-venture arrangement are seeking economies of scale in business operation and improved competitive position for accessing new or existing market sectors.

Potential disadvantages of a joint venture are many: (1) the liability is joint and several, with each party to the venture held responsible for the other's acts; and (2) although the agreement can assign responsibilities to one party or the other according to their abilities and expertise, the managerial control of joint venture can be a source of disagreement among the parties for those areas where issues are of mutual concern. To create a smoothly functioning joint venture and to minimize issues that could cause disputes, the joint-venture agreement should contain

1. *Preamble and purpose.* A preamble identifying the purpose of the joint venture.

2. *Assignment of responsibilities.* Discussion of how the work will be divided between or among the parties to the joint venture.

3. *Fiscal contributions.* Identification of initial contributions to the capital of the joint venture, future allocation of contributions, and penalties for not contributing within the prescribed time.

4. *Profits and losses.* How profits and losses will be divided and what types of accounting and audit procedures will be used.

5. *Management of the joint venture.* The joint venture is usually managed by a committee composed of representatives from each party to the agreement. Day-to-day project control is delegated to a project manager, who is selected by the joint-venture committee or board.

6. *Term of the agreement.* The duration of the agreement may be from the date of commencement to the date of completion of an individual contract (project-based agreement); or the agreement may continue in force for multiple projects over time (business market or program-based agreement).

7. *Termination or withdrawal.* If one of the parties to the joint venture has not complied with the terms of the agreement or is in default, the agreement should stipulate how the coventurer will be replaced; or if the venture is to be terminated (such as after the completion of a contract), how the remaining assets will be allocated or disposed.

8. *Dispute resolution.* The parties can draft a formalized procedure for resolving disputes. The procedure often provides for settlement of the dispute through referral of the matter to the next higher level of management; and if this is not successful, the matter is moved beyond informal settlement conferences to (in order) mediation, arbitration, minitrials, and formal court proceedings.

9. *Insurance, liability, and indemnification.* As the joint-venture partners negotiate their agreement, discussions should take place to determine the amounts and types of general liability insurance, professional liability insurance, automobile, workers' compensation, and other coverage that will be carried by the joint venture or by the individual parties. For questions of liability, each party to the joint venture shall be jointly and severally liable for any loss arising out of or relating to the performance of the design-build contracts. Within the joint venture, any liability would normally be assigned in accordance with the respective party's contributions (e.g., finances, resources). Each party to the joint venture will indemnify and hold harmless the other from and against any losses, damages, or expenses from any negligent or wrongful error, omission, or act of its employees or agents.

10. *Accounting and finance.* A system of accounting should be selected by management for regular updates (monthly or quarterly) to publish and review periodically the profit and loss statement and balance sheet of the joint venture. According to generally accepted accounting principles, the joint venture will have a full and current set of books. Each coventurer will also maintain a separate set of books, and either party may request a certified financial statement or an audit of the joint venture.

Favorable aspects of a joint-venture arrangement are similar to other structural forms of design-build: single source of responsibility, increased speed of delivery, and lower-cost growth over the life of the project. Some owners have expressed support of the joint-venture arrangement because of the perceived ability to have simultaneous attention from both the professional designer and the constructor (without having to ask questions through one of the disciplines to obtain answers as in designer-led or constructor-led design-build). For whatever reason, clients who do not want to place control directly into the hands of either the design professional or the constructor may employ the alternative structural arrangements of "joint venture" or "integrated firm."

8.2.2 Structural variation 2: owner and constructor-led design-builder

The constructor-led design-build entity is currently the most prevalent variation, because of its ability to manage schedule and costs, coupled with financial capabilities that often allow such firms to be bonded for sizable sums. Under this variation, an owner contracts directly with the constructor for all design and construction services necessary to fulfill the requirements for completing the project. To obtain professional design services, the constructor hires a design consultant through a subcontract arrangement.

The subcontract arrangement descending from the constructor-led design-build entity to the architect-engineer can take one of two forms. In the first form of arrangement, the A/E has no financial interest in the outcome of the project, nor does the A/E have a financial interest in the prime design-build company. In the absence of financial interest (other than typical fee for services and reimbursibles), the relationship is largely the same as the A/E would have with the owner in a traditional design-bid-build project. Under its traditional duties, the A/E provides the owner with services that are reasonable and prudent, but that may be inexact and/or imperfect. According to the *Architects' Handbook of Professional Practice,*[1] A/Es acting in their traditional advisory role to a client "are not legally required to guarantee that a building will function perfectly, that its roof will not leak, or that the lowest bid will be within budget."

In the second form of design subcontract arrangement, the A/E has a financial interest in the outcome of the project, or has partial ownership of the design-build entity itself. With this arrangement, the A/E will have incentives that are closely allied to those of the design-build prime. Those incentives may include improving on the delivery schedule to participate in an early completion bonus, increasing the project profits by meeting the facility program with innovative systems or products that reduce construction expenditures, or employing a particular structural system or finishes system that accords with the constructor's (or subcontractor's) special expertise to enhance the constructibility of the project, thereby lowering overall labor and material costs or allowing the owner to begin operating the facility sooner.

---

[1]Edited by David Haviland (AIA Press, Washington, DC, 1994).

When an A/E moves beyond performing services for a fee (first arrangement) to being a profit-sharing member of the design-build entity (second arrangement), its risks and responsibilities are significantly altered. For example, if the A/E is involved in the project at the outset with the constructor-led prime, it is likely that the A/E has been involved with feasibility issues, writing of the owner's program, or together with the design-builder, preparing a design in response to an owner's request for proposals (RFPs) that asks for a design concept to be submitted. This A/E will be in a position to accept responsibilities that the design will meet a prescribed budget. If, on the other hand, the A/E of record (subcontractor to the constructor-led design-build prime) is brought on board after the design concept is completed (this design concept being created either by the owner's design professional or by another designer on behalf of the design-build prime), then the A/E of record will likely decide not to accept full responsibility for bringing the ultimate design into line with a prescribed budget and will participate in the project only on the basis of a traditional fee-based professional services provider.

Among the most challenging issues for architects and engineers engaged in constructor-led design-build is the traditional A/E's reluctance (and traditional lack of incentive) to take risks on new, innovative, or untried design and construction technologies. In its proposal, the design-build entity submits a design solution or technical response to the owner's criteria. The proposal includes the design-build entity's best judgment and an implied warranty that the design and construction will function as proposed. The constructor and designer will have worked diligently on the proposal, will have arrived at joint goals for the project, and will want to maintain their professional reputations and avoid any claims or litigation (to ensure follow-on work, and to reduce the costs of insurance and bonding). This parallel and combined mission and focus on behalf of the constructor and the designer under a design-build format are powerful forces toward accepting reasonable risks and producing a quality product.

The A/E owes a duty of care to the design-build entity and a duty to the public to protect health, safety, and welfare. The owner may also require that the A/E of record (part of the design-build entity) take responsibility for certifying quality control documents and progress payments. In reality, the professional responsibilities of the design-build entity and the architect-engineer are very closely aligned. The owner looks to the constructor-led design-build entity as providing competent, professional construction and design services under a single legal agreement. Because licensing and registration is individually granted, the A/E must maintain professional care and judgment regardless of employer (government, academia, consulting firm, design-build entity, etc.).

To insist that there must be an *independent* A/E consultant to the owner in *every* project delivery may be, according to a former chief counsel of the National Society of Professional Engineers, inappropriate and contrary to the intent of the professional licensing laws. However, any owner who wishes to have third-party independent advice during the design-build process may, at its option, hire an independent A/E or design-build consultant for advice and support. Owners who may not have the time or expertise to participate in design-build often rely on an A/E owner's representative or construction monitor (CM) to facilitate the process.

A number of attorneys and A/Es have criticized the constructor-led design-build process for its perceived loss of the *independent* design professional. However, this concern is mitigated because the owner may freely choose, when it deems it necessary and/or is willing to pay the additional consulting fees, to hire an independent advisor. The question that each owner has to answer is whether the additional value of the independent consultant is worth the additional cost; or whether reliance on the A/E-of-record design professional is sufficient.

Positive aspects of constructor-led design-build may include an emphasis on adherence to schedule and cost goals of the owner. Constructor-led design-build may also provide for an assemblage of the best possible "virtual company" for the project, with custom expertise added to the team as needed, and with the party assuming the greater part of the overall risks (i.e., the constructor) still in control as leader of the design-build entity. In-house capabilities of constructor-led design-builders often include knowledge of state-of-the-art construction technology; concentrated business and financial management resources; thorough estimating and scheduling data and expertise; ability to understand and manage all of the factors that influence quality, time, and price; and availability of surety bonding if required by the terms of the contract.

8.2.3 Structural variation 3: owner- and designer-led design-builder

Since the late 1990s there has been a great deal of interest on the part of architectural and engineering practitioners and some owners (usually public) in designer-led design-build. Owners frequently have close and trusting relationships with design professionals. These often long-standing relationships permit the design professional to become aware of the owner's long-range capital facilities plans earlier than other potential design-build entities.

With designer-led design-build, it is the designer that signs a design-build contract with the owner and the designer prime then engages a constructor under a subcontract. The A/E prime becomes responsible not only for providing design services but also for maintaining construction cost and schedule, and for overseeing the means and methods of construction. Some architects and engineers have contended that design-build quality may suffer unless they are leading the design-build team. To date, there has been no study either to add credence or to refute this contention. What may be true, however, is that design professionals have been trained to identify the owner's performance objectives, and their ability to sustain performance intent throughout the delivery process may provide some reassurance to infrequent users of design-build services.

According to a project delivery manual from the American Institute of Architects[2] the concern over project quality that is perceived by many architects is eliminated when the architect heads the design-build operation. Whether this is a grounded concern or simply a job-protection bias will continue to be debated until design-build becomes as commonplace as traditional design-bid-build. However, by stating that designer-led design-build can provide high-quality projects, architects seem to be admitting that their concerns over having a separate independent professional advising the owner on every project may no longer be important or even necessary, when it is clearly affirmed that the A/E of record is part of the design-build team.

A potential disadvantage of designer-led design-build is the lack of direct communication between the owner and the constructor. This limitation places a greater emphasis on the estimating, scheduling, and project control capability of the design professional. Where these skills or resources are not available within the designer-led design-build prime, the owner and the project may suffer. This indirect and tenuous relationship between an owner and the constructor subcontractor for the portion of the project that will consume the overwhelming portion of the funding may be unacceptable to some owners.

When the design-build prime contractor is a design professional, a constructor subcontractor with no responsibility for design will try to make clear in the contract that its role is limited to construction tasks. An informed owner is likely to object to contract terms wherein, if faulty work occurred on the project, the owner would be forced to "reach through" to the constructor subcontractor. Seasoned drafters of design-build contract forms usually strive for clarity in the responsibilities language, thereby indicating that the designer-led design-builder is not any less responsible when subcontracting construction than if the designer were performing the construction work itself.

Job-site safety is an issue that most design professionals have traditionally excluded from their responsibilities. Under designer-led design-build, the prime design-builder is going to be contractually responsible for project compliance with Americans with Disabilities Act (ADA) and Occupational Safety and Health Act (OSHA) requirements. Those responsibilities will have direct impacts on the designer's insurance coverage and cannot be overlooked.

Owners expect the design-builder to have the authority to manage risks. One side of the design-build coin shows the design-builder with greater responsibility and more exposure to potential liability. Turning the coin over reveals that the design-builder possesses the ability to manage and control overall risks that could affect the design-build project, the owner, and the design-build entity itself. Although cautionary about the increased risk exposure, one California design professional claims that fees for her firm can triple when engaging in design-build projects versus accomplishing design services alone.

Another practical issue confronting designer-led design-build entities is whether the state licensing statute will allow professional corporations to

---

[2]Published by the AIA California Council in 1996.

engage in activities outside the scope of their professional discipline. If the design professional is not violating the state law by assuming responsibility for the construction process, then the designer should proceed to obtain a contractor's license to perform construction services associated with design-build delivery.

Beneficial aspects of designer-led design-build include a continuation of the owner's trusted advisor in an expanded project role. This arrangement may encourage greater attention to aesthetic design issues, and could be most appropriate for what are called *foreground structures* (those structures, facilities, and buildings where excellence in aesthetic design is the overriding objective). Unfortunately, according to one prominent architect, only about 1 in every 20 buildings is a "foreground" building.

Satisfying clients and their needs remains a primary goal of the design professional. Design-build is an alternate route to that satisfaction—a route that may have greater risk, but at the same time offering a more comprehensive embrace of the facility development process and an opportunity for more complete relationship-building along the way. As with any other major project delivery system, design-build can lead to design excellence (AIA's *Architects Handbook of Professional Practice* reminds architects that design excellence is attainable with design-build, regardless of whether the designer is in the lead). Designer-led design-build can also be the path to fully satisfied owners who return again and again for comprehensive design and construction services.

8.2.4 Structural variation 4: Integrated firm design-builder

Perhaps more than any other structural variation, integrated firm design build has tremendous potential for growth. For owners seeking true "one-stop shopping" under a single company roof, this format is ideal. With the demand-pull of the marketplace seeking integrated design and construction services, many proprietors of design professional firms have been hiring construction management expertise to position themselves for the growing design-build market. Some have even purchased construction firms and attempted to merge them with their design-oriented culture. The same approach has held true for construction firms seeking to differentiate into integrated services.

Within the global design and construction marketplace, since the late 1890s there has been a tradition of integrated design-build firms. Examples include the Austin Company (Cleveland, Ohio), operating since the 1880s and Stone and Webster (Boston, Massachusetts), in business designing and constructing process and power plants from the 1890s until being purchased in 2001. It is these venerable market-specialized firms that have kept the flame of professional design-build alive in North America despite the massive inroads made by design-bid-build in the 60 years between 1918 and 1978.

An integrated design-build firm truly furnishes the single source of responsibility that many owners seek. The integrated entity will provide clients with direct access to the design professional and the constructor, with no apparent diminution in the role or status of either. An employee at a culturally integrated design-build firm is likely to tell clients that "We are design-builders," contrasted with an employee of a firm whose primary mission is either design or construction, who may state "We do design-build," but whose allegiance is to a single discipline.

Potential disadvantages of integrated firm design-build include the challenge of having in-house personnel and resources that can provide direct experience to the needs of a given project. If the design-build project requirements are beyond the skills and capabilities of the design-builder, it will need to contract with qualified designers, constructors, and subcontractors outside the firm. For example, critics of one-stop shopping at integrated firms have pointed to the inflexibility of an integrated design-build firm in responding to an RFP for conventional steel bridges because the firm's core business was designing, casting, and installing concrete segmental spans. Rather than regard specific in-house expertise as a limitation, many integrated firms claim that their tested techniques, innovations, and efficiencies far outweigh those that can be provided under ad hoc or project-by-project design-build teams.

The greatest advantage that an integrated design-build firm can offer owners is the ability to blend the cultures of a design firm and a construction firm into a seamless whole. For professional licensing purposes, the firm can be structured to function under NCARB (National Council of Architectural Registration Boards) or NCEES (National Council of Examiners for Engineering and Surveying) state-adopted rules, depending on the professional registration of the principals. The integrated firm will also maintain a contracting license as a business entity within the state.

The regular and close internal communication between designers and constructors can help improve delivery efficiencies through the respect and empathy that will grow amongst hardworking professionals. The contrasting business approaches of a design firm (driven by hourly multiples and viewing the work as a series of iterations) versus a construction firm (lump-sum or guaranteed maximum driven and seeing the work as a unified whole) are theoretically overcome within an integrated setting. The integrated design-build firm will have a proven track record of delivering design-build projects. For some owners, this reassurance provides the integrated firm with an edge over the temporary alliances found under the other structural variations.

Over time, the number of integrated design-build firms is expected to grow significantly. The busy owner of tomorrow will be focused on its core business, and will want a total-facilities services provider to handle not only its design and construction needs, but its site selection, financing, master planning, feasibility, and other predesign services. After construction is complete, the future owner (indeed, many of today's owners) will want the design-build entity to operate and maintain the facility for a period of years, adapting the facility to new users, new production lines, or new technologies. For many in the building business, becoming a "total facilities" service provider will mean more steady revenues at higher profits.

8.2.5 Structural variation 5: developer-led design-builder

The transition from a build-to-own developer to an independent at-risk developer-led design-build entity can be a natural one. When the developer has a strong relationship with a key client, and the client asks the developer to design and build a facility for the client's ownership and operation, the developer becomes a third-party design-builder. Clients will engage the developer design-builder either as a professional agent on an hourly or cost plus fixed fee basis, or as an at-risk provider of services and products on a guaranteed maximum price (GMP) or a lump-sum cost basis. Developers who accept and execute combined design and construction contracts on an at-risk basis, regardless of the origin of the work, are design-builders. A major difference between developer-led design-builders and those that are A/E or constructor led is the developer's usual lack of in-house design or construction services. Developers usually (not always) subcontract both of these critical elements to outside designers and constructors.

Developers have benefited from the current business management philosophy of corporations, which holds that the client corporation should concentrate on its core businesses, whether they are manufacturing or food service or retail trade. While those businesses deny that they are in the real estate business, they are nonetheless invested in real estate. Developer-led design-builders are in the advantageous position of being able to add value to the transaction, such as advising the client whether it may be appropriate to acquire, lease, or build facilities in specific locations or geographic areas. It may be in the interest of the owner to structure complex deals, such as those involving subsidiary companies, multiple financial investment sources, creation of cash flows, and upgrading of existing properties.

An advantage of developer-led design-build is the knowledge of a corporation's concerns about its real estate assets. Corporations concentrate on when the returns from their assets are obtained and how the financial reports capture the revenue and expense. Not only will a corporation's board of directors be reviewing the financial reports; but shareholders, creditors, insurance firms, regulatory agencies, and others will be looking for certain indicators of periodic performance. Current return on real property assets, tax advantages of real estate, and long-term appreciation are all factors in the successful delivery and operation of facilities.

Disadvantages of developer-led design-build may dissuade a corporate owner from using this structure. The negatives may include the reluctance of many developers to participate in at-risk design and construction. Developers are ingenious when it comes to formation of a complex transaction, and earn handsome fees for their knowledge of the market and its players. However, developers are sometimes ill-equipped to contract for, oversee, or efficiently deliver design and construction. The spirit of true single-source responsibility can be lost under a fee-based developer. At the same time, there are a few at-risk developers, building facilities for their own account and for third-party owners, who have embraced design-build delivery. These developer-led design-builders are competing head to head with the traditional design and construction community in major cities throughout North America.

For major infrastructure projects, an arrangement that is akin to developer-led design-build has emerged: the *concessionaire*. In build-operate-transfer (BOT) schemes, the concessionaire is a firm or group of firms that has received a franchise to build, own, and operate a facility, and will later transfer the project back to the government. For the project to succeed, it must have an identified revenue stream and other financial guarantees necessary to attract appropriate financing. Build-own-operate packages are very complex, with the process of the project beginning with conception and transfer occurring many years later. Until there is a greater understanding of project risk and the nature of return on investment by governments at all levels, the BOT design-build-plus variation will likely be limited to the megasized projects. A further explanation of design-build-plus and engineer-procure-construct (EPC) options is shown in Chap. 18.

8.3   Operational Variations of Design-Build

Most owners agree that the choice of which *operational* variation (see Fig. 8.2) of design-build to employ on their project is of considerably more importance than what *structural* variation to employ. Generally, facility owners separate the task of *problem definition* from that of *problem solving*. Stated another way, the facility owner decides (1) whether to define its needs by resources within its organization or outside its organization and (2) when the needs or problem-to-be-solved are sufficient to hand over to a contracted entity. For some owners, reaching out immediately to a total facilities provider (such as a large design-builder or EPC firm) to develop a facility program or a project study and report is the most efficient approach. Other owners prefer to work with their traditional architectural or engineering consultant to the point of concept design before engaging a design-build entity. (See Fig. 8.3.) Both ends of this spectrum are shown in these operational variations:

Direct design-build
- With program and/or pro forma developed by design-builder
- With program and/or pro forma provided by the owner

Design criteria design-build
- With minimal (or nominal) design criteria and program from the owner
- With partial design criteria, program, and performance goals from the owner
- With extensive design criteria, program, and performance specification from owner

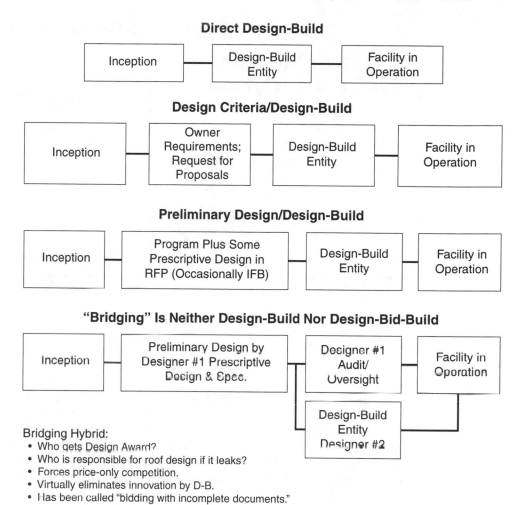

Figure 8.2 Operational variations of design-build.

Preliminary design design-build

- With design criteria, program, performance specification, and limited single line drawings
- With conceptual design and performance-based specifications
- With owner's schematic design and performance-oriented specifications

8.3.1 Operational variation 1: direct design-build

Within the genesis of every project, there are two basic tasks: defining the problem and developing the solution. Part 1, problem definition, involves data gathering, studies of site constraints, development of a facility program, crite-

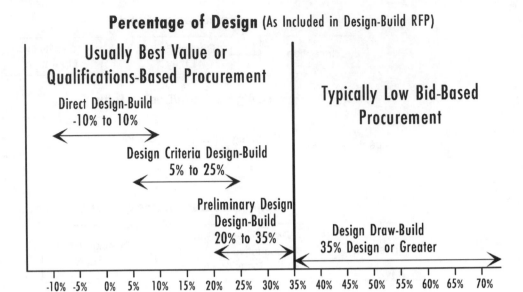

Figure 8.3 Timeline of operational variations.

ria for design, preliminary schedule, a budget estimate, and defining the responsibilities of the owner and the designer-builder. Following these definitional steps is part 2, wherein the solution to the owner's defined problem is developed by the design-build entity. The solution normally includes a technical response by the design-builder, such as design solutions (shown by preliminary drawings and outline specifications), a detailed estimate, fees or GMP, and a comprehensive schedule for design and construction.

In the private sector, both parts 1 and 2 are often accomplished by a design-build team working closely with the owner. Thus, it is not essential for the owner to contract with an independent design consultant to begin the architecture or engineering, especially where the project will be handed off to the A/E of record (who is a key component of the design-build team) at a very early stage. Some owners have compared having an independent designer prior to engaging a qualified design-builder to asking Toyota to program the needs for its new luxury car, and then giving those ideas to Honda to execute the engineering and production. While the resulting automobile may work, there is no integrated, seamless, and efficient response to the owner's needs. Other owners prefer working with design consultants (with whom they have had a long-term relationship) to work through the problem-seeking phase.

Direct design-build allows design-build project delivery to operate in its purest and most unfettered form. Except in emergencies, most government agencies have not yet gained statutory permission to employ direct design-build, although some agencies have been permitted to adopt such private-sector practices on a project-by-project basis. Direct design-build brings the entity that is in need of design and construction services together with the provider of the services at the earliest possible time during the facilities development process. *Fortune 500* companies have forged strategic alliances with facilities providers to not only provide design and construction services but also to maintain facilities inventory control; leasing, furnishings, and equipment management; space planning and development; site selection and purchase; and asset operation and disposal. Those strategic partners are direct design-builders who are in the business of serving a long-term and repeat customer.

Direct design-build (with the program and/or pro forma developed by the design-builder) is an application of the growing "integrated services" concept. If the owner requires professional assistance in preparing a program (facility needs and performance goals statement), the integrated design-build firm or team can fulfill those services. Similarly, if an owner requires assistance in developing a pro forma (budget) on which to base financial planning and capital investment decisions, integrated design-build-plus teams can usually perform those estimating, return on investment costing, and financial feasibility services.

Professional facility programmers in the employ of an integrated design-build team will initiate a project for a client by going through a problem-seeking process. The level of effort needed to identify problems that the design-build process will solve varies greatly from project to project. For more complex projects, a formal process with extensive research is necessary to uncover all the potential design problems. Very large projects or projects in environmentally sensitive areas are likely to be programming challenges. For other projects, the programming exercise is more straightforward, dictated by clearly understood performance requirements, by market competition, or by other factors that make the identification of the requirements for design more transparent.

For a design-build team to practice direct design-build *inclusive* of up-front programming and financial feasibility services, all or part of the following services would be bundled under the single contract:

1. Facility planning and requirements forecasting
2. Capital budget planning
3. Fiscal performance ratio analyses (e.g., return on assets, debt service coverage, gross rent multiplier)
4. Present value modeling
5. Risk analysis
6. Identification of facility alternative strategies
7. Acquisition planning and project delivery method selection
8. Site analysis and zoning analysis

9. Negotiating services for real property purchasing or leasing

10. Environmental analysis

11. Legal review

12. Facility programming

13. Facility performance criteria (criteria for design, materials, equipment, construction)

14. Estimating, budgeting, and cost control

15. Facility design and construction

16. Facility operation and maintenance

17. Specialized administrative services (e.g., low-voltage communication systems, marketing and leasing, security systems)

18. Database of facility information systems

19. Disposition or transfer of real property assets

Business management consultants such as KPMG Peat Marwick and Arthur Anderson have formed units that are going beyond advising corporate clients about their facility management approaches to actually absorbing the follow-on design and construction contracting within their own companies. Traditional A/Es and constructors who have worried about competition from their peers in design-build will be confronted with business rivals outside their sphere of influence. The traditional design or construction firm will be well served by moving aggressively into the integrated services model, both in response to market demand and to reassert their professional claim to the broader design and construction marketplace.

Direct design-build also may be accomplished through a "call and response" iteration, in which the owner and its facilities planning staff or consultant issue a call in the form of a streamlined RFQs (request for qualifications) or RFPs (request for proposals) containing the project pro forma and program, and the design-build entity responds with its submission. The call can include requirements for site acquisition, financing, maintenance and operation, and other design-build-plus services. The owner has chosen design-build because of its ability to place those multivariate facility delivery and management needs under a single legal agreement. The U.S. government, in its defense buildup during the Reagan years, used a call-and-response direct design-build for huge military bases, including the Fort Drum complex in upstate New York. The Army, Navy, and Air Force still have the separate legal authority to employ direct design-build, although most military projects are currently using the 1996 "two-phase" design-build methodology.

Creation of the pro forma and program allows the parties to build the project on paper, using available assumptions and data to simulate the design, construction, use and operation, and eventual sale or transfer of the facility. Design-build knowledge is extremely valuable to owners who are working through the "what to build" decisions. Determining the area, configuration, costs, revenue flow, time schedule, and aesthetics or features will determine the successful outcome of the proposed project. Direct design-build will allow the owner and its designer-builder to have maximum input at the earliest stages to influence cost, quality, and—in a word—efficiency of the facility.

8.3.2   Operational variation 2: design criteria design-build

Many owners employ a middle-of-the-road design-build approach that has become known as *design criteria design-build*. The owner, often assisted by professionals knowledgeable about facility planning and programming, sets out the criteria for the facility in clearly understood performance terms. In preparing the criteria package, the owner (and its design criteria consultant) should adhere to two basic principles:

1. Seek out and define the problem(s) and outline the parameters for design in lieu of offering design solutions.
2. Determine the minimum amount of information needed in the RFP (or other form of solicitation) for the offerors to prepare their project costs, bearing in mind that too much information forecloses on options available to the A/E of record (the A/E of record is part of the design-build team).

For design criteria design-build to be successful, the needs of the owner must be described clearly and precisely in performance terms. The performance expectations must be communicated in a solicitation that is universally understood and interpreted by potential offerors. The need for the owner to investigate the requirements of the new facility, and to define the requirements in an unambiguous manner in advance of handing the project over to the A/E of record, contrasts sharply with the traditional designer's approach of a series of design iterations with ongoing reviews and consultations with an owner. *What design criteria design-build has the potential to do is to encourage multiple outstanding design solutions from a limited group of proposers* (emphasis added). These designer-builders will, under the crucible of competition, develop individualized creative solutions to the owner's facility "problem" and requirements. The design solutions may be world-class (especially where the owner lists aesthetic design as a primary selection factor); and because they are conceived and nurtured in conjunction with input by constructors, specialty consultants, contractors, suppliers, and manufacturers, the solutions are buildable concepts disciplined by the owner's cost and schedule requirements.

Every design criteria design-build project relies on a program of facility requirements that is integral to the RFQ/RFP package. Within the RFP, the program will describe the "what" of the project, and the remainder of the key information for proposers, such as the "when" and the "how much," is delineated in the instructions to proposers, general conditions of the contract, performance specifications, and other RFP contents. What is contained in the project solicitation package varies by the type of design criteria design-build approach chosen by the owner and its consultant. Further explanations of types or variations of design criteria design-build listed at the beginning of this section are given in the following paragraphs.

Design criteria design-build with minimal or nominal project criteria.   The level of information provided under this variation is comparable to the stage of *pre-design* and *programming services* as used by the A/E professions. For projects where the contract terms are based on a GMP and the owner and design-build entity have interactive negotiations to arrive at the project scope, the design criteria approach with minimal or nominal criteria is ideal. As examples, projects that are complex or with significant design requirements, including hospitals, cleanrooms, wafer fabrication plants, and synthetic fiber manufacturing facilities have been constructed even though the issuance of minimal or nominal criteria followed by discussions between the owner and designer-builder. By contrast, some government agencies may not, under their restrictive procurement authority, conduct bilateral discussions with the offerors, and there is a need to convey greater information in identical documents that go to each of the proposers. Until their procurement laws are updated, those owners will move down the food chain toward partial criteria or extensive criteria packages.

Owners who have the flexibility to use minimal or nominal criteria plus negotiation will issue a call for design-build proposals containing some or all of this information:

- Functional efficiency, safety, and security of the facility
- Quality of systems and materials
- Design character and image
- Operating, maintenance, and energy costs/efficiencies
- Time to design and construct
- Capital cost or value

Design criteria design-build with partial project criteria.   The amount of information issued by the owner is similar to the level of "conceptual design" as used by the A/E profession. The level of "design" will likely be no more than 10 or 15 percent, although it is dangerous to stipulate aggregate level-of-design percentages because they do not convey the varying level of design in each elemental package (sitework may be at 50 percent, but structural at 5 percent and finishes at 0 percent in the RFP package). For the proposers to establish firm fixed costs for the project, these respondents must develop sufficient documentation beyond what is contained in the RFP in order to guarantee cost with sufficient confidence.

With the partial criteria variation, the RFP may contain some sketchy graphics as a way to communicate project goals to the proposers. For example, a preconcept plot plan may show the owner's idea of general arrangement of the new facility on the property. Relationship diagrams would show functional adjacencies, and a single-line recommended floorplan may be used for a preferred room layout for a hotel room, prison cell, or process manufacturing line.

As an overlay, the RFP document (when employing design criteria design-build with partial project criteria) will contain some or all of the following information:

1. General project description
2. Project-specific design criteria (minimal to allow maximum flexibility by the proposers)
3. Facility program
4. Outline performance specification
5. Agreements and contract terms
6. Schedule and milestone dates
7. Evaluation and selection criteria
8. Proposer response requirements
9. Environmental statements/analysis
10. Budget
11. Site location and zoning
12. Other information essential to the specific project

Design criteria design-build with extensive project criteria. The essence of this variation is clear and complete project criteria that treats the design problem in a comprehensive manner but stops short of offering design solutions. The extensive criteria is generated as a result of ascertaining all the owner's objectives for the project. Among the major objectives usually called out by the owner and its design criteria professional are functional efficiency, acceptable costs, image and appearance, safety and security, operation and maintenance costs, time to complete, and energy conservation.

Coupled with the statement of objectives are detailed performance requirements and performance specifications and an all-encompassing facility program. A thorough program will consist of each of these elements:

1. Mission statement of the company or agency
2. Occupant or user requirements (demographics of usage)
3. Area requirements
4. Environmental attributes
5. Functional relationship diagrams
6. Narrative statement of design goals

One reason why an owner may elect to include extensive project criteria in a solicitation is to obtain firm fixed-cost proposals (rather than fee-based or GMP proposals) from the marketplace. The use of extensive criteria will likely lower the amount of effort that a design-build consortium will need to expend to arrive at a lump-sum cost for the project. Nevertheless, where the design-build team is provided with less information, it will work within the design-build team to reconcile the design issues to arrive at a constructible solution for proposal costing purposes. There will be greater design latitude and multiple design solutions with lesser criteria. Regardless of whether sufficient project information is generated by the owner and its agents or by the design-builder, there is a point at which the two parties can move beyond a pure services for fee contract to an agreement with capital facilities project scope, overall budget, and an at-risk contract format.

In practice, the design criteria design-build *with extensive criteria* is occasionally used for *negotiated* GMP contracting, but is most frequently used where the owner has set up a much more formalized *competitive* GMP or lump-sum contracting format. In this formal call and response setting, discussions between the owner and the design-build offeror are limited or curtailed, such as for a public project where no one team can be allowed to gain preselection advantage over another.

8.3.3 Operational variation 3: preliminary design design-build

When employing design-build project delivery for the first time, many owners and their traditional designers begin with a design-build variation that has become known as *preliminary design design-build*. After gaining an understanding of the concept, and a realization that the full benefits of design-build cannot be obtained without allowing the A/E of record more flexibility, those owners often gravitate toward design criteria or direct design-build variations.

Preliminary design for a design-build project is accomplished by the owner or its retained design consultant to convey project information in, among other things, a graphic format. This variation has application for repetitive projects that need a consistent look and layout, such as a series of retail stores or fast-food outlets. For civil infrastructure projects, preliminary design information within an RFP package for projects that are more commodity-oriented, such as an asphalt overlay on a stretch of state highway, would allow traditional small and midsized engineering consulting firms and local general contractors to team together on design-build contracts. This teaming could inspire alliances that would lead to the firms becoming active design-builders, proposing in tandem on design criteria design-build highway and bridge construction projects.

The thinking behind preliminary design design-build is to have the initial architect-engineer provide the design concept and for the design-builder's architect-engineer finish the design and construct the project. The process allows the owner's design professional to perform in its traditional role: information gathering, programming, and early iterations of design concepts. Under the preliminary design design-build approach, the owner's design professional is "designer 1" for the project, and the design-build entity is "designer 2" and the A/E of record.

An owner who requires a preliminary design prior to engaging a design-build entity is likely to have either (1) a long-standing relationship with a traditional A/E or (2) a desire to work with a particular design professional (such as a signature architect) on the project. At the appropriate time, design duties are shifted from designer 1 to the production professionals (designer 2). Some owners will elect to retain the initial designer throughout the project, and others will rely on their own oversight of the design-build team. The choice is predicated on the owner's confidence in the design-build process and willingness to rely on the professionalism of the design-builder.

Preliminary design design-build with design criteria, program, performance specification, and limited single line drawings. Facility owners often care deeply about specific project elements, but have little concern about the remaining facility elements. For example, a *Fortune 100* company president may demand marble for the facade of the new corporate headquarters building, and will leave the structural wall supporting the marble cladding entirely to the architect-engineer.

When the owner's A/E details the marble curtainwall with drawings and prescriptive specifications within the RFP, this action represents the issuance of a design solution, rather than the posing of a design problem. The preliminary design design-build process can absorb some degree of prescription, such as for the specific elements about which the owner cares passionately. Successful implementation of the process depends on the design-builder having the latitude to make other design judgments and construction decisions while adhering to the project goals and objectives.

The most flexible preliminary design design-build approach is the one that is the least prescriptive. Limited graphics, such as single line drawings, are used to convey design intent for discrete elements of the project. Information about selected subsystems is communicated through conceptual drawings or sketches. Among the elements or subsystems (for a building project) that may be graphically delineated with owner-provided concept drawings are

- Site planning
- Form and massing
- Access and circulation
- Subsurface conditions
- Structural system
- Walls and fenestration

- Roofing type and slope
- Layout of key spaces or typical spaces
- Horizontal or vertical circulation systems
- Mechanical systems concepts
- Electrical systems concepts

An astute A/E working on behalf of the owner will describe those project elements listed in narrative performance terms, accompanied by single line drawings when it is necessary to communicate a design idea. Preliminary design design-build does not require that every element be shown in two-dimensional or three-dimensional representations; however, the owner may want to have key concepts drawn up to clear away ambiguities and to stipulate the use of particular materials or equipment.

Preliminary design design-build with conceptual design and performance-oriented specifications.   The conceptual design phase carries a project beyond a narrative scope or project definition into a format that is graphically documented. In previously described operational variations, each of the design-build proposers would have slightly different interpretations of the program. Owners who are looking to maximize creativity and innovation want to see the distinctions between proposal A and proposal B to determine which proposed design solution will add the most value. Other owners prefer to define their needs by issuing design concept drawings and accompanying project information. The conceptual design package (for a building project) provided by the owner may include

- Site plan
- Elevations
- Floorplans (for each level if multistory)
- Key sections
- Outline framing plan
- Outline performance specification

Additional information may be provided in performance-based specifications format, augmented by conceptual design information as necessary:

- Grading and sitework
- Foundation system
- Roof system selection
- Wall systems, exterior doors, and windows
- Room design and layout
- Major interior finishes

- Circulation systems and exits
- Elevator and escalator requirements
- Mechanical systems, service, and distribution
- Electrical systems, service, and distribution

Before assembling an overly large design-build information package that will be conveyed from the owner to the designer of record (the design-build entity), it's important to take stock of the evolution of design within a design-build context. Design starts with the "twinkle in the eye" of an owner, and begins to take form through the process of discovery and programming. Design isn't the sole province of the A/E. Owners design. Constructors design. In fact, design is a process that continues throughout the life of the facility, as those who operate and maintain it make adjustments and modifications.

The standard form of agreement between an owner and an architect (the American Institute of Architects B-141) is perhaps less distinct about the phases of design as they apply to alternative project delivery systems. First, the B-141 assumes that the owner will provide (or contract separately for) a facility program before conceptual or schematic design begins. However, either an A/E or a design-builder (A/E of record) can use the owner-provided program to explore alternative concepts to arrive at a preferred option. Because preliminary design can move rapidly beyond the first conceptual stages, it is advantageous to involve a multidisciplinary team as early as possible to exchange ideas, analyze design schemes, and coordinate the entire project in an open and efficient manner. For more complex projects, the earlier that all design and construction disciplines are brought together, the higher the probability that best-value solutions will surface.

Facility owners should exercise caution when considering the inclusion of any detail beyond conceptual design in the project information package. Owners who cross the Rubicon by prescribing aspects of the design in specific and exacting terms are reassuming responsibility for the adequacy of design (placing themselves in an implied warranty position indicating that the owner-provided design documents or specific design elements are complete and free from error).

Preliminary design design-build with owner's schematic design and performance-oriented specifications.   The CEO of one of the nation's leading design-build firms have described the preliminary design design-build variation with owner-issued schematic and early design development drawings as being on the ragged edge of design-build. Because many of the bedrock design decisions have been made throughout the programming, conceptual, and schematic design process, the design-build entity tends to become more of a drafting service. The design-builder is placed in the position of completing the design, constructing the facility, and taking responsibility for the entire design package. There is a tendency for architects and engineers working as owner advisors to immediately begin offering design solutions, rather than concentrating on thorough discovery, programming, and design criteria writing. Keeping one's fingers away from the lead holder and vellum and the CAD keyboard requires self-discipline and professionalism. The owner's design professional must always keep in mind that the design-build information package will be handed off to the A/E of record.

For a new building, an owner's design-build information package with schematic design and performance-oriented specifications may contain part or all of these elements:

General
- Facility program
- Schedule of significant delivery dates
- Contract terms
- Employment of construction manager or owner's A/E throughout the project (optional)
- Evaluation and selection criteria
- Allowances, labor rates, selection criteria, other instructions

Site
- Site plan
- Zoning analysis
- Topographical survey
- Geotechnical report
- Stormwater management system
- Environmental assessment
- Grading and landscaping plan
- Pavement requirements

Architectural
- Exterior and interior aesthetic requirements
- Major dimensions
- Required standards (space per person or function, clear heights, etc.)
- Special requirements (laboratories, cleanrooms, hazardous-materials handling, etc.)
- Floorplans (for circulation, adjacent spaces layout)
- Elevations (for wall materials, windows, doors layout)
- Sections (for heights, structures, types of construction)
- Life safety analysis and plan
- Roofplan (with slopes and drainage pattern)
- Preliminary finish schedule
- Preliminary door schedule

Structural

- System selection
- Bay spacing
- Special loading requirements
- Preliminary foundation plan
- Preliminary framing plan

Mechanical—plumbing

- Preliminary plumbing plan
- Special requirements (grease separators, on-site water treatment systems, etc.)
- Equipment

Mechanical—HVAC

- Energy conservation requirements
- Preliminary service, zoning, and distribution plan
- Equipment (capacities; locations of major equipment)
- System selection analysis

Fire protection

- Preliminary fire protection plan
- Equipment required

Electrical

- Performance requirements (lighting and power)
- Preliminary electrical service and distribution plan
- Equipment (special loads, telephone and sound systems, etc.)

8.4   Delivery Variations That Are Not Design-Build

A certain project delivery mutation attempts to position itself between design-build delivery and "traditional" design-bid-build. The system has alternatively been called *design-design-build, draw-build, detail build,* and *bridging.* Under this approach, the owner contracts with a design professional to prepare partial design documents. The owner issues the partially complete design documentation (30 to 80 percent) to the marketplace, and requests proposals. Because of the extensive design effort by the owner's designer, the system sets up the classic triad with two separate contracts: the first flowing from the owner for design and then, the second for design detailing and construction.

Under draw-build or bridging, the owner asks the bridging contractor to

1. Complete the design documents
2. Value-engineer the bridging design

3. Provide firm costs for all of the work

4. Obtain permits

5. Construct the facility

Because the transfer of design with draw-build or bridging occurs during design development, the process does not necessarily allocate risks to the party in the best position to assume those particular risks. The owner is again in the position of warrantying the completeness of the design to the bridging contractor. In reality, draw-build or bridging is neither fish nor fowl. It skirts the essence of true design-build, which is single source responsibility for design and construction, and it avoids appropriate risk assumption by obscurely diffusing responsibility. This process is more akin to traditional design-bid-build than design-build, and it is beginning to be classified as a tripartite method not unlike the sequential design-then bid-then build process that has dominated the U.S. construction industry since the very early 1900s.

8.5   Summary

Project definition is a special art. It can be minimalist when the owner's idea is grasped immediately by an experienced designer-builder, such as for an owner who simply asks for a "class A office building." On the other hand, the project definition can be extensive. An owner may want to stipulate the use of particular finishes or equipment and will employ comprehensive design criteria or schematic design drawings to convey project needs and design intent.

The proportion of design-build project information generated and issued by the owner is a variable. When less project information is provided to the design-build entity, there is an opportunity for iterations between the design-builder and the owner to develop an optimal solution or at least a preferred option. When more information is issued by the owner (or its design consultant), some of the opportunities for alternative design solutions, creative problem-solving and innovative systems, materials, means, and methods may be foreclosed.

Generally, the direct design-build and design criteria design-build variations are malleable for projects of high to medium complexity. For projects of medium to low complexity, the design criteria design-build and preliminary design design-build operational variations are applicable. For straightforward commodities purchases, such as for an asphalt overlay of a bridge deck where the design has been dictated by state standards, use of a traditional design-bid-build process wherein construction is treated as a commodities purchase is understandable.

By contrast, if the state DOT wants a new bridge replacement, a process that requires a significant design component, then the use of design-build would be ideal. The crux of true design-build, however, is that the *design* of the

DOT bridge will be accomplished as an integral part of the combined design and construction contract by the design-build entity; and not by the owner or its separately retained consultants. Like the transition to modern management theory, it is sometimes difficult to change from a command and control culture to one of entrusting responsibility to others and expecting them (indeed, contractually requiring them) to follow through.

Procurement and Selection of Design-Build Entities

9.1 The Difference between Project Delivery and Procurement

Acquisition of a new or renovated facility is a significant step for a facility owner and the owner's organization. The amount of resources devoted to this task in both staff time and money is substantial, and the potential for problems (time delay, claims for omissions, unforeseen conditions) contribute to the risk and managerial anxiety that may accompany the decision to build a new facility.

After weighing the options, however, owners may determine that designing and constructing a new facility offers a better return than leasing, purchasing, or other forms of acquisition or may decide that a new or renovated facility will provide the special amenities or unique features needed by the facility users. The decision to use a form of project delivery requires the engagement of, at minimum, design services and construction services, replete with the provision of land, labor, materials, and equipment and the management know-how to combine these disparate pieces into a finished facility.

In brief, a *project delivery system* is the process by which the components of design and construction—including professional services, labor, materials, and equipment, as well as responsibility for cost, schedule, quality, and management—are combined under an agreement that results in a completed facility. (See Fig. 9.1.) *Procurement* represents the purchasing steps that the owner or its representative must take to gain the services and commodities required under the chosen project delivery system. For example, the state of North Carolina wants to build a new office building in Raleigh, and the state public works board considers three project delivery methodologies: design-bid-build, CM (construction management) at risk, and design-build. For the office building project, the state officials decide to use traditional design-bid-build as a project delivery system. To implement the delivery system, state officials must purchase design services separately and construction services separately. Under state law, for contracts that are primarily professional design services in nature, the state uses a qualifications-based selection procedure [request for qualifications (RFQs)] and negotiates with the firm deemed best qualified to provide professional design services for the project.

Note that qualifications-based selection (QBS using an RFQ) is *not* a project delivery methodology, but a procurement/purchasing methodology. (Qualifications-based selection is used in many jurisdictions for purchasing design-only services, but may also be used to purchase construction management and design-build services in some jurisdictions.) Continuing under traditional procurement laws, after the state of North Carolina receives the finished design package from the architect-engineer, it will issue the design for bids by general contractors. To obtain construction services under traditional design-bid-build, the state is then required to purchase construction services via low-first-cost bids from construction contractors who meet the state's bonding requirements.

Public and private owners, through their peers in owner organizations, are beginning to discover the variety of options available in project delivery, procurement/purchasing, and commercial terms of contracts. Procurement/purchasing options are presented in this chapter; options for various forms of contracts are presented in Chap. 14.

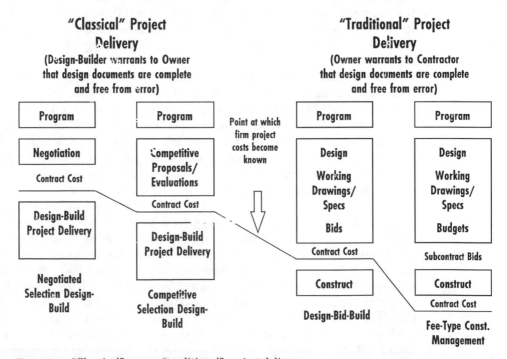

Figure 9.1 "Classical" versus "traditional" project delivery.

9.2 Purchasing (Procurement) Options for Implementing Project Delivery

Purchasing options for various services required under combined design and construction contracts range from sole source to open competition; from direct selection to low first cost (see procurement spectrum in Fig. 9.2). Sole source and direct selection have typically been within the province of purchasers of personal or professional services, where relationships and reputations are more important than other considerations. At the other end of the procurement spectrum is low cost, where buyers consider the purchase of commodities primarily on the basis of initial price.

Design-build, the project delivery system that reunites the synthesis of design with at-risk cost and schedule performance, can be procured through most, if not all, of the purchasing methodologies on the continuum. However, the choice of procurement method should be adjusted to the delivery variation and type of contract format that is chosen for the contract. Common procurement/purchasing procedures for design and construction include the methods described in the following paragraphs.

9.2.1 Sole-source selection

Private owners who have established long-term relationships with vendors and services providers often use sole-source or direct selection in the procurement of

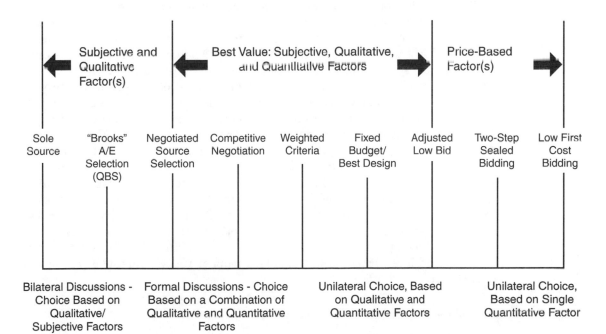

Figure 9.2 Procurement/purchasing spectrum.

design and construction services. Using selection factors such as past performance, reputation, technical and managerial qualifications, and long-standing or prior association, a private owner has the latitude to bestow contracts on the firm or individual of choice. Public owners may also employ sole-source selection when there are no other potential offerors or bidders; or when an emergency (such as an earthquake, hurricane, or civil disturbance that may endanger the health and safety of the public) allows the waiver of strict procurement rules.

The purchasing vehicle or format used for sole-source selection is a *direct purchase request* or *sole-source contract* agreement.

9.2.2 Qualifications-based selection

The purchasing process of QBS is codified in federal and state law for acquisition of specific types of professional services. In response to the owner's RFQ, entities or individuals submit their qualifications statements for review. The owner will review the qualifications statements and rank the firms on the basis of their past performance, technical competence, capacity to accomplish the work, and geographic location (if applicable). Usually, previous experience or specialized experience in the type of project named in the solicitation is critical to being ranked as one of the top three "most qualified." During the review, the owner will rank the firms according to their qualifications (which may include discussions with offerors about the services required, but not a discussion of price), and will then begin negotiations with the number 1 (most qualified) ranked offeror to reach a "fair and reasonable" price for the services required. If the owner is unable to reach a satisfactory agreement with the most qualified offeror, it will cease discussions with number 1 and begin negotiating with the second most qualified firm. In 99 percent of the cases, the public owners are successful in negotiations with the most qualified firm.

In public projects, agencies normally identify a set of criteria in the project announcement against which each qualifications statement will be evaluated. For design-build projects, a selection scoring matrix may include (1) past experience with integrated services delivery; (2) past experience with other members of the team; (3) approach to problem solving, creative strategies, and innovation; (4) quality assurance planning and management; (5) key personnel dedicated to the project; and (6) financial solidity and management skills. The evaluation matrix will often be a point of departure for final judgment by a balanced selection panel made up of procurement specialists, design and construction professionals, owner or user representatives, and other project-related experts.

The purchasing vehicles or formats used for qualifications-based selection are RFQ and qualifications statements. Firms or individuals who respond to an RFQ are called *offerors*.

9.2.3 Negotiated source selection with discussions

Generally, negotiated source selection with bilateral discussions involves a procedure of inviting proposals from services/goods providers, permits "bargaining," and usually affords an opportunity to revise offers before the award of a contract. Bargaining (in the sense of discussion, persuasion, alteration of initial assumptions and positions, and give-and-take) may apply to technical requirements, schedule, price, type of contract and its provisions, and other project-related issues. Within the *Federal Acquisition Regulations,* FAR Part 15 negotiated procedures employ a solicitation for competitive proposals, responses from proposers, and proposals being evaluated by the contracting officer or a duly constituted selection and evaluation committee. Discussions with proposers deemed to be within "the competitive range" are held, and those within the competitive range may submit best and final offers. The final proposal packages are considered, and award is made to the proposer who is providing the "best value" to the government.

Owners who procure design and construction services via negotiated source selection would use a request for proposals (RFPs). Respondents to an RFP are known as *offerors* or *proposers.*

9.2.4   Source selection with formal review (no discussions)

Owners may award contracts on the basis of the most favorable initial proposal without discussions. The use of discussions is advantageous for projects that have little definition; however, the owner may not have the time or inclination to conduct protracted negotiations about the scope and terms of the contract. In these situations, the owner will solicit proposals on the basis of information that is contained within or appended to the RFP. Relying on the guidance contained within the RFP, plus the factors for evaluation of the proposals, the team will generate their response documents. Award is made (without discussions) to the proposer submitting what is deemed to be the most advantageous offer to the owner. The federal two-phase selection procedures (Public Law 104-106) generally follow the source selection with no discussions model (although discussions would be permitted under two-phase selection, they are rarely employed).

Implementation of negotiated source selection with unilateral evaluation and selection of the successful proposal by the owner requires the use of an RFP.

9.2.5   Fixed budget/best technical response or design

Adoption of the fixed-price/best-design approach to procuring integrated design and construction services is growing rapidly. This procurement method relies on a contract price that is fixed by the owner and stated within the RFP. Since the budget is set prior to the RFP announcement, proposers work to develop qualitative and technical proposals only, under the stipulation that all price offers are equal. Fixed budget/best design is regarded as a competitive procedure wherein the proposers are competing in terms of scope and quality, rather than cost. The proposer that can fill the owner's shopping bag with the most value through their technical submission or concept design will become the awardee.

Fixed-budget/best-design procurement is implemented through an RFP.

9.2.6 Weighted criteria

The solicitation for proposals will include a requirement for the submission of a separate qualitative proposal and a price proposal from each offeror. The qualitative proposal contains technical information from the design-build team that will be rated on the basis of points, frequently with the use of a matrix that scores each proposer's response to each of the evaluation factors. After consideration of each of the qualitative proposals (a process that may include oral presentations by the teams), the price proposals will be opened. Maximum price points are assigned to the lowest dollar bid, and all others are scaled inversely proportional to that amount. The firm with the highest total points will gain the award. (See Appendix 9A.)

Weighted-criteria procurement is usually implemented using an RFP.

9.2.7 Adjusted low bid

Purchasing of design-build services through the use of adjusted low bid is similar to the approach used under weighted criteria. Following the receipt of qualitative proposals and fixed-price envelopes, the qualitative portion is scored and totaled on a scale of 1 to 100, which is expressed as a decimal (e.g., a score of 92 is shown as 0.92). Price submissions are then opened, and each price is divided by its respective qualitative score to yield an adjusted price. The adjusted low-bid amount is used to determine the successful proposer; however, the actual price submitted in the original proposal is used on the contract.

Because the adjusted low-bid process includes submission of a technical portion along with a price, the procedure relies on an RFP format for implementation.

9.2.8 Low first cost

Since 1900 or so, the use of bidding (as at an auction) has been the fashionable way to buy construction services. This method of purchasing, using the single attribute of price, is applicable to commodities and other goods that are available in multiple quantities. Design-build, however, combines architectural and engineering services for projects that are typically custom prototypes, or at minimum, must be adapted to fit a specific site and adhere to local codes and customs. Owners who employ low first-cost bidding to purchase design-build services are sending a signal to the marketplace that cost is their overriding criterion, and that the creative value of a unique proposal (solution to their facility problem) is not wanted or desired. The bidder's role amounts to completing the partial design and constructing the facility. Many design-build practitioners are uncomfortable with bidding under these conditions, because the low-bid process eliminates the creativity and innovation of the team and indicates that the owner regards the design-builder as a commodities supplier rather than an industry professional. Low first-cost procurement is usually applied by state or local governmental agencies who, under anachronistic procurement laws, must use bidding for contracts with a majority of the dollars destined for construction costs.

The format or vehicle used to implement low first-cost procurement is the *invitation for bids* (IFBs). Those who respond to IFB procurement are *bidders*.

9.2.9   Other variations

A number of other procurement, evaluation, and selection methodologies are used in design and construction, but these are not regarded as "good practice" for use with design-build contracts. Under *equivalent design/low bid,* the proposer responds to the owner's RFP with a technical submission and price submission. After review and critique by the owner and its consultant, the proposer is given a deadline to respond with specified design changes and corresponding price amendments (either additive or deductive) in order to arrive at a group of proposals that are technically equivalent (this is sometimes referred to as *technical leveling*). Revised designs and pricing are then evaluated by the owner using both base cost and amendments. Award is normally made with overwhelming emphasis on costs, because the initial review process should have created nearly equivalent designs.

A related purchasing variation is known as *meets criteria/low bid.* This format is very similar to the traditional low-bid process, except that the drawings and specifications are not absolutely complete (i.e., the drawings and specifications are typically completed to the late design development phase but not carried to the later construction documents phase). The proposals from design-build teams are evaluated and deemed to meet the base criteria, and award is made to the low-cost bidder. Except for the most utilitarian structures, such as package solid-waste transfer stations or storage buildings, this process is inappropriate for design-build delivery services.

9.3   Using Competitive Negotiation and Best Value for Design-Build

For many owners, the ability to use integrated services delivery for their projects was a significant step forward. But many have been even more intrigued by the procurement/purchasing alternatives that have come about as a result of innovative project delivery implementation. The purchase of combined design and construction services through best value rather than low bid is the single most important part of the late twentieth/early twenty-first century realignment that is occurring in the industry, according to some owners.

The goal of an owner using competitive negotiation to procure design and construction is to structure and implement each purchase to ensure that the contract is awarded to the competitor with the most favorable proposal. First, to elicit the best responses from the marketplace, the owner will want to conduct a fair and unbiased competition. Then, to arrive at an agreement that best meets the owner's needs, the owner will want to evaluate the technical and management capabilities of the team, past performance, costs, and other key factors. It is in the owner's interest, predicated by the project's goals and available resources, to find a solution that represents the best value. The best value in design and construction may not mesh with the low-first-cost solution.

The key elements of competitive negotiation and best value selection are described in the following paragraphs.

9.3.1   A source selection plan

Planning for a competitive negotiation allows the facility owner to think through a strategy for managing the acquisition. An informed owner will have gauged the market to ensure that the solicitation is formulated in a way that will attract qualified proposers and that the goals of the project are embodied in clear functional and performance specifications. The source-selection plan, the solicitation notice, and the RFP should all contain a consistent "statement of need" summarizing the technical and contractual aspects of the procurement, plus applicable conditions surrounding the acquisition such as cost, schedule, and capability performance constraints, and need for compatibility with existing or future systems or programs.

9.3.2   Obtaining proposals

In brief, the RFP package from the owner serves as an advertisement to attract worthy service providers to the project. The RFP will contain a minimum of four sections: (1) a technical description of the goods and services required, (2) an explanation of the quality and complexity of the project, (3) the location of the project and time for completion, and (4) the standard contract clauses that will govern the relationship. In addition to the technical and legal information contained in the package, a well-constituted RFP will spell out the basis for evaluation and selection. The inclusion of evaluation and selection factors is of particular importance, because it conveys what the owner considers to be important about its project to the design-build teams. When an owner includes selection criteria and weights the criteria according to their importance to the project, there is high confidence in the evaluation process and a feeling of openness and teamwork between the owner and the individual proposers.

9.3.3   Evaluation of proposals

Offerors deserve to know whether cost is secondary to quality or whether the procurement is intended to achieve a minimum standard at lowest cost.

Competition is not served where proposers are not provided with direction about the relative value of technical excellence versus price. Some courts have found against government agencies who have incorrectly stated that the award under an RFP would be made to the entity whose technical-cost relationship was most advantageous to the owner and that neither cost nor technical excellence would be controlling, when, in fact, cost was an overriding consideration in the evaluation.[1]

Evaluation starts upon receipt of the responses to an RFP, and may consist of scoring or ranking each proposer against the stated requirements. It is not appropriate for the owner to judge the merits of a proposal on the basis of criteria that are not published in advance of the submission date. Conversely, when a factor has been included in the RFP, it is improper for an owner to ignore the feature (such as stating that past performance is an important factor, then ignoring the proposer's history of outstanding performance or poor performance on a similar project). Technical scoring for best-value evaluations is usually done through numerical or adjectival ratings for each of the major factors and subfactors. A few public owners have used color coding to differentiate between proposals, but teams putting together proposals seem to prefer evaluations with numerical scoring. After the technical evaluation has been completed, owners review costs to determine the reasonableness of the cost proposal, including allowances for contingencies and cost trends. The owner must judge whether the offeror clearly understands the requirements of the RFP and whether the promissory language in the proposal has incorporated all the elements of the project that the owner deems essential. Finally, the owner must conduct the crosswalk between technical provisions and the price to verify the proposal's overall value to the owner.

9.3.4 "Competitive range" decisions

Following the evaluation of competitive proposals, an owner must decide whether to award the contract or to proceed to negotiations through oral discussions. If award is not made on a unilateral basis (i.e., without discussions between the owner and design-builder), the owner must decide which proposers will be selected for direct negotiations. Proposers who are elevated to the competitive range are those teams that the owner decides has a reasonable chance of being selected for an award. In government contracting, all proposers who are placed in the competitive range will have an opportunity to negotiate the contract with the public agency, a requirement that tends to keep the number of participants lower rather than higher. Design-build industry spokespeople have encouraged the federal government to place the competitive range at a minimum of three competitors and a maximum of five, allowing the firms to make a sound business decision about the competitiveness of their proposal within the limited pool of finalists.

---

[1]Nash, *Competitive Negotiation*, p. 13.

9.3.5   Oral or written discussions with proposers

A competitive negotiation with bilateral discussions is a process that is destined to culminate with a best and final offer. During the discussions, a public owner will reveal areas where the proposal falls short of the government's expectations. Any errors, omissions, or deficiencies in the proposal are unfolded during the negotiation as points of discussion. Proposing teams often worry about the disclosure of their innovative or ingenious solution to other teams. The federal government agrees that "technical transfusion" is unfair, and could lead to the illegal practice of technical leveling.

In the private sector, negotiations concentrating on the proposed design concept may be held, with the purpose of familiarizing the parties with each other's expectations, and providing the design-build team an opportunity to expand on information contained in its proposal. The discussions will reveal how the team members work together, and can show how responsive the team will be to the project's needs and objectives. Private owners do not operate under the constraints placed on their public counterparts; however, the conduct of negotiations will reveal the private owner's intentions for fair dealing and good faith to the design-build teams.

9.3.6   Use of best and final offers

At the conclusion of discussions with proposers within the competitive range, a public agency will issue a request for best and final offers (BAFOs). A common cutoff date is established that allows a reasonable opportunity for submission of best and final offers containing technical and cost information. After the BAFOs are evaluated, a best offeror is selected under the condition that negotiation of a definitive contract must take place within a prescribed time period. It is not uncommon for the proposer to withhold its lowest price until the best and final offer. The danger of the BAFO procedure is where an owner reopens discussions after receiving a best and final offer. The Department of Defense has now prohibited multiple rounds of BAFOs without the approval of a higher authority (DFARS 215.611) because it can turn the procurement into an auction. No such brake on abuse of the BAFO process exists with the General Services Administration (GSA) or the other civilian agencies.

It is important to remember that an owner may review initial proposals only, and a unilateral decision made by the owner on the merits of the initial proposal without resorting to a BAFO. The best and final offer process is a step beyond the unilateral RFP process, with bilateral discussions over the content of specific proposals leading to a revised final proposal.

9.3.7   Award of successful contract

At the conclusion of discussions (or at the end of BAFO evaluation, if that procedure is used) and selection of a successful proposer, a public agency will revise its documents to reflect the agreements reached during negotiations. The effort can be considerable, because the agency will be tasked with revising its scope of work, contract terms, special requirements and clauses, and government estimate to support the result of the negotiations. A record of negotiations is necessary to defend the governmental entity in case of a protest or investigation. The record serves notice that the procurement was conducted in a competitive manner, with fairness to all parties involved. Generally, the award is provided to the successful proposer in an award letter with final contract documents to follow within a reasonable period, or in an entire package containing all documents from the owner about the project. On receipt of the package, it is prudent for the design-builder to be alert for any agreements that were not included in the original documentation or discussions, as well as for additional data or requirements that may have been added.

9.3.8   Debriefing unsuccessful proposers

As soon as award has been made, fairness dictates that the owner notify the unsuccessful offerors and provide them with an opportunity of a debriefing to learn why they were not selected. It is in the owner's interest to inform the proposer about shortcomings or less desirable approaches that were contained in the unsuccessful proposal to allow the design-build team to correct such items in future competitions. The debriefing can demonstrate the owner's rationale for choosing one proposal over another with the intention of clearly showing the advantages of the successful proposal and avoiding protests from the other competitors. Timely notice and debriefing also will release resources of the unsuccessful design-build teams so that they are free to pursue other work. Some owners fear that a debriefing that is too thorough will provide ammunition for a protest of the award, or that it could inadvertently disclose proprietary information from another proposal. It is important for owners to balance these legitimate concerns, but does not override their obligation to conduct a timely and informative notice and debriefing to the unsuccessful finalists.

9.4   Brief Overview of Congressionally Authorized Design-Build Procurement Laws

After World War II, the U.S. Navy continued its success with design-build delivery for facilities by applying the method to military housing. With the exception of large-scale housing projects, the use of design-build for defense and civilian agencies almost disappeared in the 1950s, until NASA requested design-build authority after President Kennedy's historic speech promising to send a man to the moon and return him safely before the end of the decade. By the 1970s and 1980s, HUD, GSA, and the Department of Defense were implementing design-build delivery under a variety of procurement statutes.

In the 1990s, a number of legislative initiatives were undertaken to modernize government acquisition procedures. Many of these initiatives, such as past performance considerations, shortlisting, and other commercial practices, were collected into the Federal Acquisition Streamlining Act (FASA). During the mid-1990s, GSA and the Corps of Engineers began drafting language that would allow shortlisting of design-build proposers for federal projects. Working with a group of design and construction industry associations, the agencies developed what has become known as *two-phase selection* procedures for design-build.

Provisions authorizing two-phase selection were actually embodied within the Federal Acquisition Reform Act (FARA), which became a section of the fiscal year 1996 Defense Authorization Act that was enacted on February 10, 1996. The new two-phase procedures enabled contracting officers to use design-build on any directly funded federal project whenever the situation merited use of the method. Two-phase procurement is straightforward; design-build proposers submit qualifications and other requested information in response to a request for proposal. In this first phase, no cost or pricing data may be included in the response. The proposers are evaluated according to the criteria included in the solicitation and the agency selects three to five of the most qualified proposers to compete in the second phase. Evaluation of proposals in the second phase is based on key factors such as technical response, design solution, management plan, past performance, price, or other criteria of importance to the owner that is identified in the RFP.

In addition to the source selection and two-phase procurement procedures used by the federal government for design-build, the Department of Defense retains its turn-key authority, authorized by Congress under the FY 1992 National Defense Authorization Act. Congress also placed design-build authorization in the Surface Transportation Authorization Act (TEA-21) for federal-aid highway and bridge projects. Current interest in applying design-build to civil infrastructure projects is burgeoning. States and local governments, led by pioneering design-build legislation in Florida during the 1980s and 1990s, are following suit with procurement and licensing law reform measures.

9.5   Owners and Vendors/Service Providers: Procurement Proclivities

Within the business world, design and construction teams compete for contracts to earn a profit and to provide a meaningful living for their employees. Fundamentally, these business practitioners would prefer to devote their energies to core business activities (designing and constructing facilities) rather than marketing and selling their services. Given a choice, principals of these firms would prefer to be chosen directly under a sole-source contract. Sole sourcing saves the vendor or service provider most funds that it may have expended for prospecting, feasibility analysis, estimating, conceptual design, and proposal development.

Sole-source procurement may be chosen by public owners for a design and construction contract if there is a rationale for waiving other statutorily required procurement procedures. Waivers are permitted under federal and state law for emergencies such as devastating earthquakes, hurricanes, tornadoes, or civil disturbances that may endanger the health, safety, and welfare of public citizens.

For the design and construction practitioner, the second most favored procurement/purchasing methodology after sole source would likely be qualifications-based selection. Implemented through a request for qualifications (RFQs), qualifications-based selection (QBS) enables the owner to select the most qualified offeror for a one-on-one negotiation. QBS is an ideal procurement approach for direct design-build wherein little is known about the project and the owner wants the design-build entity to do programming or study and report services. Design-build contracts procured using QBS are likely to include cost-plus-fee commercial terms, which may be converted to GMP or lump sum after program definition and schematic design have been completed.

At the other end of the procurement spectrum, owners who insist on commodities-based competition (and who regard any contract that is predominantly construction as commodities-based) continue to use low-bid procurement. This auction-based purchasing methodology relies on a single measure—cost—and does not permit comparisons by the owner of other attributes that vendors or service providers may bring to the equation. Cost, by itself, is an incomplete indicator of quality and performance.

Interest in best-value procurement has grown enormously since the late 1980s. For services as complex as design and construction, many owners have discovered the power of multiattribute review of proposals based on their distinct needs and desires, followed by a realistic examination of cost or budget, to arrive at the best possible combination of value.

A day-to-day example of best-value procurement is the method that most of us use to purchase an automobile—if we had an unlimited budget and wanted ultimate quality, we would likely purchase a Rolls Royce with a lifetime warranty, or a Lamborghini and have a mechanic on retainer. If low price were our only criterion, we would buy a 15-year-old Yugo or perhaps a Ford Festiva. But many of us have multiple criteria that must be balanced, such as size (whether it can carry our family comfortably), styling (whether we will be embarrassed or pleased to be seen in the company parking lot), reliability (whether we have read consumer reports), gas mileage (whether we can afford to drive the vehicle when gas prices top $2.00 per gallon), color (whether it reflects our taste), performance (whether we can corner without body roll, drive offroad, tow a small camper trailer), and so on depending on our major selection criteria. If American consumers, every day, can make best-value judgments about the buying of cars, then surely owners can apply the same principles to objects as important as our buildings and bridges.

Appendix 9A: Examples of Criteria and Cost

See Figs. 9.3 and 9.4.

Proposer	Qualitative Score (60 Maximum)	Price Proposal	Price Score (40 Maximum)	Total Score (100 Maximum)
Firm " A"	51	$1,629,000	37	88
Firm "B"	53	1,546,000	39	92 *
Firm "C"	44	1,510,000	40	84

* Award to proposer with highest total score.

Figure 9.3   Weighted criteria and cost.

Design Build Team No.	Design Solution (Weight 10)*	Financial Capability (Weight 6)	Project Approach (Weight 8)	Management Plan (Weight 10)	Total Points
A	4 / 40	5 / 30	4 / 32	3 / 30	132
B	5 / 50	5 / 30	2 / 16	5 / 50	146
C	4 / 40	4 / 24	2 / 16	3 / 30	110

* Criteria Weight   **Where:**   **5 = Excellent**
                                             4 = Good
                                             3 = Fair
                                             2 = Poor

Key: Factor Score / Subtotal Points

Figure 9.4   Evaluation matrix with criteria and weights.

Communicating the Owner's Requirements in Developing Design-Build RFQs and RFPs

10.1 Introduction

The use of a competitive, design-build selection process for the procurement for public, institutional, and corporate facilities represents a considerable departure from the design-bid-build procedures normally employed by most such organizations in North America (Fig. 10.1). The design-build process is increasingly selected by those owners because it offers the flexibility to respond to their urgent physical, financial, legal, and political needs and restraints to a greater degree than can traditional procurement procedures. Consequentially, to take full advantage of the process, each request for qualifications (RFQs) or for design-build proposals should be drafted to fit the owner's unique situation. Standard forms and contracts for other types of project delivery are of little value in a design-build process, except for general guidance, or as a checklist. This chapter (1) discusses the various elements of the competitive, design-build selection process, (2) suggests what information should be provided to the proposers, and (3) identifies the specific facts and technical data that should be submitted to the owner for the proposals to be evaluated and considered for award of contract.

An invariable prerequisite of design-build competitions is that the owner's needs must be described precisely, and in a manner that will be universally understood and interpreted. This aspect of design-build requires the owner's staff to conduct adequate research and investigations to determine the facility requirements, and to document them unambiguously. The necessity to determine needs in advance contrasts sharply with the traditional project delivery process where needs may be loosely defined to the designer, and through the consultative

- Stipulated sum/best value
- Variable sum/best value (with cost as one of several criteria)
- Equivalent design/low bid
- Meets criteria/low bid
- Emergency/qualifications only (negotiated contract or force account)

Figure 10.1   Types of design-build selection procedures for public projects.

process of design and review, the needs and solutions are more fully defined, explored, and resolved simultaneously between owner and designer. This chapter highlights the types of information that should be determined in advance, and methods commonly used to communicate the information to the prospective proposers. In addition to the facility's physical needs, the chapter suggests that the owner know its legal, financial, and political limitations; the project's objectives, schedule, and costs; and the design-build marketplace, and discusses the wisdom of identifying the project's stakeholders and their concerns.

A complete design-build solicitation program will include the components discussed in Secs. 10.1.1 to 10.1.6.

10.1.1   Request for qualifications and competition prospectus

In a traditional design-bid-build bidding process, all responsible licensed and bonded builders must be considered as equals by public agencies evaluating their bids. While this consideration allows for an equality of opportunity among bidders, it falls short of meeting the public owner's needs on large or complicated projects because the design-bid-build process fails to recognize builders' unique skills and experiences. More importantly, those traditional public procurement methods ignore the contribution that builders can make to the creative design process. The premise of the design-build process is that all design-builders, or teams of designers and builders, are not created equal, and that it is in the public's interest, and ultimately in the design and construction industry's interest, for the owner to select only the best qualified design-builders from among those who wish to be considered. This puts double importance on the RFQ: (1) it must attract the best-qualified design-builders to the project and the selection process, and (2) the request for proposal documents must inform the potential applicants of the project objectives that the proposers are expected to meet and also describe the information needed and the criteria that will be used to determine the best-qualified proposers (finalists). The *request for design-build qualifications and competition prospectus* is the most effective, and in some cases, the only instrument that a public agency can use to advertise the project and prequalify the proposers.

10.1.2   Request for design-build proposals

The request for design-build proposals (RFPs) is the primary contractual document between the proposers and the owner. It must be supplemented by the owner, with a program of facility requirements, technical performance specifications, a form of proposal, and the applicable forms for various bid and contract bonds. The proposers, in turn, are expected to supplement their proposal form with the required and voluntary submittals describing their proposal in sufficient detail for evaluation and to ensure contract compliance (to protect the owner), and define the proposal's limitations (to protect the design-builder). The RFP represents the owner's last convenient opportunity to broadly to define its needs, requirements, and limitations before engaging in the competitive process. After selection of a design-build proposal, and execution of a design-build contract, changes in the final design are limited to the owner's responses to the interim design submittals, if any, and to owner-initiated requests for scope changes, change proposals, and subsequent change orders.

10.1.3   Instructions to proposers

The *instructions to proposers* describe how the selection process will be conducted, and lists the owner, criteria professional, technical consultants, and members of the selection panel (jury). The instructions outline the selection criteria, specify the competition schedule, and detail the submittal requirements and the rules of conduct and communication for both the owner and proposers. It also describes all important or unique conditions and requirements of the process to select the successful design-build proposal.

10.1.4   General conditions of the design-build contract

The guidelines describe some of the modifications to the traditional general conditions of contract that are necessary to accommodate the design-build methodology. Those include responsibilities of the owner, responsibilities of the design-builder, responsibilities of the design-builder's architect (including engineers and other design professionals), responsibility for permits and fees, and the role of the design and construction documents, indemnification, insurance, liquidated damages (if any), and other business-related issues.

10.1.5   Program of facility requirements

The program describes what, as opposed to the how, when, and at what cost requested elsewhere in the RFP. In several separate sections, the program should specify the owner's intentions for the character or image of the facility, the activities to occur in the facility, the functions of the individual units housed in the facility, and their interactions with each other and with visitors to the facility. For a building project, the program should specify the exact or minimum amount of usable floor area required, and the environmental conditions (power, light, heating, cooling, ventilation, etc.) required in each programmed space. Depending on the specific needs of the project, facility programs can include design directives, and design configuration criteria developed from functional and programmatic needs. Similar functional and dimensional requirements should be listed for civil infrastructure projects.

10.1.6   Performance specifications

Performance specifications (as opposed to traditional prescriptive specifications) are included in the RFP to indicate how the completed facility is expected to perform. These specifications prescribe the strength, quality, and operating criteria of the building systems and materials. They use industry and government standards, whenever practical, to denote the various levels of quality and performance desired, and to specify how those standards will be measured. Performance specifications avoid the selection of specific materials or systems, leaving the choice of, and responsibility for, such to the proposers.

10.2   Know Your Needs: Programs of Facility Requirements

10.2.1   Program narrative

This consists of a completely subjective discussion of the image that the owner wants the new facility to portray to the employees, public, and others. The specific design goals are meant to reinforce the broader issues of image, character, and so on. Many cities and institutions have adopted design guidelines for special districts, campuses, or for special land uses. These latter design guidelines should be included in the program by reference.

10.2.2   Mission statement

Most large organizations, including many state and federal agencies, have written material describing the functions of each unit, department, section, or other entity to be housed in any given facility. If not available, the owner or its criteria professional must develop descriptions of the organization's functional units and indicate how they carry out their work from day to day. This includes how the units interact with the public and the type, number, and frequency of all visitors and their reasons for visiting. Needed also is a description of all materials arriving and departing, including paperwork, and the unit's end product, whether it is a physical object or an intellectual property.

10.2.3   Staffing

In addition to the existing and projected workforce requirements of each unit, this section must include a description of each individual position's job function, including any special equipment or furniture necessary to perform the job's tasks.

10.2.4   Area requirements

Determine the net usable floor area for each workplace and support facility (use the organization's authorized space standards, where available), and summarize the total net area requirements. Use the organization's existing space standards, or develop and recommend new space standards on the basis of industry trends for similar organizations. Wherever possible, reduce the number of different-sized workstations or office spaces to as few as possible to ensure greater flexibility of the completed facility to accommodate organizational changes in the future, without extensive physical changes.

10.2.5   Physical and environmental requirements

Determine the physical and environmental requirements (temperature, light, humidity, air-changes, etc.) for each workplace and type of facility support space, including equipment requirements. This section serves as the baseline environmental conditions that must be met in the finished and occupied facility. It is a very important part of the performance requirements in the contract between the owner and the design-builder. The information needed by the designer includes the acceptable summer and winter temperature and humidity ranges, the minimum amount of air-changes, lighting levels and color ranges needed, power needs, required equipment (specify if owner- or design/builder-furnished), and security or privacy requirements. Also include a general description of the surrounding finishes (hard, soft, waterproof, special, etc.) and the acoustical environment required. (Specify the ambient allowable noise-level range, and the sound transmission class of the surrounding enclosure.)

10.2.6   Relationships

This information, the functional relationship among all programmed spaces in the facility to each other, can be indicated graphically in a "bubble diagram" or in a proximity matrix with weighted values shown. Alternately, the program document can simply describe the critical adjacency relationships and those that may be desirable but not mandatory. This section expands on the work-pattern information in the individual unit's mission statement.

Other reference documents.   The American Institute of Architects (AIA) and other similar professional organizations make available several publications on facility planning and programming.

10.3   Know Your Legal Limitations

10.3.1   Public contracts law

Does design-build procurement require specific legislative authority, or is it allowed as one of several project delivery methods available to your agency?

Research the public contracts law in your area as it applies to design-build procurement. Some states have public contract laws that address the design-build procurement technique directly, others allow the design-build process to be adopted by individual jurisdictions within the state, and some require a broad interpretation of a combination lowest responsible bidder requirements and professional services selection procedures. In many instances the term *lowest responsible bidder* has been interpreted to mean other than lowest initial cost, such as maximum value, lowest life-cycle cost, or lowest cost per value point. In either case, consult the appropriate legal authority for specifics on exactly how to structure the RFQ and RFP to comply with the applicable public contract laws.

10.3.2 Professional licensing

Do the state and local professional laws apply to design-build procurement? Research the architect and engineer licensing laws as they may apply to design-build procurement and contracting. Design-build competitions for significant public facilities often attract design professionals from a wide geographic range, as opposed to the responses to normal architect-engineer (A/E) selection procedures. It is important to communicate the requirement that it is the design-builder's responsibility to meet all the requirements of the state's professional licensing laws in the design and construction of the facility. The normal requirement in the RFQ requesting that at least one design professional in each discipline be registered (or be capable of being registered) in the state is only a part of professional licensure, and represents the typical extent of most public owners' due-diligence process. The owner should avoid undertaking the responsibility of enforcing the state's professional licensing laws.

10.3.3 Contractor prequalification

Is contractor prequalification allowed and, if so, under what conditions? The preparation of a design-build proposal will require an extensive amount of work by all members of the design-build team. To be fair to those teams participating, and to encourage the maximum competitive effort, the number of prequalified finalists (proposers) should be limited. The normal number is three, but generally does not exceed five. Complex projects requiring extensive research and development on the part of the finalists, such as courthouses, stadiums, or convention centers, should limit the number of proposers to the smaller number. Projects for simple, well-understood types of facilities, such as parking garages or generic, class A office buildings, may reasonably allow the larger number of finalists to participate.

10.3.4 Honoraria

Can a public agency pay for goods and services not received, that is, proposals not accepted? If allowed by law, we strongly recommend that honoraria be offered to the unsuccessful proposers. The honorarium or stipend serves several purposes: (1) the provision of reasonable compensation will encourage the more sought-after design-build teams to apply and, if shortlisted, to make an extra effort in the preparation of their proposal; and (2) the provision of an honorarium will tend to diffuse the potential for criticism from the unsuccessful teams that they "did all that work for nothing." Honoraria rarely cover all the design-build teams' hard cost, but they should cover a reasonable estimate of the out-of-pocket cost to prepare the minimum submission requirements. The amount of the honorarium should be proportional to the complexity of the facility, the submission requirements, the time allowed to prepare the proposal, and other considerations.

Contract authority.   What authority does the agency have to solicit proposals and enter into a contract? Because of the extraordinary effort required to prepare a priced, design-build proposal, the owner should assure itself, and then assure the potential proposers, that it is authorized to solicit design-build proposals and has the authority to accept a selected proposal. The use of the design-build proposal process for speculative efforts, such as preparation for a bond election, represents a misuse of the process and will not produce the best possible proposals. Design-builders understand the public authority's need to maintain the right to reject any or all proposals, but expect a good-faith effort on the part of the owner to engage and enter into a contract with the successful design-build proposer.

10.4   Know Your Financial Limitations and Restrictions

10.4.1   Funding in place

Is the project's funding in place, or does it require a public or legislative vote? In addition to the need to communicate the owner's authority to solicit and accept design-build proposals, it is necessary to show that the funds for the design and construction are in place, or that the funds reasonably can be expected to be available by the time it is necessary to execute a contract with the successful design-builder. Before a design-build team can be convinced to make the considerable effort to design and price a comprehensive proposal, they must feel secure in the fact that the owner has, or will have, the funds to proceed, should they be successful in their bid.

10.4.2   Funding limitations

What funding limitations are contained in bond issues or in enabling legislation? Often, bond issues and/or enabling legislation providing for the authorization and funding of major public projects contains specific language related to the size, site, or other aspects of the proposed project. It is important to include the specific language of such qualifications in the RFP.

10.4.3   Limitations of "tax-free" financing

If the project is to be a public-private, mixed-use facility, the U.S. Internal Revenue Service limits the percentage of such projects that can be intended exclusively for private use yet still enjoy the benefits of tax-free municipal bonds or similar financing. If the limitation is applicable to the project, it should be mentioned in the RFP.

10.5   Know Your Politics

10.5.1   Local professionals and builders

Learn the positions of local design professionals and builders on issues relative to design-build procurement. If the application of the design-build procurement process in the public sector is new in your area, some members of the design and construction industries will be wary of, if not outright hostile to, the procedure. To develop an effective procurement strategy that will attract the best-qualified designers and builders to the project, you will need to know their concerns and objections to the process. Many of those concerns can be addressed in the structure of the RFQ and RFP, and in the published selection criteria. Presenting and explaining the process in presubmittal conferences and at industry meetings may lessen other concerns. It is important to show that the owner is sensitive to those concerns, and has made, or will make, changes in the procedures to address them, while maintaining the advantages the design-build process offers public agencies.

10.5.2   Political rationale

Learn the political rationale or motivation to initiate the project. All except the most routine public projects will likely have a series of complex political reasons as to why the project was supported (or resisted) by various members of the applicable governing body and its political constituents. Learn the background for the decision to proceed with the project, and incorporate the rationale into the structure of the RFQ and RFP, and into the design criteria, as appropriate.

10.6   Know Your Schedule

Time is essential in most, if not all, design and construction contracts. The amount of time available for team organization, preparation of qualification statements and design-build proposals, and the schedule necessary to design and construct the facility, is critical for those contemplating participation in the selection process. Because the time allotted to the various tasks in the selection process and in the subsequent design-build contract can be adjusted reasonably to balance this resource, the critical date is the owner's absolute beneficial occupancy deadline. Once established, all other activities likely can be scheduled to fit within the elapsed time allocated. The degree to which the activities are compressed, or can be generously scheduled, will influence both overall cost and the market's response to the RFQ/RFP.

10.7   Know Your Stakeholders

10.7.1   Official chain of command

Know and communicate to the proposers the official chain of command, from the owner's project manager up to the board of the appropriate governing authority. If different from the normal chain of command, you must also communicate the "decision tree" as it will be applied to the project. The design builders will want to know, even before submitting their qualification statement, who decides the major issues in the selection process, but more importantly, who decides on the final award. The owner should make every effort to ensure that the decision process is as objective and nonpolitical as possible. A clear understanding of the official chain of command by all participants, and the owner's project administration team, is an important element of this endeavor.

10.7.2   Community and business leaders

Most large public projects have a series of interest groups and individuals who promoted the project, or who have an extraordinary interest in its successful completion. It may be wise to keep those groups and individuals informed of the project's progress. It may also be appropriate to keep those same groups and individuals involved in the project, as advisory board members or as members of evaluation or selection panels.

10.7.3   Design review committee, or planning commission

By the act of accepting a design-build proposal, most governing authorities will have inadvertently bypassed a design review or similar board, if such board must review the design of public structures prior to permitting. To avoid legal and/or political conflict with such boards, committees, or authorities, it is advisable to review the design-build selection procedure and the RFP's design criteria with the board prior to publication. If the site is known in advance of the issuance of the RFP, the owner prior to RFP publication can resolve most planning and zoning issues. Compliance with all applicable zoning ordinances and design guidelines should be made a contract requirement in the RFP.

10.7.4   Users

Know the facility's users (department heads, supervisors, etc., and public employee unions). If the facility is to house an existing organization, there should be some means for representatives of the employees to provide comment on the facility program, as it may affect their working conditions. In most design-build selection programs it is possible to include a staff committee in the design proposal review process. Similarly, review the facility program and the design criteria with representatives of the appropriate public employees unions.

10.7.5 Unions and other industry groups

Review the design-builder (contractors and their architects and engineers) prequalification process and selection criteria with the executive boards, or with the business agents of, the applicable building trade unions, contractor associations, and professional societies. If possible, address their concerns in the RFQ.

10.8 Know Your Objectives

10.8.1 Image and appearance

Subjective qualities of a facility's design program are difficult to define and put into words. However, the degree to which a design-build proposal is able to meet such design criteria for image and character are often the deciding factor in many design-build competitions. The design image is likely to evoke an emotional response from the evaluators, as opposed to a rational response, and this reaction can color their evaluation of the remaining factors. This reason makes it important to define for the designers, the owner's expectations in this regard. Interview all major stakeholders individually to arrive at a majority consensus on "what the facility should look like" or "what it should not look like."

10.8.2 Functional efficiency

Functional efficiency is not the "net to gross" efficiency of which most building owners speak. Rather, it is the ability of the facility staff to perform its day-to-day tasks efficiently, and the adequacy of the design to facilitate convenient and secure circulation of both people and materials. To the degree that this efficiency can be measured and standards established, they should be stated in the building program, and quantified in the selection criteria. That may mean the distance on one floor between functional units; it may also be the distance that materials, paperwork, or waste products have to travel within the structure, or the number of times they have to be handled. For buildings with many visitors or a need for security, the floorplan layout and exit paths can significantly affect the building's staffing requirements.

The preceding notwithstanding, if the traditional measure of building efficiency (i.e., net-to-gross floor area) is important to the owner, the RFP should clearly define how that efficiency would be measured. Both the American Institute of Architects (AIA) and the Building Owners and Manager Association (BOMA) have publications describing standard methods of measurement for building area.

10.8.3   Health and safety

For almost every type of facility there are issues of public and employee health and safety, from pedestrian security in a parking garage, to indoor air quality. Those issues of health and safety that go beyond the requirements of the applicable building codes must be assessed and specified in the RFP.

10.8.4   Value for money

All public procurement must stand the test of value received for money spent. This "value" is often a measure of the usable floor area, quality and durability of materials and systems, and the facility's operating costs (staff and utilities). It is the obligation of the owner, in a design-build competition, to inform the design-builder as to how and by whom these criteria will be evaluated and the relative priority of each in the selection process.

10.8.5   Local labor content

All other factors being equal, most public agencies would like to encourage local participation in the design and construction of their facility, at all levels. It may even be the raison d'être for the initiation of the project. If the local labor content, of both design professionals and building trades, is to be a consideration in the selection, it must be listed in the selection criteria.

10.8.6   Disadvantaged business enterprises

This should be recognized as an evolving area of law. As of publication of these guidelines, most government agencies have some form of minority contract participation guidelines that apply to the procurement and bid evaluation process. Because design-build procurement may be considered a combination of both professional services and construction, the application of the agency's disadvantaged business enterprise (DBE) guidelines must be carefully considered. At the time of award, most designs are not complete enough for the design builder to obtain firm bids from most subcontract trades. For this reason, it is not practical to ask the proposer to name its DBE subcontractors at the time of bid, except possibly for those participating on the design team. Those and other aspects of design-build procurement should be discussed with the agency's contract compliance and legal officers before drafting the RFQ and RFP.

10.9   Know Your Costs

10.9.1   Reasonable cost to design and construct a project

Prior to the publication of the RFP, the owner must be confident that the facility program and performance specifications can be met for the fixed lump-sum proposal amount (or within the owner's maximum allowable proposal amount). If the owner's expectations exceed the reality of the design and construction marketplace, the level of competition will be reduced significantly, and the likelihood of receiving design-build proposals with significant exceptions either to quality or to quantity will be almost certain. In the first instance, the most qualified teams of design-builders will not apply for the project if they perceive that the request is unrealistic. In the second instance, a qualified proposal may not be legally acceptable, or is certain to be challenged by the unsuccessful proposers.

10.9.2 Contingencies

It is important to illustrate to the design-builder that the contingency and other costs of a project have been budgeted and financed. This gives the proposer the confidence that reasonable changes to the contract (by the owner, or related to circumstances not the responsibility of the design-builder) can be funded, and that the owner will not attempt to make the design-builder absorb the costs for such changes. This confidence will translate to a more competitive attitude on the part of the proposers.

10.10   Know the Market

10.10.1   Qualified and interested design-build teams

Before developing a strategic plan for the procurement of design-build proposals, it is important to know the number of qualified design/build teams in the area (architects, engineers, and contractors) and their interest in responding to a design-build RFP. To attract a sufficient number of qualified teams, you may need to go outside the area (region or state) for one or more of the required disciplines. Sometimes, announcement of the owner's "intent to request design-build proposals" will get enough responses to determine the level of interest and the skills available in the region.

10.10.2   How to attract design-build teams

Review with the individuals and firms responding to the "intent" advertisement, either a draft of the RFP, or an outline of the design-build procurement process. Be sure to include all responding firms in any communications or invitations to briefing or informational meetings. Consolidate the responses and consider the inclusion of all suggestions not likely to diminish the quality of the design-build proposals, or that do not put the owner at any additional risk for design integrity, cost exposure, or schedule completion.

10.11   Develop a Strategic Plan

From the data and other necessary information assembled, develop a plan of action that appears most likely to produce the results described in the owner's list of goals and objectives. Review the plan with your legal advisers, procurement specialists, and others who may have experience with the design-build selection process. This is the creative part of design-build program management, and probably its most critical element. It is not necessary to write out the strategic plan, but it is important to communicate the essential elements of your plan (see list in Fig. 10.2) to those in the decision matrix, and to get their concurrence. This concurrence is particularly necessary from those responsible to develop or approve the selection criteria, and from those individuals who will be accountable for the technical and design evaluations.

10.12   Request for Design-Build Qualifications and Competition Prospectus

10.12.1   Project purpose

The RFQ is a formal request for the necessary and desirable qualifications from design-builders wishing to be considered for the competitive proposal preparation phase of the selection process. But just as important, it is the initial document in a strategic plan to "sell" the project to the most qualified teams of designers and builders. In a best-value or similar competitive selection process where price is not the sole criterion, the proposers will not simply respond with a price "bid" as they do in the traditional design-bid-build process, but will compete on a combination of important factors. The proposers will compete first on the basis of experience, talent, past performance, resources, design, construction, and management skills. Later, in the following phase, they will compete on the basis of design innovation and excellence, value-added to the project, and responsiveness to the owner's objectives, including cost. The project outline is the owner's first opportunity to attract the targeted design-builders, and to encourage them to assemble teams that can offer the subjective attributes the owner's needs for the design and construction of a successful facility.

10.12.2   Summary project description

The summary project description should contain these descriptive elements that will indicate to prospective design-builders regardless of whether the

■ Identification of owner	■ Honoraria
■ Description of project and scope	■ RFP requirements
■ Building type and size	■ Summary of proposal selection criteria
■ Estimated cost	■ Basis of award
■ Project schedule	■ Identification of jurors
■ Selection process	■ Minimum requirements of D/B team
■ Type of design-build competition	■ Submittal requirements
■ Key dates	■ Prequalification selection criteria
■ Presubmittal conference	■ Submittal deadline and address
■ Number of finalists	

Figure 10.2   Elements of a design-build request for qualifications.

project is of any interest to them. The following information should be included:

- Owner, including organizational component(s) to occupy or use the facility.
- Building type, including any special areas (auditoriums, cafeterias, laboratories, etc.) that may require specialty design or construction experience.
- Net or gross floor area and number of stories, if known.
- Site location and site area.
- Estimated contract award amount or limits, and financial constraints (if any).
- Estimated construction start and completion dates.
- Services necessary, such as design, construction, demolition, renovation, hazardous-material removal, and, if necessary, the need for the design-builder to provide a site and/or financing and/or facility operations. The latter services, while not rare, are not strictly a part of the design-build process as described in these guidelines, but rather are part of what is normally referred to as turn-key or developer competitions.

10.12.3   Summary of selection process

In a single paragraph, describe the two-phase, invited, design-build proposal selection process (including interviews, if appropriate) that the owner intends to use for this project. If necessary, reference the appropriate state or municipal regulations that prescribe the selection process. Include the following information:

- Owner's contact person, name, title, address, and phone/fax numbers, and the restrictions or limitations on direct communications
- Deadline for submission of design-builders' qualification statements, and number of copies required
- The maximum number of finalists the owner intends to invite to participate in the next phase of the selection process
- Any compensation to be offered to the finalists for the preparation of design-build proposals

10.12.4   Project objectives

Summarize the owner's project objectives, in general order of importance. That will give the prospective design-build applicants an indication of the owner's values, and some guidance in the assembly of the design-build team. The design-builder will attempt to match each objective with staff members and/or team member firms with suitable talent, experience, skill, and reputation. The objectives, at this point, may be the simple, comprehensive qualities the owner seeks for its facility. Most design-build projects will have similar, broadly stated objectives. Therefore, it is the owner's rank order of project objectives that will be most important to the design-builder at the prequalification stage. Examples include

- Design excellence, character, and image
- Functional efficiency, safety, and security
- Quality of systems and materials
- Operating and maintenance costs and energy efficiency
- Elapsed time to design and construct
- Capital cost or value

10.12.5 Project budget

A summary of the project budget will indicate to the design-builders the relative value of the design and construction to the entire project. It should disclose the cost of land, legal fees, furnishings and equipment, move-in costs, initial operating cost, insurance, contingency, and other costs. Generally, it will indicate that the owner is fiscally realistic and responsible, and that the project is reasonably secure, and will likely proceed. The recognition of those project cost categories by the owner will give the design-builders confidence in the project as a whole, and some assurance that contingencies have been accounted for, and won't later be taken out of the basic design and construction budget. A simple analysis of the project budget and the project scope will indicate to the design-builder whether the competition will be a "budget challenge," or if there will be sufficient funds to allow design flexibility, and materials and system quality improvements over and above a code-minimum facility. The results of such an analysis will likely figure heavily in the design-builder's team composition decisions.

10.12.6 Project financing

The most pertinent piece of information a design-builder will seek when analyzing its decision to submit its qualifications is the progress of the project's financing. Nothing is a greater disincentive to a prospective design-builder than the understanding that the project's funding is somehow in doubt. It is to the owner's advantage to delay the call for qualifications until the project's financing is reasonably secure. Prospective applicants will understand the need to get approval to sell bonds, or a similar technical or procedural financing activity, if the approval is to come from the same legislative body as the issuing agency. They will be leery of project financing that needs approval by voters, or by another legislative body. Similarly, the knowledge that project funding has been challenged by a taxpayer's lawsuit or similar legal action will greatly discourage active participation by the most sought-after design-build teams.

In some instances, when occupancy and completion dates are critical, owners will proceed with the RFQ phase, with the expectation that the project's funding will be assured before commencement of the RFP phase.

A secondary consideration is that the design-builder can be expected to hold its price proposal open for a limited time only after submittal (generally described in the proposal form and the bid bond form). An exception to this limitation is the inclusion of a construction-industry inflation index in the form of proposal to adjust for delays in the execution of the contract. If time for a voter's referendum or court action must be allowed after the receipt of design-build proposals (especially if the results of same are in doubt), the owner will not receive the design-builders' best efforts and/or prices.

10.12.7 Building area summary

Similar to the summary project budget, the building area summary will help the prospective design-builders analyze the project's design challenge. A one-page summary of the floor area requirements will indicate to the design-builder whether the owner has been realistic in allocating space for specific functions, and for supporting areas, such as circulation, mechanical, and electrical areas. This analysis will help define the design and budget challenges inherent in the project. The information provided by the project descriptions will allow the design team members to estimate the scope of the work necessary to prepare a proposal. This same information also allows the prospective design-builders, at their own risk, to get an early start on the development of a design concept and a pricing strategy, well before prequalification. In very competitive situations, an early start can be a considerable advantage to a finalist.

10.12.8 Project schedule

Of the three essential pieces of information that a proposer needs to know—needs (program), resources (money and land), and time—the latter is often overlooked in a request for design-build proposals. All contracts must state a time period in which the services and/or product are to be delivered, and the RFQ is the first place to give the proposer the owner's specific needs in this regard. Time of performance requirements in an RFQ and the RFP are best stated in elapsed consecutive calendar days from the date of notice to proceed. In this way, changes in the schedule to solicit, receive, evaluate, select, and award can be changed without affecting the project schedule. Except in locations subject to extreme weather seasons, or "drop-dead dates" for completion, the date of the start of the contract will not affect the proposer's cost. In those instances where the completion date is critical, the RFP must include a "but not later than" qualifier in the project schedule. The RFQ should summarize the project schedule for the prospective design-builder respondents.

10.12.9   Interim project schedule events

The project schedule is the place to specify the timeframe for interim events. These events may be submittal requirements of the design-builder, such as design development drawings or construction documents. They may also include requirements of the owner, direct or through a third party, such as site availability, completion of an environmental impact report, or the delivery of owner-furnished equipment or materials. Again, the interim deadline requirements should be stated in elapsed days and may be an obligation of the design-builder or owner. The obligation of the owner to complete specific submittal reviews within a specified period may also be included in the project schedule. The project schedule may reference the contract general conditions' time limitations on certain notices and actions.

10.12.10   Competition schedule

The competition or RFP schedule is the schedule of the entire selection process, and should include all activities from initial announcement to notice of award. The schedule should be stated in specific calendar dates, and it should clearly identify the time allotted for the preparation of qualification statements and for design-build proposals. An outline of the competition schedule should be included in the announcement, and a detailed schedule included in the request for qualifications, and included or referenced in the request for proposals. Owners should consider carefully any changes to the competition schedule after it has been initially published. The proposers will view a commitment to the selection schedule as an owner's commitment to the project and the specifics of the RFP, such as selection criteria and performance requirements. Unexplained changes to the schedule may make the design build teams suspicious that other changes may also be contemplated. Owners must also understand that to compete successfully in a design-build competition, team members must make specific time commitments for preparation of the proposal. Schedule changes in the middle of a proposal preparation period discourage the competitors and make it difficult for them to focus their efforts on the proposal, all to the ultimate detriment of the owner.

10.12.11   Basis of contract award

Prequalification.   The owner's decision process for selecting the most qualified design-builders is the most critical aspect of the design-build prequalification process. The owner must establish requirements (selection criteria) to measure and determine the most qualified firms to invite to participate in the proposal preparation phase. For the evaluation and selection process to succeed, both the owner and the prospective proposer must convey to each other clear, accurate, and complete information. Only firms that have been qualified by the owner may submit a proposal, and the owner may select a proposal only from a prequalified firm. The prequalified design-build team must remain exactly as described in the design-builder's qualification statement, unless the owner has authorized a change or substitution. It is customary to announce to all applicants and prequalified firms the complete list of qualified firms and their component members.

Complete proposal.   A valid and complete design-build proposal meeting all the requirements of the RFP must be submitted. Proposals may not be conditioned or qualified in any way, except as may be allowed in the RFP (see Sec. 10.14.7). Proposals with such limitations or conditions are subject to disqualification and rejection. Award must be based on the receipt of a complete proposal.

Functional facility.   Proposals must provide for a complete and functional facility. In the absence of any description of function in the RFP, the owner has the right to expect the proposal to provide for a facility that will be fully operational in every significant aspect, and for the purposes for which the facility was intended, in the same way that a purchaser of an automobile expects the vehicle to operate when the purchaser provides fuel and a competent driver. Award will be made on the evidence and on an unqualified assumption that the proposal provides a completely functional facility.

Jury's authority to recommend.   The owner must state the authority of selection panel (jury) to prequalify applicant design-build teams, and to recommend a specific design-build proposal to the owner. Separately, the owner must also describe its limited authority to accept (or reject) the jury's recommendation and to enter into a contract with the successful proposer.

Owner's right to reject jury's recommendation.   The "basis of award" section of the RFP includes the owner's right to reject any or all proposals, including a proposal that has been recommended by the jury. That right of the owner to reject a recommended proposal should *not* include the right to subsequently select from among the other (not recommended) proposals. In the instance of the rejection of a recommended proposal, the owner must pay the recommended proposer the minimum contract amount specified in the RFP.

10.12.12   Design-build competition process

Overview.   It is strongly recommended that all design-build competitions be conducted in at least two stages. A prequalification phase is followed by one or more additional stages to narrow the applicant field down to the few that will prepare and submit complete design-build proposals. This section of the RFQ is to inform the potential applicants, in advance, of the process that will be used by the owner to select the successful proposal.

Stage 1: request for design-build qualifications. Under the best-value design-build procurement method, considerable effort is placed on identifying and qualifying the project delivery firms (or teams of firms) who have the greatest potential to design and build the facility in question. This is not the case under the traditional design-bid-build process used by most public agencies.

Prequalification occurs under stage 1 of the design-build competitive selection process in which the owner issues a request for design-build qualifications (RFQ) document. In this document, the owner must establish clear, objective standards or criteria that the jury will use to determine the most qualified firms to be invited to prepare and submit a design-build proposal. The two essential parts of an RFQ are the owner's requirements, conditions, and criteria, as well as the submittal requirements. Both the owner and the applicants need a thorough understanding of the other's intentions. Consequently, owners must explain all significant parameters relating to the qualification, and subsequent selection of a successful design-builder and its proposal.

A request for qualifications is a well-accepted procedure in the design industry, but much less so among construction contractors. Most design firms have prepared materials and standard forms available in advance to respond to the normal requests for qualifications by public agencies under various selection procedures.

All that notwithstanding, owners also may request additional and more specific information, such as design-build experience, building type experience (similar to the owner's project), narrative statements on design and management approaches, organizational charts, construction management and quality control plans, safety records, change-order records, firm brochures and photographs, and/or slides. (Arrangements must be made to return applicants' slides.) Some public agencies also require the applicant design-builders to submit information on the number or percentage of minority and women-owned businesses (or specifically to identify such firms) it expects to use in the design, and in the construction of the project. Generally, prequalification submittal information formats that have been designed to elicit specific or project relevant information are more successful than predetermined or printed-form formats.

Interim stages: interviews and interviews-with-concepts (optional). It is common for project sponsors to use the interview process to narrow the list of potential proposers down to the few (at least three, not more than five) to be invited to prepare and submit design-build proposals. In such interviews, the design-builder and its team of architects, engineers, and specialty design consultants expand on the information provided in their qualification statement, and respond to questions from the selection panel. If the design-build team-member firms, individuals and/or their projects are well known to the selection panel, interviews add very little further insights.

Less common is the practice of requiring, or allowing, the design-builder to discuss with the jury one or more preliminary concepts they may pursue, if invited to participate in the proposal preparation stage. These "concept" presentations may take the form of sketches, mass models, or floor- and siteplan overlays on thin paper, all created during the interview meeting among the jurors and the interviewees. The purpose of the "concept" interview is not to select a design concept, but to allow the jurors an opportunity to see how the individual team members work together, and to determine how responsive the team will be to the project's needs and objectives.

A disadvantage to the interview-with-concepts procedure is that either the owner may expect to receive, or the proposer may feel some obligation to pursue, the "concept" in the next phase of the competition, even though the concept was developed with limited information about the project, and in a very compressed time schedule. On the other hand, an advantage to the process is that the design-build team has an opportunity to explore with the jury members several preliminary design concepts, and to get collective and individual reactions to various and wide-ranging design suggestions. These insights into the selection panel's "mind" can be very valuable in a design competition.

In the case of interviews-with-concepts, the owner must make an effort to keep the preparation and presentation as simple as possible. It is also critical that all such information presented during the interview be treated confidentially. Because original and potentially useful work is conducted on the owner's behalf, an honorarium is appropriate. The amount offered should be proportional to the time and effort required, and should compensate the design-builder for the travel and other out-of-pocket costs incurred by the team members.

Stage 2: design-build proposals.   This is the essential part of every competitive design-build selection process. The preparation of a design-build proposal by a prequalified team of designers, builders, and subcontractors is an intensive, sometime hectic, and hopefully, a very creative process. For most major public projects, the team will include both new and previous participants, some as generalists within their disciplines, others as specialists in one aspect or the other of the building type or its design and construction solution. The design-builder must assume the role of strategist, design-process manager, logistics planner, scheduler, and cost estimator. In a competitive situation, the process is almost never routine, and each team member is expected to provide an extraordinary contribution. Facility owners should recognize the exceptional and unique character of the process the designers and builders are undertaking, all in the owner's behalf, by supporting the process and lessening the negative impacts on the teams as much as possible. When received, evaluated, clarified, and selected, the proposal represents the "other half" of the contract between owner and design-builder, the owner's request for proposal documents constitutes the first half.

10.12.13   Selection panel (jury)

General.   It is the owner's responsibility to provide a selection panel or "jury" that is knowledgeable of the needs of the owner, and able to understand the nuances and complexities of the design solutions to be submitted. The owner must assign to the jury the sole responsibility of evaluating the eligible proposals (with the possible assistance of a technical evaluation committee) and selecting the proposal that best meets the owner's published objectives. The owner must limit its authority to accepting or rejecting the jury's recommendation. The owner may not select any proposal other than that recommended by the jury, but may reject the recommendation and start over with any other legal selection process. If the owner (board of directors, council, commission or authority, etc.) wishes to reserve the selection of the successful design-build proposal to itself, then it must make clear in the RFQ that it intends to be its own "jury," and to refrain from appointing an "advisory only" panel. The proposers can only reasonably be expected to target their proposal to one, and only one "jury."

Technical evaluation committee.   On large and complex projects, the jury can benefit from the use of a technical evaluation committee or panel. This group of technical and functional "experts" is normally composed of those responsible for the compilation of the facility requirements and performance specifications in the RFP. They may be employees of the owner, representatives of the facility users, and the owner's technical and building type consultants. Their responsibility is to evaluate each proposal in the area of their expertise and experience. They may score or rank those portions of the proposal relative to the same area of competing proposals, and relative to the minimum requirements of the RFP. They are not expected to evaluate the broader, subjective areas of the design proposals, such as building image or impression. They may not give the jury any overall ranking or scoring of the proposals, only sectional evaluations. Their reports to the jury should be both in writing and in person. Technical evaluation committees function as technical staffs of the selection panels. The evaluation reports serve as supplementary documents in the jury's final report to the owner, and can assist in the owner's debriefing to the competing design-build teams, after the selection process has been concluded. On less complicated projects, one or more of the jury members can assume the role of the technical evaluation committee.

Number of jurors.   The number of different, competing interests that have to be addressed by the proposals most often determines the number of jurors on a design-build selection panel. The dynamics of such groups suggests that five to seven members works very well. With that number of people on a jury panel, it is difficult for any one individual to get enough support to argue against the will of the majority. Most jurors in small groups will try to find and support a common ground. On larger jury panels (more than 12 members), jurors with divergent opinions can often find enough support to make a unanimous recommendation difficult. In larger discussion groups, individuals are less inclined to seek accommodation with the majority's opinions.

Independent jurors. For a public facility, half or more of the jury members should be financially independent of the owner's organization; that is, a majority of the panel should not be employees, nor should they be officers or members of the organization's board of directors or other governing body. The owner's professional consultants are generally considered capable of expressing independent opinions, and may serve on a jury. However, technical and operational consultants may best be used on the technical evaluation committee. To the extent practical, design evaluation jurors should be selected from among leaders in their respective professions, businesses, and/or communities, national, regional, or local, as appropriate. The ability to read and understand architects' and engineers' models, drawings, and other documents is essential to fair and equitable selection.

Jury composition. Owners should consider these categories of juror types when organizing a selection panel:

- Owner or representative(s) familiar with the owner's operations and objectives
- Individuals in similar business or operations, but not direct competitors
- Employees or facility user representatives
- Two or three design professionals, such as architects, engineers, or landscape architects, at least one of whom should be from outside the immediate area, but well known and respected professionally
- Noncompeting builder or other individual familiar with the construction industry in the region
- Community representative(s)

Jury chair. If the owner intends to participate on the jury panel, then the owner should serve as the jury's chair. Otherwise, the owner should appoint a chair from among those individuals with previous experience on a design selection panel. The chair should be a voting member of the panel.

Some jury panels are organized by owners to represent a broad spectrum of community, employee, and other interests. In such instances, there may be a natural suspicion that one member or a special-interest group may have more influence on the final recommendation than another may. To alleviate that situation, we recommend that the owner appoint the competition's *criteria professional* as the nonvoting chair of the selection panel. The chair's role would then be to facilitate a fair and even discussion of the merits of each of the proposals, and to arrive at a recommendation the entire panel will support.

10.12.14 Design criteria professional

Neutral party. If the preparation of the design-build proposals is expected to require any significant effort on the part of the design-builders and their architects, engineers, and consultants, the participants will expect the owner to appoint a neutral third party to be responsible for the administration of the selection process. The impartiality and objectivity of the competition administrator is best maintained if the individual is not an employee of the owner and has no financial interest in any of the competing design-builders or in any of their team member firms.

Duties and responsibilities.   Duties may include

- Preparation of all or part of the RFQ and RFP documents
- Communications with design-builder applicants and prequalified proposers
- Certification of the completeness and timeliness of the submitted proposals
- Assisting in the technical evaluation of proposals
- Assisting the jury in its review of the proposals
- Recording the jury's comments
- Certification of jury's recommendation and final report of the jury
- Debriefing for the competing design-builders

The criteria professional has the additional responsibility to advise the owner of any actions of, or information concerning, the competing design-builders that may disqualify them from the RFP process.

10.12.15   Owner's right to change competition schedule and/or terminate RFP process

Changes in the competition schedule are often unpredictable. They should be avoided whenever possible after the announcement and publication of the initial RFQ. Changes after the commencement of the proposal preparation phase are particularly disruptive to the process and discouraging to the participants. Schedule changes should not be made unless there is a clear and obvious reason for doing so. Most participants in a competitive selection process will assume, in the absence of any legitimate reason, that the delay or change was made for the convenience of another competitor. Unsubstantiated changes in the schedule weaken the proposers' enthusiasm for the project. The assumption is that if the owner can change one aspect of the selection process in midcourse, then it is likely the owner will feel free to change other aspects of the project and/or the selection process.

Termination of a design-build selection process can occur, and all public owners have to reserve the right to do so, when necessary. If the process is canceled prior to the commencement of the proposal preparation phase, then compensation to the prequalified proposers is optional. If cancellation occurs after the midpoint in the proposal preparation phase, the owner should compensate each proposer with the full honorarium amount. Earlier cancellations may be compensated in proportion to the time expended.

10.12.16   Request for design-build qualifications (stage 1)

Advertisements.   The official announcement of the project should be placed in the "legal announcement" newspapers serving the agency's jurisdiction. Additional legal advertisements should be placed in other areas from which the owner expects to draw design-builder proposers (see Sec. 10.11, on developing a strategic plan). If the project is large, or the timeframe for the selection process is short, a "notice of intent to request design-build qualifications" may be placed early in the RFQ document preparation phase. If the owner has targeted national or international design and construction firms, then display advertisements in national and international publications may be warranted. Display advertisement requires longer lead times and greater expense than the normal legal announcements.

For most design-build prequalification procedures, the legal advertisement cannot contain enough information for the potential proposers to adequately assemble a design and construction team, and to complete a qualification statement. For this reason, the advertisement should contain only basic identification and project scope information. Also include the nature of the design-build selection process, number of stages and key dates in the selection process, and the offer of compensation (or no compensation) for the finalists. Send interested parties the request for design-build qualifications and competition prospectus document (RFQ).

Registered RFQ holders.   In addition to jurors, technical evaluators, and others in the owner's "official" project family, only firms and individuals requesting an RFQ through official channels should receive one. Writers of inquiries received prior to the advertisement may be sent a copy, or may be sent only a copy of the advertisement. A complete registry of RFQ holders, their contact person, phone and fax numbers, and other pertinent information must be maintained. The registry is necessary to be able to send RFQ addenda to registrants. The registry is also a good indication of the type and level of interest the project has generated.

The mailing list of registered RFQ holders should be distributed to all holders from time to time. This facilitates team building, particularly among specialty design consultants, and heightens the level of competition among potential proposers.

Questions and answers.   Questions concerning a new design-build selection and contracting process are inevitable. Addenda are the only way to issue additional written information and answers to questions, and make the information and instructions binding on both the owner and the proposers. Questions, beyond requests for the RFQ document, must be limited to written (letter or facsimile) questions from registered RFQ holders. Answers must be published in addenda to the RFQ. Draft copies of the questions and answers can be distributed to the RFQ holders by facsimile, if all holders receive the questions and answers (Q&As) at the same (or approximately same) time. Addenda should be distributed in the same manner as the initial RFQ. Owners may allow RFQ holders to request overnight courier delivery of addenda and other documents, at the registrant's expense.

Presubmission meeting(s). For significant projects where there is considerable interest, or where the submission requirements are complicated, the owner should schedule one or more presubmission meeting(s) not later than 2 weeks prior to the deadline for submission of the qualification statements. The meeting(s) allow the owner's staff and consultants to answer potential proposers' questions in person (to be followed up by written addenda), to "sell" the project, and to gauge the level of interest among the targeted design and construction firms.

Competition schedule. In addition to the obvious listing of deadlines for submissions, interviews, and selection, the competition schedule is important because it allows the design-builder to obtain schedule and performance commitments from the design-build team member firms prior to committing to the team composition in its qualification statement. It is important that the schedule list the specific calendar time set aside for the preparation of the design-build proposals, and the schedule of in-person presentations, if any.

Minimum requirements. The minimum requirements for consideration for prequalification should be as objective as possible. These may include

- Design and/or construction experience in the facility type and scope
- Minimum experience of design-builder in design-build method of contracting, and similar experience of the design team (together and separately)
- Minimum level of bonding capacity, and "proof" thereof
- Minimum insurance requirements
- Limitations on the geographic location of the design-builder and/or the architect/engineer (A/E) of record, if any
- Minimum design team composition, including some specialty consultants
- Minimum DBE participation on the design team
- Required license, registration, and/or tax status of design-builder and team members

Selection criteria for prequalification. Selection criteria are, by nature, a combination of subjective and objective factors. The owner should list only enough criteria to allow the jury to make reasonable evaluations sufficient to recommend the desired number of prequalified firms to participate in the next stage of the selection process. Criteria that require extensive, detailed supporting information or analysis will not likely be used in a meaningful way by the jury.

The criteria should be listed in order of importance, and weighted. Typical selection criteria may include

- Design and construction excellence, including owner and industry references
- Specific experience of the design-build team members in the subject building type
- Actual design-build experience similar to that requested in the RFP [as opposed to quasi–design-build projects, such as developer projects, guaranteed maximum prices (GMPs), and construction management (CM) or CM/ general contractor (GC) contracts]
- Design-build experience of the team members working together on the same project
- Financial strength and stability of design-builder
- Performance records of design-builder and design team, including fees, recent comparable costs, value engineering savings, change-order record, and on-time performance
- Quality of technical and managerial organization proposed
- Degree of local professional participation, including DBE or other business firm classifications
- Quality of individuals proposed for key positions
- Design approach or philosophy, or preliminary design concept
- CM plan, including time, cost, and quality control
- Resources (staff, equipment, and capital) available to the project

However, the selection panel should be instructed to limit its evaluation of the respondent teams only to the selection criteria listed in the RFQ. To do otherwise may invite a legal challenge to the selection process.

Receipt of qualification statements.   Indicate the usual when, where, and how the qualification statements are to be delivered, labeled, and so on.

No compensation for preparation of qualification statements.   It is not normal to offer compensation for qualification statements, and the potential applicants should be notified of this in the RFQ.

Required format of qualification statements.   To determine eligibility, and to make quick and accurate comparisons among competing applicant teams, a consistent format of information is desirable. There are several forms available from government agencies, professional organizations, and trade associations that would include most of the typical information requested in a design-build RFQ. Use of those forms, to the extent possible, allows for convenient comparisons by the jurors, and reduces the cost to the applicant firms to prepare the information. To further simplify the application and evaluation process, the owner can specify the sequence and format of the qualification statements, limit the number of pages of each section (particularly useful to limit the number and/or size of individual résumés), and encourage the applicants to submit printed brochures and other prepared materials, in lieu of additional narrative.

Number of copies. Applicant design-build teams prefer that the jurors and other evaluators review "original" copies of their qualification statements, rather than "copied" versions. Therefore, the RFQ should specify a requirement for a sufficient number of copies of the qualification statements to supply each reviewer, and a record copy for the file.

Photographic slides.   Photographic slides (35-mm, color) allow for a quick and convenient review and comparison of experiences among competing design-build teams, particularly if the jury includes members that are in neither the design professions nor the construction industry. Most design firms maintain a slide library of completed projects. If used, request a single set of slides, normally a maximum of 20 or 40 (20 slides can fit into a single slide holder sheet). All slides must be identified by facility and designer and/or builder, and indexed on a separate index page (to enable the projectionists to identify the individual slides while the slide is in the projector). The owner must return all slides after prequalification. However, the owner should reserve the right to make copies of all slides submitted.

References.   If references are requested, a maximum number should be specified. The reference information should include the contact person, and the project name(s), dates, and services corresponding to the reference. Applicants should be instructed to brief the reference prior to submittal, and to verify location, address, phone number, and availability of the contact person.

Number of finalists to be selected.   The number of design-build teams the owner intends to prequalify for each phase of the selection process must be stated in the RFQ. The usual range of finalists has been between three and five. Five proposals can offer the owner a broad range of design solutions without unnecessarily diminishing the level of competition among proposers. Because of the great expense to prepare a design-build proposal, and need to offer an equitable honorarium to the unsuccessful proposers, the trend lately has been to reduce the number of competing firms in the final phase to three, and sometimes four. If a security deposit or bond is required of the finalists to ensure that all prequalified proposers submit a valid proposal, three finalist teams can be adequate. If such security is not required, additional design-builders may have to be prequalified as "insurance" that the owner receives a suitable number of responsive design-build proposals.

An owner may consider additional factors when determining the number of finalists to prequalify. For instance, if interviews-with-concepts are used, an owner may feel more comfortable with fewer finalists because it has a deeper understanding of the design directions the proposers are likely to pursue than it would have with an experience-only prequalification process. Also, fewer design-build proposals allow the owner's evaluation panel the time to examine the submitted proposals in more detail and to resolve conflicts and ambiguities with a subsequent clarification process.

Notification.   The RFQ should indicate the method by which the applicant teams will be notified of the jury's recommendations, and the owner's action on the recommendations. The competition schedule will indicate the date of notification. If the jury's recommendations are to be delivered in a public forum, and the owner must act in the same or similar public forum, it is particularly important to inform the applicants of the time and place.

Qualification statements become property of owner.   Except for the photographic slides, the RFQ should indicate that all submitted material becomes the property of the owner.

Description of the agreement to prepare a priced design-build proposal.   This is a short-form or letter-type agreement between the owner and the finalists. In the agreement the owner promises to provide timely information, to employ qualified evaluators and jurors, to evaluate the proposals in accordance with the published criteria, and to compensate finalists who submit valid proposals. The finalists promise to abide by the rules and regulations of the RFP, to avoid contact with the owner, evaluators, and jurors, except as provided for in the RFP, and to pay a security deposit or bond to ensure their performance (if security deposit is required).

10.12.17   Outline of requests for design-build proposals (stage 2)

Purpose of RFP outline in the RFQ.   The design-builders that intend to submit qualification statements in response to the owner's RFQ should be informed in advance of the requirements of the RFP. This allows the design-builder the opportunity to assemble its team of designers and other subcontractors with an understanding of each team members' responsibilities, tasks, schedule, and submittals required by the RFP. Alternatively, an owner may publish both the RFQ and the RFP simultaneously as a single, comprehensive document. The RFQ should indicate the owner's intentions with respect to these items and should forewarn the prospective design-builders of the following:

Agreement to prepare a priced design-build proposal. An agreement to prepare a design-build proposal must be executed by the design-builder before it may receive the RFP and participate in the owner's briefings for the finalists. A security deposit or bond will be required to ensure that the proposer will submit a valid design-build proposal. An honorarium or stipend will be paid to the unsuccessful proposers. The agreement will specify that the proposer comply with all the requirements of the RFP, or the security deposit will be retained and the honorarium will not be paid. The RFP will be an attachment to the agreement.

Competition schedule. Indicate that an expanded competition schedule similar to the one included in the RFQ, but with additional interim events scheduled, will be included in the RFP.

Requirements for owner's briefing(s). Because of the requirement for commitments of professional personnel, and travel expenses associated with the owner's briefing meetings, it is appropriate to inform the potential proposers of the requirement for mandatory attendance at owner's briefings in the RFQ.

Questions and answers and other communication restrictions. Potential applicants must be informed of the restrictions on contact and communications at the RFQ stage, because they are applicable immediately.

Submittal requirements. The RFP's requirements for submittals (supplements to the proposal form) are a significant cost item in the preparation of a design-build proposal. For this reason, the design-builder and its associated design professionals should know the scope of work required to participate in the proposal preparation phase.

Fixed or variable price and alternates required or allowed. A critical element in the design-builder's decision to submit a qualification statement, and for the composition of its design-build team, is to know to what degree the selection process is cost- or value-weighted. The method of cost determination must be communicated in the RFQ.

Technical evaluation of proposals and selection criteria. The method of, and the criteria for, the evaluation and final selection of the winning design-build proposal will indicate to the prospective proposer the technical and professional talents it will need to compete effectively for the project. Some indication of this aspect of evaluation and selection should be included in the RFQ to be fair to the proposers. Note that the RFP does not allow the proposer to delete or change design professional firms after it has been prequalified.

Request for clarifications of proposals process. The request for clarifications, after the owner's initial evaluation of the submitted design-build proposals, is an additional element in the team's scope of work, and should be mentioned in the RFQ.

Public exhibition of models and display boards. If public exhibition of the design-build proposals is a factor in the selection process, it should be indicated to the potential proposers in the RFQ.

Requirements for in-person presentations. Because of the need to establish long-term time commitments of senior management and design principals to conduct in-person presentations for the jury and/or the technical evaluation committee, an early indication of that requirement, and the dates indicated in the competition schedule, would help the prospective design-build teams plan schedule commitments.

Jury recommendation. Potential design-build applicants should be notified in advance that the jury alone will make the recommendation to the owner of the winning design-build proposal. This fact, plus the identification of the jurors, will greatly influence the composition of the design-build team.

Contract award. The owner will have the authority to accept or reject the jury's recommendation. Under the RFP procedures, the owner will not be able to select any other design-build proposal. It is to the owner's advantage to indicate to the prospective design-build teams that they will have to design for only one jury, not two.

Final report of jury. Indicate that the final report of the jury, including the complete technical evaluation, will be available to the finalists after the owner executes a contract with the successful proposer, but not before.

10.12.18   Prequalification selection criteria

Typical criteria are listed in Fig. 10.3.

Need for selection criteria.   The ability of the design-builder to successfully complete projects efficiently and on time is a key attribute that an owner must consider when evaluating prospective design-build proposers. Design and construction management capabilities vary from one organization (or team) to another. While most firms are inherently eager to compete for different types and sizes of projects, some firms are better qualified in terms of management skills, design quality, quality control, finance, experience, and personnel than other firms for a specific type and size of project. For this reason, minimum requirements, combined with a screening process using published and weighted evaluation criteria, are essential. Collectively, those requirements and judging factors are referred to as prequalification selection criteria.

- Builder's financial and bonding capacity
- D/B team's building-type experience
- Record of design and technical excellence
- Staff experience
- Design-build experience
- D/B's organization-management plan
- D/B's quality control plan
- D/B's record of on-budget performance
- D/B's record of on-schedule performance

Figure 10.3 Typical prequalification selection criteria.

Design-builder's organization and management plan. Some organizations are constantly reshaped by management based on their concept of a winning strategy. With this in mind, it is important that the organizational structure of each applicant organization be identified clearly for the specific project for which it is applying. With this information the jury can determine, at least relatively, which organizations have the greatest potential for meeting the owner's objectives, and delivering the project successfully and efficiently.

The selection criteria, therefore, should be structured to require the applicant demonstrate that its proposed organization-management plan has the capability, special experience, and the technical competence to deliver the proposed facility within the schedule, quality level, and other existent and/or specified conditions. Capabilities to be examined and evaluated should include: client relations, design and construction management, quality control, schedule control, and risk management. An organizational chart is often provided to illustrate lines of reporting and responsibility within the organization.

Design-builder's financial capacity. The successful proposer must have the financial capacity to absorb the project start-up costs, and to maintain the day-to-day working financial aspects of the organizational structure. The submission requirements should require applicants to identify project financial arrangements, proof of ability to bond the contract, their current situation with regard to significant claims and/or lawsuits, pending or outstanding judgments of any significant size, and an indication of their financial reputation among the applicable business, banking, and credit institutions. These financial qualifications take on extra significance in light of the fact that, through the prequalification process, the owner is limiting open competition on the project. For that reason alone, the public owner must exercise extreme care and use prudent judgment in the selection of prospective proposers. The ability to provide a contract bond is not, in and of itself, sufficient evidence of financial capability.

Design-builder's management experience and capability. Design-build projects, with their combined design and construction responsibilities, must be carefully and prudently managed to meet both the owner's and the design-builder's goals. Without a balanced effort, adversarial situations may occur, either within the design-builder's organization, or between the design-builder and the owner. Strong project team leadership capabilities, and relevant management experience, for both the design and the construction sequences, are essential selection criteria.

Design-builder's construction experience. Much will be required of the construction team and, ultimately, their efforts will be judged on the quality of the work produced, and on the timely and profitable completion of the project. Design-builder applicants should be asked to demonstrate the level of construction experience and competence of the proposed team members (prime and major design-build subcontractors) and their ability and willingness to partner with the owner, designers, and others to meet their collective objectives.

Architects' and engineers' design experience. Whether the owner receives a well-designed and functional facility is greatly dependent on the team's architectural and engineering capabilities. Broad knowledge and experience in such important areas as aesthetics, client communications, facility programming, urban and site planning, environmental design considerations, design of building systems and interiors, materials, equipment, color selection, and cost and schedule control is essential. Demonstrated excellence in some or all of those areas should be required for prequalification.

Qualifications of key personnel. Essential to all projects are the specific key personnel forming the design-build team for the project. Regardless of past achievements of an organization, its ability to respond successfully on future work is dependent on the specific capabilities of key personnel assigned to the project. The RFQ should require that those applying for prequalification must identify key personnel, and provide short (one-page maximum) résumés to demonstrate the specific knowledge, experience, and capability of the individual to perform in the assigned role. This can be demonstrated by education, relevant project experience, professional or technical certifications, and peer recognition of achievement or excellence. Applicants must be reminded that, after prequalification, significant changes in the composition of the design-build team's member firms, personnel, roles, or responsibilities may not be made without the prior, written approval of the owner. The RFQ should list the minimum staff positions for which specific personnel must be assigned and identified by name, title, firm, project assignment, and résumé.

Team members' common project experience. The ability of a project team to work together effectively and in harmony is essential for a design-build team where most participants share a common goal. Owners should look for evidence of such favorable common experience among the key member firms and/or individuals, either on another design-build project or on a project of similar scope and complexity.

Design-build experience. For the owners of most large public projects, design-build procurement is a new or recent experience and consequently a relatively new experience for those local or regional firms that seek and perform design and construction for such projects. Familiarity with the process, risks, responsibilities, and types of participants (on both the owner's and the design-builder's team) is a valuable asset for a design-build team. On significant or critical projects, public owners should avoid proposers without some firm or individual experience in the design-build contracting method.

10.12.19   Qualification statement outline

Organization. The applicant (the design-build entity) should provide a clear and descriptive organization outline or chart. All components of the design-build team should be identified, along with their project roles and responsibilities, and their reporting accountability. It is important that the owner understands the type of organization, such as single entity, joint venture, or association of builders, designers, and major subcontractors. The owner must also understand how that organization will operate as one, fully integrated and cohesive unit. Lines of authority and communication should be clearly shown. Responsibilities of individuals within the organization should be listed, and the methods used to coordinate and integrate the activities and responsibilities of each should be indicated on the chart or in the narrative.

Financial capacity.   Indicate the design-builder's form of business (e.g., corporation, partnership, joint venture, or sole proprietor). Provide a copy of the design-builder's latest financial statement, and quarterly updates, if available. If the design-builder is a joint venture, provide similar information for each member of the joint venture. Disclose any unpaid judgments in excess of $1 million (or other appropriate amount). Disclose any disputed or nonadjudicated claims in excess of $5 million (similarly). If the design-builder is not a public company, this financial information will be held in confidence, and it will be examined only by the official responsible for its evaluation (such as the agency's chief financial officer, or its currently contracted certified public accountants).

Provide evidence that the design-builder's current bonding capacity is at least as large as the published contract award amount. This evidence may be in the form of a letter from a licensed bonding company (surety), or from an agent normally representing such a company.

For the design-builder (and joint-venture partners), compare anticipated gross monthly billings for the next 12 months for uncompleted work presently under contract, with monthly gross billings for design and construction over the past 24 months.

Management and construction experience.   In keeping with the submitted organizational chart, describe how the design-builder intends to manage this project, including internal, subconsultant, subcontractor, and owner performance. Provide recent (completed within the past 5 years) examples of the management of similar projects that required management organization, skills, and expertise similar to those required by this project. Identify those persons within the design-builder's organization who held responsible positions on each example project cited, and explain. Provide owner references for each project cited.

Provide recent examples of similar construction experience, including facility type, contract value, schedule, and any peculiarities that may be germane to this project. Identify one or more individuals in the design-builder's organization who held responsible positions on each cited example. Provide owner references for each project cited.

Design experience.   In keeping with the submitted organizational chart, identify all design disciplines and specialty consultants the design-builder intends to employ in the project design. List each firm or subconsultant, its area of responsibility, and the corresponding principal in charge. For each firm, give recent (completed within the past 5 years) examples of similar projects. Identify individuals proposed for this project that held responsible positions on the examples cited. Provide owner references for each project cited. Provide a matrix indicating the participation in a similar capacity, if any, of each firm and specialty consultant proposed for this project, on the projects cited.

Key personnel résumés.   Provide a one-page résumé for each individual listed on the organization chart. List the individual's firm and position therein, including project responsibility, education, license or registration, affiliations, publications, awards, and relevant experience over the past 5 years.

Brochures (optional).   If available, provide a preprinted brochure for the design-builder and each firm in the joint venture and/or listed on the organization chart. Provide one photograph (maximum) for each project cited. Identify (by attachment) project and the firms associated with its design and/or construction.

35-mm slides.   Provide photographic slides and an index (up to 20 or 40, 35-mm slides in plastic page holders, to be submitted with multiple copies of the qualification statement). Slides will be returned to the design-builder. Owner should reserve the right to make copies of all slides submitted.

10.12.20   Additional features of design-build contract

Purpose.   The purpose for listing some of the features of the design-build contract is to allow the prospective proposer (design-builder) to evaluate the risks associated with the design-build contract prior to its submittal of a qualification statement. The items listed in this section are those that may have a significant financial impact and/or those features of the contract that vary considerably from the normal public works contracts issued by the agency.

Eligibility: requirements for professional and contractor licensing.   Public owners have an obligation to ensure that the entity to which they contract for the design and construction of a facility possesses the licenses and professional registrations required by law. Beyond listing the requirement to have or to obtain such licenses when required, the owner should not attempt to administer the jurisdiction's license regulations in detail. The RFQ document should list the names and addresses of the appropriate agencies to which prospective proposers should direct their inquiries about licenses and professional registrations.

Other restrictions.   Some owners require that the architectural and engineering construction documents be produced within the state, or within a specified distance from the project site. This is to ensure that those professionals are conveniently available to the owner during the construction-document review stage, and during the subsequent construction phase. The requirement also attempts to ensure that the designers are familiar with local or regional customs, laws, codes, regulations, materials, and craft capabilities. If other objective restrictions apply to the design-builder, the designers, or any other member of the design-builder's team, they should be listed together in the RFQ.

Proposal bond requirement.   Indicate the type and amount of proposal bond required at the time of submittal of the design-build proposal, if any bond is to be required.

Contract bond must include design.   Indicate to the proposers that the contract bond (performance and payment bond) must cover the entire contract between the owner and the design-builder, including any necessary professional, architectural, and engineering services. The bond must not be limited solely to construction value and associated risks. The contract bond should cover the initial performance of design services sufficient to complete the facility and satisfy the requirements of substantial and final completion. The contract bond is not intended to protect the design-builder from claims associated with design errors or omissions discovered after completion. Those risks can be addressed by a *contractor's design liability policy.*

Liquidated damages.   If liquidated damages are to be assessed for the design-builder's failure to complete the contract within the time specified (or within such time as may be extended by change order), the duration (in elapsed calendar days) and the daily amount of liquidated damages should be mentioned in the RFQ.

DBE contracting and similar requirements.   If the agency has requirements or guidelines for a percentage of disadvantaged, minority, or women's business enterprise contracting, and/or minority or women hiring, those requirements should be summarized in the RFQ. If the DBE contracting requirements apply separately to architectural and engineering professional subcontractors, they must be specified in detail in the RFQ.

Quality control plan.   Some RFQs require the submittal of a construction management and quality control plan with the design-builder's qualifications statement. If required, it should be specified in the appropriate section of the RFQ. Most design-build contracts have specific requirements for the builder's quality control operations and reporting. This responsibility is distinct from the owner's quality assurance operations.

Responsibility for hazardous materials on site.   In the event the owner is not contracting with the design-builder for hazardous-waste remediation, the contract must state that the design-builder will not be held responsible for the removal of concealed hazardous materials that may be discovered on site during construction.

Owner-furnished information.   List, generally, the type of information that will be furnished by the owner (geotechnical, topographical, and legal) to enable the design-builder to assess the risk that must be assumed to make a lump-sum proposal.

Requirement for design-builder to conduct additional subsurface investigations (optional).   Normally, the owner will have obtained a preliminary geotechnical investigation report prior to the publication of the RFP, and that report must be shared with all finalists. However, because the specific location of the structure on the site and the extent of below-grade construction is not known, the contract must provide for an allocation of responsibility in the event that actual geotechnical conditions differ from those set forth in the preliminary report. Also, it may be appropriate for the owner to conduct additional investigations during proposal preparation.

Codes.   The contract will require the design-builder to comply with all applicable codes and building regulations in force or announced prior to the date of submittal of the proposal. However, if the owner requires additional compliance with other codes, such as with a code that may not yet have been adopted by the jurisdiction, it should specify the codes in the RFP and mention that fact in the RFQ.

Permits.   The contract should specify which permits the design-builder must obtain that are necessary for the construction and initial occupancy of the facility. This includes permits that must be applied for in the name of the landowner. It is normal to omit zoning and land-use variances from this requirement, as they are routinely applied for prior to the award of the design-build contract.

Training requirements for owner's staff.   Because most design-build contracts are based on performance specifications, as opposed to prescriptive specifications, it is normal to require the design-builder [or its MEP (mechanical-electrical-plumbing) design-build subcontractors] to operate the facility until compliance with the performance criteria can be ascertained. This provides an opportunity for the training of the owner's staff in the operation of the building systems. If this is a requirement of the contract, it would be helpful for the design-builder to know of the requirement as it assembles its team of designers and design-build subcontractors.

Required construction drawing format (CAD).   If the contract requires the design-builder to provide CAD (computer-aided design) drawings for the owner's use after occupancy, the design-builder must know the specifics of this requirement before submitting its list of design professionals.

Retainage.   If the contract's requirements for retainage are different from the agency's normal public works contracts, the design-builder should be informed in advance. Additionally, the application of retainage provisions to the design-builder's cost of professional services should be summarized in the RFQ.

Dispute resolution.   The contract documents specifically should set forth a mechanism for how disputes between the parties should be resolved. This could include a variety of alternative dispute resolution mechanisms, such as dispute review boards, mediation, and arbitration.[1]

Requirement to use firms listed in design-builder's qualification statement.   The design-builder should be warned in the RFQ that it must retain all the design professional firms listed in its qualification statement, for the entire period of the contract, and for the duties and responsibilities assigned in the same document, unless specifically authorized otherwise by the owner.

10.13   Interviews or Interviews-with-Concepts (Optional)

10.13.1   Purpose of Interviews

The intent of these optional in-person interviews between the design-builder and the jury is to enable the jurors to further evaluate the capabilities and qualities of the design-builder (and its design team), to distinguish between seemingly equals, and reduce the list of finalists to a prescribed number. The interviews should be restricted to jurors and members of the design-builder's organization listed on the organizational chart in their qualification statement. In a typical interview, no new information is anticipated, or allowed, but the design-builder is asked to introduce team members in person, expand on their qualifications, and respond to the jury's questions.

Typical areas of inquiry by the jury include

- Design-build team relationships and previous experience
- Methods used by the team members to design a facility within budget
- Techniques for conflict resolution within the design-build team
- Scheme for team leadership and management
- Project analysis (intended to reveal solution methodology)
- Quality control philosophy and implementation during design and construction

---

[1]See R. M. Matyas et al., *Construction Dispute Review Board Manual,* McGraw-Hill, New York, 1996.

The questions, proposed by the jurors, should not be revealed in advance of the interview. An off-the-cuff, candid answer is a very good predictor of how the team members respond to challenge, and how they develop a response.

10.13.2   Purpose of Interviews-with-Concepts

The aim of an interview-with-concept is to examine the design-build team's awareness of, and initial response to, the owner's objectives, program, site, resources, and the constraints therein. It should not be considered as a design concept for the facility, but rather a preliminary expression of the creative capabilities and sensibilities of the team members in response to the site, and to the owner's limited program information. Like the traditional interview, this additional insight is to allow the jurors to evaluate further the design-build teams, to make a recommendation of finalists to participate in the proposal preparation phase. However, the interview-with-concept has two additional advantages: (1) it allows the jury to see how the team members work together under competitive pressure, and (2) it allows the design-build team to meet the jurors and to get their responses to one or more preliminary design efforts. This last feature begins to alleviate some of the concerns the teams have for the lack of meaningful, private communication between designer and owner in a competitive design-build selection process.

10.13.3   Briefing for design-build teams

If the subject facility has any degree of complexity, the owner may chose to have the criteria professional conduct a common briefing for those design-build teams invited to participate in the interviews (or interviews-with-concepts). The briefing may include an explanation of the purposes of the interviews, ground rules for the conduct of the interviews, an introduction to the owner's key staff, an overview of the facility program, and a tour of the site and/or the owner's present operations. The briefing should include an opportunity for questions and answers, with publication and distribution of the attendance list and the Q&As to all invited design-build teams. The preinterview briefing process can substitute for a similar briefing normally conducted for the RFP finalists. To create a preliminary concept of any value to the jury, the owner must furnish the invitees a reasonably comprehensive facility program, including design guidelines or objectives, and maps and drawings of existing site conditions.

10.13.4   Letter agreement

Because an interview-with-concepts will require some creative work on the part of the invitees, and as the results of that work may ultimately be of benefit to the owner, it is customary to offer a small honorarium or stipend to each design-builder. The letter of invitation or agreement will stipulate the amount of the honorarium and/or the expenses to be reimbursed, and will ask the design-builder to acknowledge its acceptance of the terms and conditions of the agreement.

10.13.5   Interview-with-concepts format

To keep the cost to the teams reasonable, and to allow the jury to judge the team's creative abilities, some presentation restrictions should apply. All jurors on the selection panel must be present, familiar with the facility program and site conditions, and prepared to discuss both with the team. All graphical work must be created in the meeting with the jury, and no previously prepared graphics (other than those depicting existing conditions) may be taken into the meeting. Adequate time should be scheduled in the meetings with the jurors to allow significant communication between the team and the jurors, and to permit the team to illustrate its initial impressions or concepts with quick sketches and/or plan overlays. The owner must allow sufficient time in the interview for the team to fully develop and explain the concept(s) to the jurors, and time for the jurors to respond. This may require an interview of 2 to 3 h each; 2 h would be a minimum, and 3 or 4 h preferable for complex projects. This type of interview is often referred to as a "chalk talk." All materials are returned to the team at the end of the interview. Make a videotape of the interview for the record, and for the use of the team interviewed.

10.13.6   Videotapes

Videotapes of the interviews-with-concepts can be used for several purposes. Copies of videotapes of the design-builder's own interview can be provided to the individual design-builder shortly after the last interview. This tape will remind the design-build team of the jury's response to the concepts, and their answers to the team's specific questions. Later, after the selection process has been completed, the tapes can serve as evidence that the interviews-with-concepts were conducted fairly and evenly, and that the jurors shared no team's concepts or design innovations with any other team.

10.13.7   Communications

Communications to the design-builders, outside the briefing and the interview, must be in writing, and distributed to all invitees simultaneously.

10.13.8   Notifications

Notification of the jury's recommendations and the owner's action must be prompt, and in writing to all invitees simultaneously.

10.14   Request for Design-Build Proposals

Elements of a request for proposals are listed in Fig. 10.4.

- Identification of owner, consultants, jury, and design-build teams
- Instructions to proposers
- Eligibility and honoraria
- Communications
- Preproposal conference(s)
- Competition schedule
- Proposal form
- Alternates
- Supplements to proposal form
- Presentations
- Disqualification
- Weighted proposal selection criteria
- Basis of award
- Information provided by owner
- General conditions of contract
- Agreement and bond forms
- Program of facility requirements
- Performance specifications

Figure 10.4   Elements of a design-build request for proposals.

10.14.1   Organization

Owner's organization and consultants.   On the cover page, list the senior public official and public body issuing the RFP and requesting design-build proposals. On the inside title page(s), list the agency, department and/or criteria professional (or consultant team) responsible for development of the RFP. Indicate, by name and title, all officials in the chain of command relative to the project and this RFP. Elsewhere in the RFP, state that those individuals and their organizations may not be contacted by the proposers, concerning the subject of this RFP, except as specified in the RFP.

Qualified design-build teams and their member firms.   List each prequalified design-build entity and its member firms, architects, engineers, and specialty consultants. List the official name, address, and phone and fax numbers of the design-builder's single contact person. Elsewhere in the RFP, state that only those prequalified design-builders may receive an RFP, and only they may submit a proposal. Also, refer to the contract requirement for the design-builder to retain the designers and subcontractors proposed in the design-builder's qualification statement and documented in this section of the RFP.

Jury.   List all members of the owner-appointed jury and their organizations and positions. Some owners include a short, one-paragraph biographical sketch of each juror in the RFP. Contact by proposers is forbidden.

Technical evaluation panel.   If the jury is to be assisted by a panel of advisers or consultants responsible for the technical evaluation of the submitted design-build proposals, list their names, titles, and organizations. Contact by proposers also is forbidden.

10.14.2   Instructions to proposers

Definition and terms.   The RFP, with attachments of the RFQ, design-builder's qualification statement, proposal, bonds, and other documents, will constitute the contract between the owner and the design-builder. As such, it should include a definition of the names and terms used in the various documents. It is recommended that a single list of definitions be included in the RFP, and indicate that it applies to all contract documents, unless specifically indicated otherwise. Capitalized names and terms in the RFP and other contract documents specifically refer to the definitions listed in this section.

Eligibility to receive RFP documents and to submit a design-build proposal.   Only those prequalified design-builders listed in the title pages of the RFP are entitled to receive the RFP and addenda, and only they may submit a design-build proposal. Each design-build entity must remain exactly as indicated in the design-builder's qualification statement, unless the owner authorizes a substitution in writing. However, the proposer is free to add more design and construction resources without such permission. Duties and responsibilities of the prequalified design-build team members may not be significantly changed without the owner's permission.

Official request for design-build proposals.   This section should include a verbatim copy of the official advertisement of the owner's request for design-build proposals, exactly as it was initially published in the publication of record. Such advertisement normally includes the name and address of the owner and its contact person, a brief description of the project and its scope, minimum requirements for both design and construction, licensing and bonding requirements, description of the selection process (two or three-stage, invited competition), and the form and deadline by which to submit qualification statements.

Competition schedule.   The competition schedule lists all the critical dates in the competitive selection process, from initial advertisement for qualifications to the scheduled date of the announcement of award. Except for the deadline for qualification statements in the official advertisement, this should be the only place where specific dates are indicated. In other sections, if it is necessary to refer to a date, it should be done by reference, for example, "by the date indicated in the competition schedule." If it is ever necessary to change any date in the competition schedule, this is the only section to be amended. The elapsed number of calendar days allowed for design and construction, and any interim milestones for submittals, and other activity should be indicated in the general conditions of the design-build contract.

Information provided to proposers.   It is important to document the specific information provided to the design-build finalists, but that is not bound into the RFP. Such information normally includes topographical and boundary-line maps, legal description of the site, preliminary geotechnical investigation reports, plans of existing structures on or surrounding the site, environmental impact studies, master plans, feasibility reports, and programming or cost studies. The owner is obligated to provide the finalists with all pertinent information that conceivably may be useful to the proposer, to avoid the possibility that some significant information may be available to some, but not to others. If some of the information is old and considered not reasonably useful or directly applicable, it should be listed in the RFP by title and date, and copies made available to the finalists on written request.

10.14.3 Supplements to proposal form

General. Supplements to the proposal form are the proposal bonds, required certificates (if any), and the exhibits describing the design-builder's offer to the owner. Those exhibits describe the specific facility offered by the design-builder in response to the RFP. The RFP must indicate the minimum submittal requirements, and may also include a limitation on the number, size, and type of submittals that will be allowed. For drawings, display boards, and models, indicate scale, size, and any limitations on graphical technique, such as "all drawings must be black line on white background," or "perspective drawings may be in color or black and white, at the proposer's option."

Appropriate level of detail. The level of detail needed to require in the submittal documents should be determined by the owner, and specified in the RFP. The specifics required for each design discipline and each building or facility system will depend on the complexity and uniqueness of the project, the capabilities of the finalists, the amount of the honorarium, the time allowed, and the amount of detail necessary for evaluation and contract security. For building projects, the owner should consult AIA Document B162, *Scope of Designated Services,* for a description of the detail required for each phase of design. If the subject of the design-build selection process is civil infrastructure or other predominately engineer-designed facility, the owner should consult EJCDC Document 1910-1 or similar documents for a description of the detail required for the initial phases of design. For most design-build selection procedures, a level of detail equivalent to a complete schematic design would be sufficient. Often proposers will submit exhibits that exceed the minimum levels of detail required in the RFP. In fairness to all proposers, unlimited technical submittals should be discouraged or disallowed.

Drawings. For most design-build competitions involving buildings, the drawings can be limited to landscape siteplans, and architectural plans, sections, and elevations. Some owners ask for schematic or diagrammatic layouts of foundations, structural systems, MEP, and other building systems. Interior and exterior perspective drawings are almost always required. For civil infrastructure or predominately engineering design proposals, the documentation necessary to describe the design-build proposal will vary considerably. Owners should consult with appropriate design professionals to ascertain the kind of drawings necessary for the type of work being considered.

Model (optional).   For a major public building, a model should be required. For most other types of buildings, they are not necessary for technical or aesthetic evaluation. Models are very costly for the finalists to produce. Very often, the model must be started before all the exterior details have been determined, resulting in discrepancies within the submittal documents. The owner, in light of the purposes the model is expected to serve, should consider this requirement carefully, together with the honorarium, and the time allowed to prepare a design-build proposal. If required, smaller-scale models cost less than larger and more detailed models. White-on-white models are often specified to allow the evaluators to concentrate on the comparative forms and shapes of the different proposals. If the RFP requires a model, the document must specify scale, color, identification, and whether a base and/or a clear plastic cover is necessary. For some large, urban design–type projects, the owner will commission a base model of the areas around the site, and require the proposer to submit a building model only within a specified model "footprint" or template. This greatly reduces the cost to the proposer, but limits the number of proposals that can be evaluated or displayed "in context" at one time.

Design display boards (optional).   If the design proposals are to be put on public display, or if the jury is large, it may be appropriate to require specific landscape, architectural, and perspective drawings to be mounted on rigid boards and submitted with the proposal. Color may be allowed, if not allowed on the base submittal drawings. Size (20 × 30 to 36 × 48 in) and number (4 to 8) and the details of mounting, covers, frames, identification, text, and other elements must be specified.

Color and material samples board (optional).   Again, if the design proposals are to be put on public display, or if the jurors are predominately laypeople (not design professionals), it may be appropriate to request the finalists to submit color and material sample boards. For most building projects, a single board can contain samples of both exterior cladding materials and the major interior materials and colors. Sometimes a second board of interior-only materials and colors is requested. Exterior material samples can be heavy, and some flexibility should be allowed about the form in which these samples are submitted and displayed.

10.14.4   Other information at option of proposer

General.   Provisions in the instructions to proposers to allow the finalists to submit additional information and materials at their option should be considered carefully. To stay competitive, some proposers may feel that they have to prepare and submit substantially more technical and design information than the minimum requirements. This places a time and cost burden on the proposers, and adds considerably to the task of the owner's evaluation team. As a general rule, the RFP should request only enough information to be able to evaluate the proposals, and to have sufficient particulars to describe the facility (beyond the program and performance specifications) so as not to put the agency at inordinate risk in the design-build contract.

Additional technical information.   It may be appropriate to allow the proposers to submit additional technical drawings, specifications, calculations, and special reports (elevator studies, exit capacity calculations, energy consumption calculations, etc.). This additional information typically has to be prepared by the design-build team to arrive at a priced proposal. Normally, little additional effort is necessary to prepare those exhibits for submittal with the design proposals. This additional information can serve to protect the interests of both parties to the design-build contract by describing more precisely what is offered in response to the RFP. This additional information should be limited to the owner's technical evaluation committee, and not used with the design display materials.

Additional professional resources.   Although there are provisions in the RFP and in the contract to prevent unauthorized substitutions on the design-build team, the design-builder must have the liberty to add those professional capabilities it feels is necessary to satisfy the responsibilities inherent in its design proposal. The demands of the design challenge cannot always be determined during the prequalification stage, and this supplementary submittal allows the proposer the opportunity to document any professional or other resources it has added to its design-build team. The format and limitations of this type of submittal should follow the format prescribed by the RFQ.

Unsolicited alternates.   It is common for a design-build team, in the course of preparing its proposal, to discover alternatives to its design that technically violate the requirements or limitations of the budget, schedule, program, and/or performance specifications; but it may add value to the facility. Such alternates may be additive, deductive, or cost-neutral, and the proposer would like the owner to consider them. The design-build selection process, however, must be based strictly on the proposer's responses to the RFP and its requirements. Therefore, the jury must not consider any unsolicited alternates until at least a preliminary evaluation and scoring of the base proposals has been completed. Alternatively, and preferred, is the procedure to limit consideration of unsolicited alternates to the winning proposal, and to return the other unsolicited alternates unopened.

10.14.5   Proposal selection criteria

Typical criteria are listed in Fig. 10.5.

General. Beyond the mandatory requirements of the program and performance specifications, the "proposal selection criteria" is probably the most critically examined section in the RFP. If honestly derived from the owner's stated objectives for the project, and if accurately interpreted by the design-build teams, the criteria will hold the answer to a responsive proposal. Thus, it is important for the owner and its project staff, jurors, and consultants to make every effort to define and summarize those criteria, achieve a consensus among those responsible for evaluation and selection, and present the criteria in a meaningful way in the RFP.

Criteria. The criteria may be subjective or objective. They may include cost or value as a criterion. The number of criteria cited should be kept to an absolute minimum. Typically, the range is from 5 to 10 separate criteria. In either case, the criteria must be weighted in some manner to indicate the relative importance of the specific selection factor to the owner. It serves little purpose to list criteria with a weighting less than 5 percent of the total available. Therefore, it may be necessary to summarize several factors under one heading; for instance, *utility* may represent functional efficiency, operational cost, maintenance cost, durability, reliability, safety, and security. It would be appropriate to list the elements or factors that constitute a selection criterion, but it would not be effective to list separate scores or weightings for each subfactor.

Excluded criteria. In a public, competitive selection process, the determination of "prequalified" in the first phase of the process is an absolute. There should be no degrees of prequalification. All proposers must be allowed to prepare and submit a design-build proposal with the understanding that they are considered by the owner to be absolutely equal to the other finalists, with respect to their professional, technical, and construction capability and financial standing, each as being adequate for the subject project. Therefore, prequalification criteria should not be used along with proposal selection criteria. The proposal selection criteria should be limited to the quality, quantity, and value of the

---

- Architectural image and character
- Alternate for engineering project: technical innovation and environmental acceptability of engineered solution
- Functional efficiency and flexibility
- Quality of materials and systems
- Quantity of usable area
- Access
- Safety and security
- Energy conservation
- Operation and maintenance costs
- Cost/value comparison
- Completion schedule

Figure 10.5 Typical proposal selection criteria.

"product" proposed. The exception to this rule may be a situation where the proposal includes a very technical design element requiring specialized expertise and experience. An example may be the use of tension fabric roof over a long span, when no such structure was contemplated during the prequalification phase. The owner's request for clarification process, or the proposer's optional submittal information concerning additional technical resources, may resolve the issue and maintain the proposer's "prequalified" status.

10.14.6 Communications

It is in the owner's best interest to maintain a high level of competition among the finalists. To accomplish this, the proposers must feel that they have an equal opportunity to prepare a winning design-build proposal. Crucial to establishing and maintaining this attitude of equality of opportunity is the establishment of an unbiased method of communications among finalists and owner. The following are devices, techniques, and procedures that both improve owner-proposer communication and create an atmosphere of impartiality:

- Develop a comprehensive and detailed program and performance specification that anticipates the proposers' questions.

- Conduct a thorough, in-person briefing for the design-build teams, using the owner's operational personnel and others who are familiar with the functions and operations of the units to be housed in the new facility, then publish extensive notes of those meetings.

- Conduct multiple question-answer (Q&A) sessions for the finalists in joint meetings, then publish the Q&As in an addenda to the RFP.

- Allow only written questions outside the Q&A sessions, and provide written answers to all finalists, simultaneously by facsimile, and published addenda, with source of questions held confidential.

- Disallow contact with the owner, owner's staff, jurors, consultants, and other designated individuals, except as prescribed in the RFP.

- Consider midcourse, private meetings between owner and design-build teams, if all finalists agree to procedures, order, limitations, and so on. Meetings should be of equal duration and videotaped (similar to interviews with concepts).

- Establish cutoff date for questions and a deadline for last addenda to RFP at least 2 weeks prior to submission deadline.

10.14.7 Deviations and exceptions

Variable-priced proposals. If the RFP does not fix or cap the base proposal price, then the owner should not allow the proposers to qualify their proposal with any deviations or exceptions to the RFP and its program and performance specifications. Such qualified proposals must be rejected immediately, without further evaluation of the other design-build submittal materials. Proposals accepted on the basis of such qualifications would, justifiably, be subject to a protest from the unsuccessful proposers. Proposal qualifications are, however, not the same as unsolicited alternates. (See Sec. 10.14.4.)

Fixed-price proposals. If the RFP fixes or caps the base proposal price, the owner must allow the proposers the opportunity either to (1) exceed the fixed price or cap, under a scoring penalty in the evaluation process; or (2) submit specific deviations or exceptions to the RFP, with similar evaluation penalties. The penalty for such proposal qualifications should be severe enough to make it a last-recourse act on the part of the design-builder not able to provide a proposal within the cap. This safety-valve procedure would circumvent the necessity to reject all proposals, if no proposer was able to meet the requirements of the RFP within the price limitation. An alternative to downgrading qualified proposals would be to reject proposals with qualifications, unless all proposals contained a specific qualification, deviation, or exception. Proposal qualifications must be listed specifically or referenced in the proposal form.

The corollary to a stipulated sum/best-value RFP is the obligation of the owner to reasonably assure itself that the facility can be designed and built for the owner-determined fixed-price amount. This will require the development of a detailed program and performance specifications, and the services of an independent construction cost estimating consultant, prior to the establishment of the fixed price, and the publication of the RFP.

10.14.8 Owner's alternates

The use of owner's alternates in a design-build proposal is discouraged. It puts the design-build team preparing the proposal in the uncomfortable position of trying to determine if it is better to make the base proposal more attractive, or the alternate lower-priced (assuming additive alternates). Alternates also have the potential to create more than one highest-scoring proposal, depending on the owner's decision on the alternates. Alternates diffuse the focus of the design-build competition and create confusion in the minds of the evaluators and jurors. The owner should be confident enough of its financing and preproposal cost estimates to confine the design-build selection process to a clearly defined objective of quality, quantity, and price. An option to owner's alternates would be a prioritized list of desired but not required facility attributes in the RFP that the proposer may chose to include within the base proposal price, and its proposal would be evaluated accordingly, on a predetermined basis.

10.14.9 Presentations

For design-build competitions in which subjective design criteria are the predominant factor in selection, the design-build proposer must be given sufficient opportunity to present and defend its proposal before the evaluators and jury. The primary methods of communicating the proposal are the narrative and graphical materials attached to the proposal form and submitted in response to the RFP. A third and highly effective method of communication is the in-person presentation of the design-build proposal to the owner, its evaluators, and its jury. This type of presentation allows the jurors the opportunity to hear firsthand from the designers, and possibly to understand the derivation of the design, and the rationale for the form, character, image, and so on. It also allows the jury a chance to ask introspective questions of the design-build team just prior to the jury's final deliberations. If in-person presentations are required, the following procedures should be considered:

- The order of presentation must be determined by lottery, and the dates, duration, and venue determined by the owner well in advance.

- Only members and employees of the member design-build team firms may participate in the presentations (no professional presenters).

- Members of competing proposers' teams must be excluded from the audience.

- For public projects, the public and press should be invited to attend the presentations (but not to participate in the Q&A period).

- The criteria professional must determine in advance what presentation materials, other than the proposal submittal materials, if any, may be used in the presentations.

- Design-builders specifically should be prevented from modifying or adding to their proposal during their in-person presentations. This restriction includes technical information not contained in their initial submittal, unless it is in response to a question from the jury or technical evaluation team.

- The owner must limit, and provide, all audiovisual equipment.

- Presentations should be videotaped.

10.14.10   Disqualification

The owner's criteria professional or other, third-party, program administrator must be given the responsibility and the authority to monitor the proposal preparation phase, the proposal submissions, and the evaluation phase to ensure equality of treatment to all, and complete compliance with the RFP requirements and restrictions by all proposers. If significant and intentional breeches of the RFP procedures occur, the consultant must investigate and recommend corrective actions, including proposer disqualification, if warranted. If unintentional or unlisted discrepancies appear in the proposals, the consultant must require the proposer to certify that the proposal will meet every requirement of the RFP, or disqualify the proposal. (See Sec. 10.14.7 for intentional discrepancies.) The basis for disqualification, at each phase of the selection process, should be described in the RFP.

10.14.11 Proposal form

In addition to the normal provisions of a public owner's proposal form, the following modifications may apply to a design-build proposal form:

- Proposal amount fixed in advance by owner (optional) and printed on the form included as an attachment to the RFP.

- If the proposal amount is capped, include this phrase after the blank for the base proposal amount: "which amount shall be equal to or less than $XXX."

- The proposer must certify and warrant in the proposal form that its proposal meets or exceeds every requirement of the RFP, including, but not limited to, the program of facility requirements, and the performance specifications.

- However, if deviations and exceptions are allowed, the proposer must list and reference each and every deviation from the requirements of the RFP that are contained in its proposal.

- Proposer must attach list of documents and other exhibits that constitute its complete proposal.

10.14.12 Bond forms

Proposal bond.   Because the payment of the honorarium is dependent on the submission of a valid proposal, and, if selected, the execution of a design-build contract, some owners forgo the requirement for a bid bond on fixed-price proposals, even though the honorarium is likely to be considerably less than the normal bid bond. Proposal bonds, however, should be required for any type of variable-price design-build proposal.

Contract bond.   The form of contract bond form must specify that the bond shall secure the design-builder's faithful performance of the entire contract, including any and all professional design services necessary to complete the contract. Separate bonds, or construction-only bonds, must not be allowed to secure a design-build contract.

10.14.13 Agreement form

A copy of the form of agreement between the owner and the design-builder must be included as an attachment to the RFP. It should include, by listing, and on execution, by attaching the RFQ, the design-builder's qualification statement, the RFP and addenda thereto, the design-builder's proposal and supplements thereto, requests for clarifications issued by the owner (if any), design-builder's clarifications (if any), contract bond, and any certifications, certificates, corporate board resolutions, or other submittals required by the RFP.

**10.14.14   General conditions of the design-build
contract**

In addition to the normal provisions of the general conditions of contract, an owner must consider for addition or modification, as they may apply to the design-build procurement process.

Design-builder's proposal.   Proposal must be defined in the general conditions to include the design-builder's submitted proposal and any preaward clarifications thereto. If the proposal exceeds the requirements of the RFP, at the owner's sole option, the proposal may take precedence.

Design and construction documents.   When accepted and approved by the owner for progress, the design development and construction documents (plans, specifications, shop drawings, and equipment cutsheets) produced by the design-builder after award become a part of the contract documents. But the owner's approval and acceptance of those documents do not relieve the design-builder from the obligation to meet the requirements of the RFP and its proposal, unless specifically indicated otherwise by the owner on an item-by-item basis. Design development and construction documents are accepted by the owner for information purposes only and do not constitute authority from the owner to deviate from the RFP and proposal, unless so indicated.

Nevertheless, there can often be significant disputes between the parties as to the legal significance of owner's review, approval, and acceptance of design and construction documents, regardless of delivery system employed.

Permits and fees.   The design-builder may be responsible to apply, pay for, and qualify for all construction and occupancy permits required by the applicable governmental agencies, to the extent that the permits are related to design and construction. The owner must agree to cooperate with the design-builder to apply for and obtain all applicable permits. The owner should apply and pay for land-use or zoning variances, if necessary, but only if the owner provides the site.

Project coordination.   Include provisions to submit design and construction documents for approval by the owner at various stages, and define the stages and the level of documentation required. If appropriate, require in-person presentations to owner's staff by principals of appropriate design-build team member firms, as part of specific design document submittals.

Insurance.   It is seldom necessary to require professional liability insurance certificates from the architectural and engineering firms associated with the design-builder, because the owner does not have direct contract with those firms, and is relying on the design-builder to meet the design criteria. Consequently, the design-builder will be responsible for covering any claims arising from design.

When professional liability insurance *is* required by a design-build RFP, it is most commonly in the form of a project policy for the design team, listing the design-builder as a coinsured. Typically, a project policy for this type of insurance will cover all the design professionals associated with the design-builder on the project, and is available in higher amounts of coverage than individual policies for professional design firms.

An alternative to the project policy is a relatively new product called a *contractor's design liability policy,* which is meant to protect the design-builder from claims related to design errors or omission. Owners should discuss the issue of risk management for the project under consideration with their insurance consultant, agent, or risk management officer.

Responsibilities of the design-builder. Describe the design-builder's responsibility to provide all services (including design), labor, materials, and equipment, and the management necessary to achieve the intent of the RFP. Describe a process to adjudicate differences, if any, between the owner's design review comments and the contract requirements.

Design-builder's architect-engineer

Definition. Define the design-builder's architect-engineer to encompass the entire design team, namely, all professional architectural, engineering, and specialty consultants required to design the facility and produce the necessary design and construction documents. Describe the professional responsibilities of the design team with respect to this project, including those mentioned in the following paragraphs.

Retain listed firms The design-builder's architect engineer should be the same in-house staff and/or firm(s) listed in the design-builder's qualification statement, unless the owner has approved a change, and said staff and/or firm(s) should be retained and employed on the project throughout the contract period.

Professional responsibility. The design-builder's architect-engineer should at all times attempt to safeguard the health, safety, and well-being of the public (including the employees in, and visitors to, the facility), regardless of any instructions from the design-builder to the contrary.

Inspections. The design-builder's architect-engineer should inspect the work periodically and reject that work that does not comply with the construction documents prepared and/or approved by it.

Interpretations. The design-builder's architect-engineer should make interpretations of its construction documents when requested to do so by the owner, which interpretations shall be reasonably inferred from those documents. The design-builder's A/E should be reasonably available to the owner for such interpretations and other information related to the design and construction documents.

Review shop drawings. The design-builder's architect-engineer should review and approve all shop drawings, samples, and other items for compliance with the construction documents and the intent of the RFP, prior to their submission to the owner. (Owner does not approve, but acknowledges progress represented by the submissions.)

10.14.15   Program of facility requirements

The program of facility requirements and the accompanying performance specifications are the owner's detailed and specific expression of needs. As primary contract documents, they describe the end product to be produced and delivered by the design-builder. The owner is cautioned to prepare these documents with considerable thought and attention. Once a design-build proposal has been made, it is difficult to make significant changes to those requirements. All subsequent cost-related changes have to be made by a prescribed change-order proposal and acceptance process, generally without the benefit of competitive pricing. Some facility managers find this particular constraint a welcome discipline to their internal user needs collection, analysis, and documentation process. The component parts of a program of facility requirements are

- Program narrative discussing the image, character, and broad design goals of the proposed facility
- Mission statements of the functional units expected to occupy or use the facility
- Staffing patterns and summary job descriptions for each workplace to be housed in the facility
- Net area requirements of each workplace and support facility
- Physical and environmental requirements of each type of programmed area
- Internal and interoffice relationship diagrams
- Graphic space standards
- Sketches to communicate specific arrangements of rooms, or of components within specific rooms (such as in a courtroom)
- Space or room criteria sheets that identify all types of finishes, utilities, equipment, and other features to be included in each space (for typical and for special "one-of-a-kind" spaces)

10.14.16   Performance specifications

General. For each building system (structure, roof, exterior closure, ceiling-lighting systems, HVAC, etc.), the owner must specify the minimum performance expected at occupancy, and the applicable industry standards desired. The Construction Specifications Institute's/Construction Specifications Canada (CSI/CSC) *UniFormat,* a uniform classification of construction systems and assemblies, is recommended for the development of the performance specifications. CSI's 16-division *MasterFormat* classification system is not suitable for the owner's RFP, but may be used by the design-builders in their design proposals to describe their technical solutions. To the extent possible, the owner should avoid design or system selection decisions of any type unless it is necessary for technical coordination with the owner's other physical plant systems, or integration into existing maintenance and repair procedures.

Prescriptive specification has the effect of removing design responsibility for the specific building system from the design-builder and placing it with the owner. This act contravenes the basic premise of design-build: a single point of responsibility for design and construction. If necessary to designate minimum quality levels desired by the owner, some materials, building systems or subsystems specifically can be excluded in the performance specifications. See Chap. 12, "Performance Specifying for Integrated Services Projects."

Classifications.   The CSI/CSC *UniFormat* classifies construction systems and assemblies into eight broad categories at levels 1 and 2 with letter designations, and preceded by administrative information designated by numbers:

10	Project Description
20	Proposal, Bidding, and Contracting
30	Cost Summary
A	Substructure
	A10 Foundations
	A20 Basement construction
B	Shell
	B10 Superstructure
	B20 Exterior enclosure
	B30 Roofing
C	Interiors
	C10 Interior construction
	C20 Stairs
	C30 Interior finishes
D	Services
	D10 Conveying
	D20 Plumbing
	D30 Heating, ventilating, and air conditioning (HVAC)
	D40 Fire protection
	D50 Electrical
E	Equipment and Furnishings
	E10 Equipment
	E20 Furnishings
F	Special Construction and Demolition
	F10 Special construction
	F20 Selective building demolition
G	Building Sitework
	G10 Site preparation
	G20 Site improvements

G30   Site civil and mechanical utilities
G40   Site electrical utilities
G90   Other site construction
Z   General
Z10   General Requirements
Z20   Contingencies

Those categories can be used to arrange brief project descriptions and preliminary cost estimates. Category Z contains general information and is normally placed immediately before "project description" when *UniFormat* numbers and titles are used in performance specifications.

Section format for performance specifications

Part 1.   Performance
    A.   Basic Function
    B.   Areas of Concern:
      1.   Amenity and Comfort
        ■ Performance requirement
        ■ Criterion for measurement
        ■ Means of substantiation
      2.   Health and Safety
        ■ Performance requirement
        ■ Criterion for measurement
        ■ Means of substantiation
      3.   Structure
        ■ Performance requirement
        ■ Criterion for measurement
        ■ Means of substantiation
      4.   Durability
        ■ Performance requirement
        ■ Criterion for measurement
        ■ Means of substantiation
      5.   Operations and Maintenance
        ■ Performance requirement
        ■ Criterion for measurement
        ■ Means of substantiation
Part 2.   Products and Assemblies
Part 3.   Methods of Design and Construction

Performance standards and attributes.   Within each category or building trade, industry standards are the most common reference for establishing performance standards. Some of the industry standards organizations typically referenced are

- American Society of Testing and Materials (ASTM)
- Underwriters Laboratories, Inc. (UL)
- Canadian Standards Association (CSA)

- Federal Specifications (FS)
- National Fire Protection Association (NFPA)
- Illuminating Engineering Society (IES)
- American Society of Heating, Refrigerating and Air Conditioning Engineers (ASHRAE)

Trade and manufacturing associations also can provide generic standards and data oriented to specific building systems, equipment, products, and materials.

PerSpective.   The Construction Specifications Institute (CSI) and the Design-Build Institute of America (DBIA) have formed a joint venture to develop and distribute a performance specification guide for buildings. Named *PerSpective,* the product is a Windows-based performance specifications software for owners, design-builders, and specifiers. It allows the industry conveniently and consistently to use performance criteria for RFPs and design-build proposals.

10.14.17   Project administration manual

The project administration manual describes the procedures to be followed by both the design-builder and the owner's representative during the term of the design-build contract. It will include

- Directory of project principals and their official contact (address and phone/fax numbers)
- List of fixed events (reports, submittals, etc.) described in the RFP
- List of the design-builder's notice requirements, as specified in the RFP
- List of the design-builder's submittal requirements, as specified in the RFP
- List of the owner's fixed notice or response requirements, as specified in the RFP
- Correspondence procedures between the design-builder and the owner's representative
- Schedule of regular and special meetings, and procedures for minutes
- List of records required of both design-builder and owner's representative (submittal logs, change-order logs, etc.)
- Forms to be used by the design-builder (application for payment, etc.)

10.14.18   Addenda

In addition to the normal and anticipated changes in an RFP document, the addenda should contain proposers' questions and the owner's corresponding answers. Those questions may have been submitted in writing or asked in person by a design-build team member at any one of several preproposal meetings conducted by the criteria professional. The RFP does not allow any other type of communication between the owner and the finalists. The design-builder must be able to rely on the accuracy of those answers as a matter of contract. For this reason, they are to be published as part of the addenda to the RFP.

10.14.19   Attachments

Information provided by owner.   The owner is obligated to provide the finalists with information concerning the physical, environmental, and legal conditions relative to the site and its surroundings. It is in the owner's interest to reduce the design-builder's risk by providing or obtaining as much information as possible about the owner's needs and constraints. The typical information furnished to the finalists, beyond that described in the RFP, includes

- Legal description and property line survey of the site, including easements, deed restrictions, etc.
- Copy of legislation or resolution authorizing project and/or its financing
- Preliminary geotechnical site investigation, including hazardous materials survey
- Topographical site survey, including tree survey
- Facility master plan and/or preliminary feasibility studies
- Preliminary cost estimates
- Environmental impact statement
- Descriptions of surrounding improvements (streets, buildings, etc.)
- Applicable design guidelines, codes, regulations, standards, etc.
- Preconcept documents (sometimes called *bridging documents*) describing functional floorplan layouts, special room arrangements, and other details and requirements of the program of facility requirements not easily communicated by written text

Conceptual Estimating
and Scheduling
for Design-Build

11.1 Introduction

The valuable and specialized skills of facility planners, programmers, speci-fiers, cost estimators, and other professionals are crucial in the complex and demanding process of design-build. As projects are conceptualized, and alter-native facility elements and systems are considered, the knowledge of these individuals brings a discipline to design and a verisimilitude to cost.

The challenging process of cost estimating for a design-build project is a skill and an art that is mastered by those with the right aptitude, temperament, and experience. Because most owners have fiscal restraints that affect their capital facilities investments, there is heavy dependence on the work of the cost estimator to develop accurate cost forecasts at every stage of the project. A capable design-build cost estimator has a comprehensive understanding of the costs of labor, materials and equipment, and means and methods of both design and construction. In addition, the design-build cost estimator should understand facility programming and formulation of an owner's overall bud-get for a project. Because the owner's overall budget can be significantly high-er than the estimated design and construction costs due to such items as land acquisition, financing, or furnishings costs, it is essential that the estimator be able to distinguish between the hard and soft costs for the entire project.

The estimator's ability to analyze the productivity of the design and con-struction consortium when applied to the difficulties and risks of the particu-lar project will determine the long-term success of the design-build team. With the application of efficiency and productivity measures, the seasoned estima-tor can help a firm gain market advantage on selected projects. Productivity is a measurement of the value-adding process in design and construction wherein the outputs represent greater worth than the inputs. The more value added per unit of dollar and/or time input, the higher the productivity of the design-build team. Ways of adding value include:

- Managing and guiding information flow
- Use of efficient equipment
- Innovative production processes
- Teaming of workers
- Ferreting out unproductive steps and downtime
- Use of better or more easily handled materials
- Better tools to leverage more work output and higher-quality output
- Removal or mitigation of constraints, such as inclement weather or regulatory roadblocks
- Focus on project owner's needs to meet overall goals first (as opposed to secondary goals or goals that are not overarching)

Determination of the estimated cost in design and construction is not confined to knowledge of labor, materials, equipment, and overhead but is directly tied to the decision to use one or more of the alternative methods of production. A designer may choose to use computer aided design (CAD) over lead-on-vellum; a constructor may prefer a metal preengineered concrete forming system over site-built wood forms. Given available assets and resources, the design-builder will seek the most efficient method that results in the highest production and lowest cost when multiple resources (starting with labor and material) are combined. An estimator's knowledge of the *combined* significance of multiple inputs will allow the design-builder to make important quality-versus-cost-versus-speed tradeoffs that are integral to single-source responsibility contracts.

11.2   Estimating and Scheduling for Design-Build Projects Are Different from Traditional Project Approaches

The expression "early knowledge of firm project costs" is often cited as one of the major benefits of design-build delivery. Design-builders often guarantee the cost of the project very early in the design effort, often before any graphical representation of the facility has been developed. Estimating for design-build projects at the early stages requires estimators who are versed in the *design and construction process,* and can think in terms of facility systems and assemblies, rather than the detailed "takeoff" experts, who rely most heavily on complete drawings and specifications to demarcate each piece of material or equipment.

The design-build estimator is economically conceptualizing a completed building or other type of facility. This compilation of systems and assemblies is carefully cobbled together by recalling similar completed projects and other contemporary and historical data and then adjusting the estimate for special conditions and risks. Design-build estimates require pricing knowledge beyond construction, often including basic understanding of facility operating costs; durability of materials; cost of land and rezoning of land; financing costs of both origination and time value of the funds; design costs of architecture, engineering, and specialty subconsultants; and other costs related to planning, designing, and delivering the facility.

An experienced design-build estimator knows that there are three crucial objectives associated with every design-build project: determination of level of quality, accurate baselining and control of project costs, and forecasting and delivering the facility on schedule. In traditional design-bid-build construction, the estimator's skill in predicting construction costs from completed designs directly affects the success of the builder. Construction is a high-risk business because this "facility manufacturer" must predict its costs before actually having manufactured the product. By contrast, an industrial manufacturer can determine product prices on the basis of the engineering and production costs of the product, and prices can be adjusted *after* the manufacturing process to ensure profits. Given the commercial terms of most construction contracts, significant cost adjustments are not possible after the fact.

With the determination of firm costs not only before construction, but often prior to any graphical design, design-build is a unique and complex undertaking for the practitioner. The business success or profitability of a design-builder is directly dependent on the estimator's ability to establish and adjust facility and systems pricing throughout the development process. The qualified design-build estimator will understand how to elaborate on functional and aesthetic performance criteria by selecting appropriate assemblies and components, then arriving at a reasonable projection of facility costs. The knowledge, flexibility, and team orientation of a conceptual estimator makes this position absolutely vital to the design-build process.

11.3   Stages of Design-Build Budgets and Estimates

Management of facilities, whether buildings, bridges, or baseball fields, requires a comprehensive approach. A design-build entity will want to understand the owner's total facilities need in order to craft a package of services that will add value to the client's organization. The design-builder's estimator will become familiar with the owner's overall facilities development plan, the capital budget plan, and the operations and maintenance plan. If the client is lacking facilities audit information or requires a strategic development policy, the estimator (or other professionals on the design-build team) is in the position of adding these needed services as part of an overall professional services contract. In addition, the design-build estimator will watch for maintenance and operations opportunities that may become available after completion of design and construction. Stages of design-build estimates include

- *Feasibility estimates.* A project starts with an owner's "statement of need" for a facility. A feasibility estimate is usually done in conjunction with the statement of need to determine whether the project can be built, depending on available resources. This early estimate is also a valuable tool for obtaining needed project financing (private projects) or budget authorization/appropriation (public projects).

- *Performance criteria estimates (at the predesign or programming stage).* The scope of the project is often developed through a programming or study-and-report exercise. The creation of a scope of work narrative allows the owner to designate those facility attributes about which it "cares" the most. The resultant level of performance that is identified for the entire facility or for elements of the facility provide the design-build estimator with sufficient information to detail a formal design-build proposal.

- *Conceptual design estimates.* The estimator turns to more traditional estimating techniques when the project is further quantified through conceptual and schematic graphic representations. (Please note that in design-build, the conceptual design is often prepared by the *design-build team's designer,* instead of being issued by the owner's design consultant; because the owner wants the benefit of alternative innovative solutions.) Preliminary drawings may go through one or more iterations (with an accompanying conceptual design estimate) to adjust for changes in, for example, room size or adjacency, or chiller piping or heat-pump ducting. Estimators must "visualize" all the assemblies and components of the project that are not shown.

- *Detailed estimates.* With detailed estimating, the resulting work product is intended to show the type, amount, and quality of discrete materials and components of the project, along with the work processes (falsework, non-permanent construction, mobilization, etc.) necessary to execute the contract. Detailed estimates usually depend on relatively complete drawings and level-of-quality specifications (although the specifications may be performance-based rather than prescriptive). The American Association of Cost Engineers classes a definitive estimate at the 80 percent design level as being with −5 and +15 percent of the final project costs, barring further scope change.

At the point of the project where systematic cost control becomes an important issue, the estimate is tied to schedule. The resultant project budget has cost breakdowns that are formulated for both the design-builder and the owner. The design-builder will want to have costs broken down by work item within the schedule. The owner and the design-builder view the project in terms of functional elements. Functional elements of a facility contain components and assemblies that enable that portion of a facility to serve a particular purpose. With proper cost and schedule coding, the same data can be configured to provide detailed progress reports to the design-builder as well as the owner.

11.3.1 Feasibility estimates

Since the economic downturn in the late 1980s, revival of demand for facilities has resulted from a series of demographic and other factors. Projects from new town developments to retail, from institutional buildings to recreational facilities, from industrial plants to transportation and water infrastructure projects, have been developed in response to economic and social demands. The key drivers for growth of midsized to large-scale projects have been

- Population increases in concentrated geographic areas
- Increasing incomes
- Improvements in transportation infrastructure
- No-growth or low-growth regulations and zoning challenges
- Environmental controls and challenges
- Governmental preference for dealing with large-scale rather than small-scale development
- New financing techniques, including real estate investment trusts (REITs) and public-private partnerships

Determining the feasibility of successfully executing and operating a project requires a systematic approach: (1) the owner's and user's overall goals and objectives should be identified, (2) the quantitative demand for facilities of this type should be researched (this step is often called a *demand study* or a *market analysis*), and (3) a feasibility analysis of the specific project should be conducted. For a firm offering integrated services, it is desirable that the conceptual estimator or other professional staff have capabilities in generating the needed financial reports included with feasibility studies.

Establishing the owner's objectives is a necessary first step in all capital facilities investments. The objectives must be known before specific uses or functions can legitimately be chosen for the property, location, or right-of-way. Although most owners are concerned primarily about financial considerations, some owners are motivated by other goals as well, such as architectural achievement, public safety and well-being, or job creation. Generating a clear statement of the owner's objectives is an invaluable preparatory step because it allows the user requirements (generated during the performance criteria estimating stage) to be placed in an overall context. As examples, an industrial owner may need increased production capacity as its number one goal, with a lower unit cost for each good manufactured as a secondary goal. A shopping center owner may be interested in current investment return, tax advantages, and long-term appreciation of real property value.

Once the "spirit" of the facility initiative is stated through the writing down of goals, the "letter" of the development opportunity begins to form through a demand study or targeted market research. For a bridge project, the demand study will help determine the projected traffic counts for deciding service life and setting toll rates. Market research for an apartment complex would look at the rate of development that could be supported (absorption rates) and help determine the constraints of environmental regulation, zoning, and competitors' costs and pricing. The heart of any demand study or market research is the objective and reasoned estimates by the analyst. The factors considered in a demand study or market analysis include

Commercial projects

- What is the size of the market and what percentage of the market would be attracted to the site?
- What are the indirect constraints that would affect the demand (or supply) of the development? Among the concerns will be zoning laws, potential local opposition, licensing requirements, and other social (community) and physical (site) challenges.
- What type of facility is justified by the market demand? For example, will the geographic area support an indoor shopping mall, or is a shopping center strip better suited given the demographics?
- How much square footage can be leased annually? What is the lease rate? What is the mix of space? A first-floor walk-in insurance office may pay $30 per square foot, whereas the top five floors of the office building that is leased to a state agency may pay $10 per foot.
- What is the expected revenue of various types of tenants who may inhabit the facility? A high-technology service center will be generating greater income than a pet store, and a good market study will acknowledge a mix of activities and businesses.

Industrial and process projects

- What are sizes and trends in the industrial market, and what segments of the market could be attracted to the site?
- What are the indirect constraints that could affect the industrial project? In addition to zoning regulations and environmental requirements, the availability of transportation is a critical factor in industrial development.
- What quantity of acreage or square footage could be sold or leased to industrial users annually? Much of the demand for industrial land is national, rather than regional or local, in scope. Site location depends on inputs (labor, materials, capital) and outputs (ease of distribution, proximity to wholesale or retail markets).
- What is the projected selling price of the improved industrial property? Are there additional services or features that would add to the salability? Additional features may include tax inducements, industrial development bonds, state-supported training programs, government-funded off-site improvements, and special security measures.

Civil infrastructure projects

- What are the costs of the land or rights-of-way?
- Is there an identified revenue stream to support the construction investment? The revenue may come from gasoline taxes, utility fees for water or wastewater, tolls for transportation projects, or other appropriated or nonappropriated source.
- Is there sufficient demand to justify the investment? Depending on the types of projects, factors such as traffic counts and trip generation, population growth, economic development, extent of service area, peak-usage projections, and environmental requirements must be considered.
- What are the results of the environmental assessment? Addressing the environmental impacts will likely require multiple permits from various federal, state, and local agencies, along with mitigation measures that may result in additional costs.
- How will the sponsoring government fund operation and maintenance for the project?

The feasibility estimate serves as a tool for managers who must decide between two or more courses of action. Frequently, the chief goal of the organization is to develop needed capital facilities at the greatest cost savings, or to invest in projects with the highest rate of return. Undertaken during the statement-of-need phase of the project, the feasibility estimate enables the owner to make cost/benefit comparisons in the absence of extensive project data.

Capital expenditure and project selection decisions are based on knowledge of investment objectives, funding inflows and outflows, design/economic life of the facility, investment risk, and related capital budgeting analysis issues. When combined with construction and project design knowledge, the analyst will be able to synthesize the data to form a reasonably accurate feasibility estimate. For private-sector commercial real estate projects, *investment objectives* will include

Current income. A regular cash flow from the investment is the goal of many holders of commercial real estate. Given the risk of real estate development, the investor is usually seeking a higher current return than from other investments.

Future appreciation. A future return is expected by most investors when the property is leased or sold to new users. Because of their present tax situation, some investors prefer future income that is subject to long-term capital gains taxes rather than paying higher short-term rates.

Hedge against inflation. Capital facility investments generally retain their value as the nation's price index spirals upward. In fact, in some geographic areas commercial real estate values not only keep pace with but actually exceed the increases represented by the general price index.

Tax treatment. Whether the investment will cause the investors to pay additional tax or will allow them to show a tax loss is of foremost concern. Implicit in this observation is whether the tax imposed will be based on ordinary income or long-term capital gains. The owner or investor will weigh the investment objectives against the *investment risks* of the capital project.

Risk of market changes. Changes in supply and demand can help or hurt the commercial development project. The abandoned hotel standing by the old state highway route that has been bypassed by the new interstate is a poignant reminder of importance of location. Demand can also change with new technology, such as what may happen to all the videocassette rental stores when the capabilities of home-based systems tied to Internet-based providers cause the cassette tape to become obsolete.

Legal risks. A major legal risk is having clear title to the property, without encumbrances, liens, environmental liabilities, easements, and other restrictions. Risks are often manageable through insurance coverage that can mitigate the exposure. Additional legal and administrative risks include those associated with building and fire code requirements, mortgage terms, tax requirements and contracts with tenants, facility operating companies, and other facility subcontractors.

Investment risks. The prudent investor in capital facilities will gauge the investment risks against the likelihood of those risks impacting the specific project or facilities program. Among the risks are inflation or deflation, either of which can reduce the value of holding an investment over time. Some economists argue that a real estate investment can be a good hedge against inflation where the value of the property is increasing at a pace that equals or exceeds the rate of inflation. Similarly, an upward movement of interest rates can affect the value of capital facilities by lowering (through the formula of net operating income divided by rate) the appraised value of the property. Additional investment risks include natural hazards, which may result from development in an earthquake zone or on a coastal plain, and the illiquidity of capital facilities, which does not allow the investor to depend on timely conversion of capital assets into cash.

Nonnatural and natural hazards. Depending on the location of the project, there may be risks of dam failure, earthquakes, flooding, civil disturbances, or other events that would expose the owners to increased liability or that might otherwise affect the ability of the property to attract tenants, customers, or users. In many cases, insurance is available up to certain limits to help minimize the impact of many of these risks.

To determine the value of a commercial or industrial facilities project, the knowledgeable owner will ask for the creation of a cash flow model. Although there are many components to the cash flow model, the entire exercise starts with the estimated price of acquisition; that is, the cost to acquire land and other improvements, plus the total cost to design and construct the new facility. A feasibility estimate for a new shopping center will undoubtedly be constructed using a dynamic cash flow model that recognizes discounted rates of return. Whether the project is a shopping center or an office building or a hotel, the pro forma will contain similar items and the feasibility analysis will use similar techniques. Generally, the variables will be grouped into three areas: investment inputs, cash inflows and outflows, and property disposal (liquidation, trade, or other disposition of asset).

For the proposed shopping center complex, the design-build estimator and the project investor will likely want to determine the project's feasibility through a discounted cash flow method to determine the facility's net present value. The idea of present value relies on the commonsense maxim that the expectation of receiving a dollar in the future is worth less than the dollar received today. Further, the return on investment must be weighed against what could be earned from the dollar if it were invested in other securities, such as stocks, bonds, or treasury bills. Within the feasibility estimate, the time-value-of-money concept will be underscored.

The *investment input variables* group of the pro forma estimate will include

1. *Cost of the land.* The value of the land is treated differently from improvements for tax and finance purposes. The cost of the land may be tied to an option that must be exercised within a required period of time, and its value is also predicated on zoning and density regulations. Land and its attendant bundle of rights is usually one of the more predictable variables in the feasibility estimate.

2. *Facility design and development costs.* The initial cost of the facility design and development can be determined from the design-build estimator's historical data, cross-checked against published costs from for-profit and nonprofit sources, such as Cushman and Wakefield, the Urban Land Institute, and R. S. Means Company. Facility managers may use a replacement index method based on a percent value of the facility that must be replaced annually or an appraised value incorporating a condition value multiplier. For new structures, a level-of-quality index coupled with geographic and market adjustment factors will be tools of the prudent design-build estimator.

3. *Equity investment.* The amount placed "down" (down payment) on the project by the facility investors is the equity. For international BOT (build-operate-transfer) projects, the design and construction group often provides some of the equity, but looks to pure investors to provide most of the initial funding. Design-build consortia are normally more highly leveraged and lightly capitalized when compared with venture capitalist firms. For companies whose core business is design and construction, there is a preference to move on to the next development contract rather than settling in for long-term operation of the facility. After the equity investment is determined, the remaining costs of the project are usually supplied by debt financing in the form of a mortgage.

4. *Mortgage.* From the French middle ages came the concept of a "dead pledge" for securing a loan against property. The mortgage specified the contract terms for repayment of the debt to a creditor. The key mortgage terms are the interest rate and the number of years for repayment. Both rates are subject to the condition of the economy and the competitiveness of the market.

5. *Operating expense.* The percentage of gross income that will be used for operating expenses for the property is a vital part of the feasibility estimate. The percentage amount can be derived from the owner's historical data, or an average figure from similar properties in the industry can be used.

6. *Depreciation and economic life of the facility.* The tax laws dictate the form of depreciation according to the type of property and the tax status of the owners. The accelerated depreciation rules for real property were tossed out by the U.S. Congress during the late 1980s. Most commercial property today is limited to a straight-line, multiyear (27+) cycle, pushing the investment decision more toward cash flow and appreciation rather than the availability of a tax shelter. The life of the facility is determined by physical design, functional use, economic considerations, and related concerns; however, the number of years is usually none other than those shown in the Internal Revenue Service (IRS) tables for depreciation.

7. *Income tax percentage.* Depending on the tax brackets of the investors, the amount of income that may be sheltered from tax is adjusted upward or downward. The arbitrary cutoff or phaseout levels in the tax law have very real practical effects for the individual. For specific types of property, the occupation of the investor also has a bearing, with the tax code currently favoring full-time real estate professionals with maximum depreciation schedules.

8. *Capitalization rate.* The capitalization rate, also known as the *debt constant,* can be found in or derived from standard mortgage rate charts. For example, a 30-year mortgage with a market value of $100,000 and with an 8 percent interest rate would result in a capitalization rate of 1.64.

9. *The owner's expected rate of return.* The owner/investor's required return on investment is a rate set at a level where cash flows are discounted and the rate reflects both the minimum acceptable return and the owner's view of the marketplace. This discount factor is carried forward into the cash flow pro forma. A higher expected rate of return will show a lower investment value, whereas a lower expected rate of return will show a higher overall investment value.

Included in the *cash flow pro forma* would be:

1. *Owner's required rate of return.* This discount factor is carried forward from the investment inputs variables group and applied to the cash flow pro forma.

2. *Gross income.* The gross income figure is the bedrock (or quicksand) on which the entire development venture is dependent. An overstated gross income amount could result in an undercapitalized undertaking. Due diligence by the estimator, the owner, and other consultants is necessary to determine a gross income amount that is achievable.

3. *Occupancy or absorption rates.* Rates that are sustainable given the experiences in the local area are usually acceptable. This information can be taken from the market study that has been done for the project.

4. *Effective gross income.* To calculate effective gross income, multiply the gross income by the occupancy rate. The gross income figure as listed in item 2 above does not include potential vacancies that must be included for every commercial property.

5. *Total operating expenses.* Operating expenses are a percentage of effective gross income. To calculate total operating expenses, multiply the amount of effective gross income by a percentage determined with industry or historical data within the investment input variables. When the cash flow model incorporates the total operating expenses with the mortgage interest and the depreciation allowance, the aggregate of the three is the "sum of tax deductibles."

6. *Interest on borrowed funds (interest paid on mortgage).* The interest paid on funds that have been secured for the property are an important part of the cash flow analysis. The mortgage interest rate declines as the principal is amortized (the amount of principal as part of the level payment slowly rises over the years, and the amount of interest as a portion of the level payment declines).

7. *Book value of the property.* The pro forma is based on determining a book value, which is calculated by subtracting the cumulative amounts of depreciation from the original value (of real property improvements, not the land itself), and then adding new capital improvements.

8. *Straight-line (or other type of) depreciation.* Depreciation is the theoretical or actual loss in value of real property through physical deterioration, functional obsolescence, or economic change. The annual depreciation allowance available to the property owner is included in the cash flow analysis. Under current tax law, the depreciation term varies according to type of property. For some types of properties, there may be opportunities to look at building components as depreciable assets, such as for specialized furnishings or other elements of a structure that may have a differing useful life (regular tax code review is appropriate).

9. *Net income before tax.* The *net operating income* (NOI) is obtained by subtracting the *sum of tax deductibles* (operating expenses, mortgage interest, and depreciation) and total debt service from effective gross income.

10. *Taxable income.* To determine taxes owed on income from the investment for the year, the investor will subtract the sum of tax deductibles from

effective gross income. Whereas a property may have net income for the year, because of interest payments and depreciation allowance, the project may operate at a loss for the year, which may result in significant tax advantages for the investors (ability to shelter other income).

11. *Income taxes.* Federal income taxes that may be incurred through the operation of the property are inserted at the end of each operating period, which is typically one year. In many pro formas, the income tax entry is zero for the first few years, until interest payments and allowed depreciation are exceeded by effective gross income. When actual taxable income is realized, income taxes are included for each operating period.

12. *Tax benefit.* Because the government recognizes the advantage in stimulating various forms of investment, certain policies give advantages to capital facilities investors. To determine the tax benefit of an individual investor (or a class of investors), multiply the amount of taxable income for a period by the income tax bracket of the investor(s). The so-called tax shelter declines over time as the amount attributable to deductible items decreases and the taxable income increases.

13. *Total annual debt service.* This entry in the pro forma is determined by multiplying the capitalization rate (debt constant) by the amount of money borrowed to finance the investment. The annualized amount includes total principal and interest paid during the year.

14. *Annual mortgage principal.* When the interest payments are subtracted from the year's debt service, the amount of the principal payments shown as the remainder will allow investors to see their degree of equity in the project.

15. *Cumulative principal.* Calculating the cumulative principal for future years in constant dollars allows investors to see future equity increases for any given year.

16. *Remaining mortgage principal.* To determine the principal remaining to be paid on the property, subtract the annual principal payment from the previous year's remaining principal amount.

17. *Annual cash flow after tax.* The amount of annual cash flow for the purposes of the feasibility study is an after-tax dollar amount that includes current net operating income (net income before tax and tax benefit for a given year, less the required tax on that income).

18. *Discounted after-tax cash flow.* This entry shows the discounted value of current and future cash flows after tax. The discount factor is taken from the owner's required rate of return or from market data to determine the financial viability of the project for the specific investor(s) involved.

19. *Discounted cash flow when compared against initial equity.* The rate of return on initial equity when viewed in terms of annual discounted cash flow will reveal a declining rate of return over the life of the project since cash flow amounts received in the future will have a lower present value.

20. *Cumulative discounted after-tax cash flow.* The piece de resistance of the cash flow pro forma is the cumulative after-tax cash flow on a discounted basis. This portion of the pro forma contains the present value of cash inflows at the end of each time period. This information will help determine whether the asset should be held for an additional period of years, or whether it may be advantageous to retrieve present equity and invest in another project with potentially higher returns.

Part 3 of the project pro forma is concerned with *disposition of the capital asset:*

1. *Estimated sale price.* The sale price is derived by looking at property/facility appreciation rates for similar projects in the geographic area. A market analysis of current and future demand and supply of similar properties will add to the professional judgment as to what a conservative selling price should be at some defined future date. A hoped-for or inflated selling price should not be used because it will unreasonably skew other aspects of the project pro forma.

2. *Capital improvements.* An estimate of future capital improvements or major repairs is an important component of a capital facility wherein the asset will be held for multiple years. The amounts for capital improvements or significant repairs will likely increase after the facility has been in use for a few years, and these amounts will add to the book value of the project.

3. *Book value of the facility.* At the end of each time period (generally, a month or year, depending on the breakdown of the pro forma), the remaining book value is derived by subtracting cumulative depreciation taken from the original book value of the improvements plus land, and then adding capital improvements for the period.

4. *Capital gains.* The capital gains exposure in any given year is determined by subtracting the remaining book value of the facility and land from the estimated selling price.

5. *Amount of sale proceeds subject to capital gains.* In some instances, a portion of the sale proceeds may be subject to ordinary income tax rates (property not held for the required period of time, or prior use of accelerated depreciation—a method that is no longer allowable under the tax law). This entry in the pro forma reminds the analyst to check for tax consequences that may be higher than capital gains rates.

6. *Total tax at sale.* The tax liability of the investor(s) at the sale of the property would include any recapture of depreciation at ordinary individual rates plus the amount subject to capital gains.

7. *Net proceeds to seller of the facility.* Calculating the net amount available to the seller in current dollars is determined by subtracting the remaining mortgage principal plus the total tax liability from the estimated selling price.

8. *Discounted net proceeds.* Within the pro forma, the discounted net to the investors shows the present value of the net proceeds at the time of the analysis. The discount rate used once again is the investor's required rate of return or capitalization rate.

9. *Total discounted return.* To calculate the total discounted return, add the discounted net to seller plus the cumulative discounted after-tax cash flows.

10. *Total return to equity.* This figure represents the total percentage return to the original equity when compared to the total discounted return for any given year. To determine total return to equity, divide total discounted return by initial equity.

11. *Net present value.* The difference between the original equity investment and the total discounted return for a given period provides this informational number. If the figure for this line of the pro forma is zero, the analysts know that the project is meeting the investor's required rate of return. If the figure is less than zero, the investor understands that the project may earn less than the required rate of return. If the figure is positive, the investor will receive (in theory) a higher rate of return than the required rate.

12. *Investment value.* The culmination of the financial feasibility analysis is the bottom-line investment value of the project based on the inputs of the at-risk investor(s). The investment value of a project depends directly on the date at which the investor will sell or transfer the asset. For example, if the investor sells the property in the twelfth year because of lower tax advantages, instead of selling the property in the sixth year, the increase in market value over investment value will likely result in a higher rate of return.

The financing entity, however, bases the amount that it will lend on the project from the standpoint of appraised market value and not on investment value. The loan:value ratio on the appraised market value may force the investor(s) to arrange a higher down payment on the property. The higher down payment requirement can have a ripple effect in the pro forma, making the internal rate of return less than acceptable. The key to investment value is the determination of the present value of all the cash flow benefits coupled with benefits derived from disposal of the asset.

Even without having a particular investor in mind, the conceptual estimator can develop a cash flow model to test various assumptions. A range of assumptions based on "typical" investor inputs could be instructive, or may serve as a way to attract bona fide investors to the project. The averages needed for the "typical" cash flow package would include

- Required rate of return on equity
- Tax brackets of "average" investor(s) for the property
- Expected investment holding period

- Mortgage interest rate
- Required loan:value ratio
- Operating ratio for the specific type of project
- Expected appreciation rates

The estimator for an integrated services team will need three skills to produce a realistic feasibility estimate. First, the ability to conduct a market and feasibility analysis of the subject project is necessary, together with an understanding of accounting principles and tax laws. Second, a grounding in facility design of the type contemplated for the investment is critical. With very little hard information about the final appearance or individual material quality in the facility, the predesign estimator will be asked to develop an overall realistic cost based on the functional requirements articulated by the investors. Third, the integrated services estimator must possess macroknowledge of design and construction pricing to arrive at the second most important figure (after financing) in the pro forma. A traditional estimator within a general contracting firm who has no design or accounting-financing aptitude will be ill-prepared to assume a lead estimating role for a design-build entity without further training, experience, and motivation.

11.3.2 Performance criteria estimates (scope estimating at the predesign or programming stage)

A facilities *performance criteria estimate* is conducted in the absence of design development or construction drawings and without the details of prescriptive specifications. A performance criteria estimate may be used to communicate with an owner about the adequacy of the budget to support the scope of the project. On the other hand, the performance criteria estimate may serve as the basis for a cost-plus fee or guaranteed maximum price (GMP) proposal. At best, the estimator is furnished with a detailed narrative program describing the owner's needs, wants, and aspirations for the project.

For estimators employed in the design-build business, it is important to have the ability to build the project in their heads. A capacity to visualize in three dimensions, and to see products, components, assemblies, and systems that are embedded within the structure, are sought-after traits. For example, the makeup of a laboratory room along the exterior of a building that must also meet energy and sound-deadening performance requirements will consist of a combination of materials that must bear loads, insulate (both sound and thermal), repel water, handle wind loads, and allow for attachment of interior finishes. The room may include hoods and ducting for volatile organic compounds, oxygen piping, fire-suppression systems, raised-access flooring, in-floor ducting for fiberoptic cable, suspended air-compression equipment, and other necessary amenities. The design-build estimator will anticipate the structural, safety, aesthetic, and other needs generated by these systems and include appropriate quantities and prices.

The secret to performance criteria estimating is not simply the number crunching but is the process for developing the bases for the estimate. It is a valuable design-build service to assist the owner in confirming and validating the various assumptions that have been made about the project. Does the concept make sense when the expected functions and quality of the facility are laid alongside the cost and schedule expectations? Or is it prudent to go back and adjust (increase or decrease) the scope, quality, cost, or schedule to make this a realistic project before going into schematic design?

To assist and confirm the owner's assumptions, the project expectations and performance criteria should be applied to the specific site or sites:

1. The design-builder will want to see whether the owner has control of the site. For a road project, has the right-of-way been acquired? For a commercial development site, does the owner have an option that may be exercised, or is the property owned in fee simple?

2. Has the zoning been analyzed for the type and use of facility that is being proposed? Among the important issues are setback requirements, building height limitation, floor:area ratios (FARs), and adjacency to incompatible uses (as defined by the zoning code).

3. Have environmental concerns been reviewed? Environmental issues include soil types, wetlands, hazardous materials, and archeological or historic artifacts. No owner or design-builder wants to have surprises after the fact; rather, it is better to leave no stone unturned early in an effort to speed delivery of the project later.

4. Have building code analyses been based on the facility scope, use, and function? Meetings with the building code department and the fire marshal with regard to facility type, occupancies, travel distances, separations, and fire-suppression requirements will enable the design-builder to place reasonable brackets around the costs for these systems.

5. Given the specific site with its attendant benefits and constraints, the development of an order-of-magnitude estimate for each of the major systems for the facility will allow an owner to see whether the design-builder's proposed systems are matching the owner's expectations. Is the proposed precast exterior panel system an acceptable alternative to the owner's original wish to use oversize buff-color brick?

An estimator who is working at the performance criteria level will approach a project with a comprehensiveness that embraces the fully functioning facility. By contrast, a traditional estimator who is bidding work for a public owner is usually concerned only with what is clearly delineated on the plans or spelled out in the specifications. The integrated services estimator in a design-build environment is ascribing to a new standard of care; that is, thinking of the facility's ultimate use and occupancy, rather than simply responding to inevitably incomplete blueline or CAD representations of how the facility goes together.

Using performance criteria, historical costing data, and judgment, the conceptual estimator can employ a variety of techniques to establish a value for the proposed facility or project. The overall facility quality, as indicated by the program and performance criteria, enable the use of comparable data from previous projects, appraisals, commercial valuation, and cost services and other sources.

Performance criteria estimating methods.   For new facilities, performance criteria and program estimates typically rely on one of the following approaches:

Base unit/order-of-magnitude estimation.   By identifying outputs from a plant or factory, or occupancies of a hotel or school, and dividing the inputs or occupancies by the historical costs for the related facility, a determination of the cost of the base unit can be approximated. For example, with knowledge of a university system's year-by-year costs per student it is possible to multiply the number of students to be served by the instructional cost per student to calculate the cost of the proposed classroom building; or by the housing cost per student to approximate the cost of a new dormitory.

The problem with using the costs of previous similar facilities is that allowances have not been made for changes in facility quality, size, location, or other significant factors. The estimator must adjust predictions for design and construction costs in accordance with the performance standards set forth for the facility. A speculative suburban office building will have a set of performance drivers different from those of a downtown signature class A office facility. The accessibility of an industry database of costs, such as is increasingly possible through information technology systems, will assist the estimator in making informed adjustments to the design-build firm's historical estimating information.

An order-of-magnitude estimate may also be derived from appraisal information and techniques. Appraisal methods include the following:

- *Replacement value.* Calculate the current replacement value using comparables, tax records, and commercially available appraisal data. A lending institution is more readily swayed by comparable real property and capital asset values than by theoretical pro forma projections.

- *Life-cycle method.* The life-cycle estimating method for design and construction brings current and future costs in alignment as present values. The initial and ongoing costs of the design alternatives are examined under this method, including projected operating and maintenance expenditures, program operating costs, repair and replacement costs, and finally, the cost of disposal, transfer, or salvage. A simple straight-line projection of ongoing costs for a building or the use of a sum-of-the-digits method will help the performance criteria estimator incorporate future renewal funds for the facility.

- *Cost or income approaches.* Sales data on properties with attributes similar to those of the proposed project may provide a starting point for the valuation of the land and improvements. However, it can be difficult to find property and capital assets that approximate the unique conditions of the subject project. Past sales prices or valuations of similar properties seldom include disclosure of the financing, unique locational qualities, or physical differences between various properties. Further, there may be a difference between the appraised value of a property and its maximum investment value. A traditional approach to estimating the value of income-producing property is to divide the net operating income by the capitalization rate to determine value. Although useful, the use of net operating income alone in the numerator tends to exclude other sources of property value, such as appreciation and tax shelter. The same problem exists for the denominator, where the capitalization rate is based on mortgage rates and are not placed higher to reflect the required rate of return of the investors (to compensate, an average between the mortgage interest rate and the investors' required rate of return can be used).

Square-foot estimating.   The programming package for a facility can be translated into a predicted size and volume. For example, if the town's airport facility program shows a need for a terminal of 150,000 ft^2, a 1.5-mi runway and associated taxiways and aprons, and associated landside improvements, the cost can be determined by using data from other airport developers and operators by multiplying the comparable costs by the determined predicted costs. If the airport facility is reduced in size, it will cost less; if the town wants an expanded facility for future growth, it will have to find more funding.

The other important variable in unit cost estimates is the quality of the proposed construction. An intensity of facility functions, such as the number of bathrooms, kitchens, or high-tech meeting rooms, can drastically affect the squares-footage costs of buildings. Facilities with more subdivided space will cost more than those with open spaces, except where large clear-span open spaces drive up structural costs. Facility complexity is another influencer of cost, where the increased use of special piping, venting, and washable materials causes a medical facility to be priced higher than a simple retail facility.

Durability of facility materials is a third quality factor that should be included in unit cost estimates. The use of stone or tile floors in the airport lobby versus composition tile over concrete is an example of a materials selection where the former materials may outlast the latter by a ratio of 3:1. A vital skill of the design-build estimator is the ability to translate a facility program into square-footage and quality assumptions that are synchronous with the owner's expectations.

Parameter and factor estimating.   For more than a generation, McGraw-Hill Construction Information Group has been collecting and distributing information that is related to costs of the work as performed by various trades. Parameter estimating involves the assignment of lump-sum costs to individual trades; then dividing the total cost by the appropriate parameter measure (e.g., amount of store front aluminum and glass in square feet) to arrive at a unit cost for the work put in place by that particular trade, as compared with the aggregate cost of the facility.

The application of parameter estimating is based on the mathematical definition of parameter: a variable appearing within a mathematical expression or distribution, such that all possible values of the variable correspond to different expressions and distributions. The approach of McGraw-Hill's *Engineering News-Record* (ENR) to data collection for parameter costing is broken down into building type; standard parameter measures for most buildings; characteristics of design, including story heights and floor areas; and design ratios, such as core area to rentable space, or heating BTU (British thermal unit) requirements per square foot.

To apply the collected parameter data, the ENR methodology relies on allocating the data collected to the various trades found on the typical building project. The parameter subtotals are combined with unit cost data to determine predicted in-place cost by trade area. The sum of the subtotal costs adds up to an estimated cost for the facility. The parameter estimating method is occasionally employed by small- and medium-sized real estate developers to prepare a price during financial negotiations or for examining project costs with construction managers and subcontractors before signing contracts for design and construction.

Factor estimating is a useful method for developing costs for projects that may have a predominate piece of equipment or functional components that contribute heavily to the overall costs of the project. For example, the equipment within a carwash facility is often more important than the surrounding walls and roof. Similarly, when considering an ice rink, the refrigeration equipment and water systems are more costly than any other single component of the facility.

Factor estimating is commonly used for industrial and process facilities such as food processing plants or petrochemical facilities. The accuracy of a factor estimate depends on the amount of data collected from previous projects and the degree of similarity between the proposed project and the previous work. If the design-build estimator was using factor estimating to arrive at a program-level cost for an ice rink, it is likely that the refrigeration equipment purchasing and installation costs would be the primary factor, followed by piping, rink floor structure, clear-span roofing structure, and heating and air conditioning. For the proposed ice rink, the estimator would review previous data to discover, for example, that the clear-span roof structure represents 60 percent of the cost of the refrigeration equipment. This relationship between the factor for the roof structure and the cost of refrigeration equipment is expressed as 0.60, and the other secondary factors are expressed with a similar percentage factor appropriate to their presence within the project. Factors are carried forward for use in future projects; *not the costs themselves.* If, for the proposed facility, the factors collected from previous projects are reliable and the factors for the proposed project have a corresponding cost ratio as was exhibited in previous projects, then the estimate should be valid for most predesign or programming purposes.

Range estimating. At the predesign or programming stage of a design-build project, it may be advisable to apply a simulation exercise to a budget estimate to determine its degree of certainty or uncertainty. Range estimating can be applied at any stage of estimating to simulate the probabilities for various costs that make up an estimate for a facility. For reasons of simplification, range estimating is a method of establishing an expected value, and then an optimistic and pessimistic value, for each line item in an estimate. Added to the expected cost/lowest cost/highest cost assumptions are confidence limits (or level of probability expectations) for each item.

Range estimating is an important tool for looking at the high-risk line items and work packages in an estimate. To apply range estimating, the project is divided into divisions of the work or facility components, and the estimator details the degree of variability for each item. Using computer modeling, it is possible to generate a series of variable facility estimates on the basis of the potential costs of each line item. It is unlikely that the project will have a cost that is the sum of the highest of each line item; nor is it realistic to expect that the lowest costs will be realized. Instead, the range estimate may point out a specific direction to take with design value analysis, and it will serve as a management tool for work package focus during the project execution process. Design and construction organizations can also employ range estimating as a way of determining the amount of contingency needed to undertake the project.

Performance criteria estimating as applied in process projects. Much of the body of knowledge about engineering estimates that are derived from early stages of project definition have been generated in the industrial/process industry. The American National Standards Institute classifies estimates according to ANSI Standard Reference Z94.2-1989. Performance criteria estimates would be categorized as "class 4" or "class 5" estimates:

An ANSI class 5 estimate assumes a level of project definition of 0 to 2 percent. Class 5 estimates are prepared for strategic business planning purposes, according to ANSI, which would include market studies, assessment of project success, location studies, evaluation of resource needs, and long-range capital planning. An ANSI class 4 estimate assumes a level of project definition of 1 to 15 percent (note overlap with class 5). Class 4 estimates are used for project screening, alternative scheme analysis, confirmation of technical feasibility, and approval for proceeding to design development. The American National Standards Institute places a degree of accuracy on class 5 estimates as minus 50 percent to plus 100 percent and on class 4 estimates, minus 30 percent and plus 50 percent.

According to Hollman,[1] an effective estimating methodology consists of

---

[1] John K. Hollman, Eastman Kodak.

1. *Cost estimating relationships (CERs).* CERs are estimating algorithms that may appear as (cost resource) = (factor × parameter). A *factor* is a unit cost factor in terms of resource and unit of measure. A parameter consists of the units of measure of the estimated item. The resultant cost resource is dollars (labor, materials, or total combined) or time (labor-hours, rental hours, or similar cost amount). Cost estimating relationships are either *stochastic* in nature (based on conjectural cost relationships and statistical analysis) or *deterministic* (based on conclusive and definitive cost relationships). The outcome of stochastic approaches are typically called *parametric estimates,* and the outcome of deterministic approaches are usually termed *detailed unit cost estimates.*

2. *Cost estimating database.* For accurate estimating, cost factors should be kept in a standard reference format. The data may be created in-house or may be obtained from third-party sources. Among the types of cost factors are detailed unit cost (derived from labor-hour factors and material unit costs), unit cost assemblies (created from compilations of labor-hour factors and assembly unit costs), equipment factors (developed by either equipment type, trade discipline, or process plant type), and parametric cost models (created from statistical analysis and historical data and through professional perspective). Additional cost factors may be generated through ratio factors that are actual or de facto standards from industry experts or through gross unit costs from third-party sources. Data sources usually fall into three categories: published reference manuals, evaluation of unit cost models from standard engineering designs, and regression analysis of historical cost data. Potential problems with the data are out-of-date cost factors (especially in a volatile labor or material market), internal consistency (recent data do not match up with historical data because of differences in project approach or changed conditions), and the individual estimator's point of reference. (Do the firm's estimators see the work in the same way?)

3. *Cost updating.* Maintenance of the cost estimating data enables the estimating team to incorporate recent project experience and closeout information. The project record will include summary cost and schedule results that can be compared to original project estimates. Technical information embedded in the cost updates describe project deliverables and scope of the work, enabling the data users to calibrate the cost information with adjustments for site-specific factors. Estimators have higher confidence with data outputs from the average results of a large number of projects, which helps iron out the discrepancies that may exist from one or two projects. Using project control data with cost estimating data with parallel account data from industry performance, it is possible to benchmark cost performance against other firms.

4. *Cost estimating tools.* Forms and software are the manual and electronic tools available to the design-build estimator. For modest projects, a basic organizational form may be all that is required (database searches and extensive recalculations are not required). For larger or more complex projects, estimators employ various forms of spreadsheets which allow retrieval of data and calculations through the software's built-in analytical functions. A step beyond the basic spreadsheet is the relational database where data are captured in

cells, and functions and calculations are performed as commands permitting customized manipulation of the data. Commercial estimating software contains extensive cost coding capability and allows the user to create and store unit cost assemblies. Owners and practitioners often select their cost estimating software according to how it interfaces with other software applications, including finance, job cost accounting and scheduling for project management, computer-aided design, or combined computer-aided engineering and cost modeling systems. For parametric estimating processes at the performance requirements level, some software packages come already equipped with prebuilt unit cost models for some types of facilities. Others allow the user to create customized unit cost models.

Estimators working with complex design-build projects at the performance criteria stage have been using complex parametric cost models and parametric unit cost models. A *complex parametric model* can yield more output information than simply costs. Users of these models often refer to themselves as *parametric analysts* rather than *cost estimators,* because they are generating simulation studies, facility optimization studies, cost-of-quality alternatives, and life-cycle cost analyses. The U.S. Army Corps of Engineers and the U.S. Naval Facilities Engineering Command have funded software development that employs regression models based on internal and external cost data for the prediction of costs for proposed federal facilities.

A *parametric unit cost model* is a hybrid combination of unit cost modeling and complex parametric cost modeling. Each line item in a unit cost model has an algorithm with fixed design parameters and cost factors. By contrast, a parametric unit cost model has variable parameters and/or factors in the algorithms of the cost model's constituent line items. For example, a storage building in coastal Virginia may have components of footings, foundation wall, concrete floor, CMU walls, and wood truss roof. The basic building is designed for 90-mph (90-mi/h) wind load. But rather than creating fixed models for changing the components under wind loadings for 60, 110, or 135 mph or another variation based on expectations of hurricane conditions, a mathematical relationship could be developed that modifies the original building's wall thickness and roof anchorage depending on wind loading and uplift for each 1-mph increase or decrease. After the functional relationship is determined, the estimator could use a single algorithm to estimate the cost of a similar building with design adjustments for different wind conditions. The parametric unit cost estimating process may be used for projects that have only basic project scope information up to and including preliminary design.

11.3.3   Conceptual and schematic design estimates (estimating from the conceptual design drawings stage up to the schematic design phase)

Experienced design-build practitioners are universally familiar with conceptual estimating as a practitioner's tool for responding to an owner's design-build request for proposals (RFP) package. The RFP typically contains a program of facility requirements and a description of design intent. Using an RFP, the owner frequently asks the design-build entity to *respond* with a conceptual design, project management plan, and price. (Some owners or their consultants will also include concept drawings as part of the package, with disclaimers that the concepts are only suggestions and are generally not to be relied on as design directives.)

The estimator in the design criteria design-build process works with the design-build team's designer to develop enough in the way of design solution sketches (concepts or schemes) to estimate the team's proposal. (However, when the owner provides additional detail, such as a concept drawing for a typical classroom in a school building, the estimator may rely on portions of the owner's package rather than the team's designer. The design-builder will clearly indicate this reliance on owner-provided information in its proposal submittal.)

To fully appreciate the contrasts between estimator's role in design-bid-build versus design-build, think of two weather forecasters. Assume that the first forecaster has a good comprehension of meteorology, plus radar to show the locations of clouds and rainstorms and that the second forecaster possesses a good knowledge of meteorology, and a passion for weather and climate issues, but no radar to tell exactly what is happening in a specific geographic area. The second forecaster must rely on less exacting instruments, accumulated experience, and historical data, plus a good dose of dead reckoning. For a traditional lump-sum design-bid-build project, the estimator organizes items of work and their costs into the CSI 16-division format. The breakdown of items of work is detailed both on the completed drawings and in the specifications, and the estimator "takes off" quantities, multiplies these quantities by costs, and arrives at subtotal costs for each division and subdivision of the master format.

Because the details of completed drawings are not available to the conceptual estimator, the traditional "takeoff of quantities times installation costs" cannot be used to yield a project estimate. Instead, the estimator must know facility systems, design, and construction.

Elemental estimating is one method of determining projected costs for a facility that is directly applicable to design-build at the conceptual design stage. The elements may be derived from the CSI/ASTM UniFormat outline:

A.   Substructure

B.   Shell

C.   Interiors

D.   Services

E.   Equipment and furnishings

F.   Other building construction

G.   Building sitework

Z.   General

For example, the building envelope in UniFormat (Section B, "Building Envelope") would encompass multiple divisions from the 16-part MasterFormat:

Division 1.   Scaffolding

Division 4.   Masonry

Division 5.   Metals

Division 7.   Insulation and caulking

Division 8.   Curtainwall, glass, doors, and frames

Division 9.   Painting, finishes, and coatings

Division 11.   Specialties

Some estimators are using the Means systems estimating information, which can be organized around building systems. Similarly, the Canadian Institute of Quantity Surveyors has published an elemental estimating format that includes nine "hard cost" elements (substructure, structure, exterior cladding, interior partitions and doors, vertical movement, interior finishes, fitting and equipment, services, and site development) and two "soft cost" elements (overhead and profit and contingencies).

When a design-build team assembles for the purpose of responding to an owner's RFP, the disciplines making up the team will analyze the RFP to determine the project's objectives. The goals signify functional, aesthetic, and longevity guidelines for the team; which makes tradeoffs and compromises between budget and quality, speed of delivery and environmental protection, or innovative materials versus energy efficiency. The owner's goals for the facility are embodied within the selection criteria and weights, and further amplified in the performance requirements for the project.

With a shared understanding of the owner's performance goals, the design-build team launches into developing a technical solution for the design problem. The design concept that is evolved may be based on typical functional floorplans for that type of facility, or the concept for a bridge may rely on a particular image gleaned from owner comments or discussions with the local constituency about historical river crossings. Working with requirements and possible solutions, the design-build team narrows the alternatives down to a manageable set of achievable concepts. The concept(s) can be evaluated by the team through an issue-by-issue comparison to the owner's objectives.

Knowledgeable teams will often list the deliverables that are expected to result from the evolution of design concepts into a schematic design suitable for an RFP response. The owner may not expect or want all the deliverables; therefore, the concept-to-design process may be done only for purposes internal to the team. According to AIA's *The Architect's Handbook of Professional Practice,* the process of schematic design establishes the general scope, conceptual design, and scale and relationships among the components of the project. The primary objective is to arrive at a clearly defined, feasible concept that meets quality, cost, and schedule objectives of the owner.

Documentation of the scheme will be accomplished by the expanded design-build team, which will include professional engineers for selection and conceptual engineering of building systems; specialty contractors and manufacturers for systems and components selection, plus early systems design and costing; and other team specialists to solve the design challenges of the project. The team's documentation output may include

1. Site plan (although this may be issued by the owner to the design-build teams)

2. Floor plans (showing overall functional layout and typical spaces)

3. Elevations (fenestration, wall finishes, expansion joints)

4. Key sections (showing structural support systems, weatherproofing, floor-to-wall transitions)

5. Outline performance specification

6. Summary of the design metrics with a brief explanation of how it meets the program

7. An estimate based on the conceptual design

8. Deliverables sought by the owner (model, CAD layout, renderings, drawings)

Conceptual estimates done for large-scale design-build projects by experienced design-build firms usually rely on two methods:

1. *Price indices* consisting of charts and graphs that organize data from previous project experience are produced to show percentage of change relative to scope of the project, time to complete based on average productivity (and variations above and below that average), and costs of principal inputs (labor, materials, and equipment for the element of work).

2. *Progressive elemental cost estimating,* which continues in parallel with the conceptual and schematic design process, is done to cost the evolving design and design alternatives. For large-scale civil works and process projects, the British Chartered Institute of Building describes operational estimating techniques in its *Code of Estimating Practice.* These estimating techniques forecast the cost of completing an entire construction operation, rather than coming out with unit rates for each item of work. The estimator takes into account all resources necessary to complete the design and construction of a facility element by analyzing the project program and management plan in conjunction with conceptual sketches and schematic designs evolved by the design-builder's design team.

The advantages of progressive elemental cost estimating are

1. The estimate is based on an overall understanding of the designer's intent for that portion of the work, including the most recent modifications or adjustments made during the consideration of alternatives.

2. The estimate encompasses entire elements and assemblies. When the project is part of a larger process or has a civil-type infrastructure, it is difficult to apportion the costs among a traditional group of measured items because of the specialized nature of particular sites or pieces of equipment. Systems and assemblies estimating enables the project team to address the macroeconomic issues of the project as consensus is gained for various design concepts.

3. The productivity rates are based on the overall program or major elements within the program instead of standard outputs. The estimate is able to take into account the combination of inputs involved, such as combinations of materials and resources and multitrades relationships (e.g., two-way hauling of materials, cross-skilling of trades for critical-path components, sequencing of the work).

4. The progressive conceptual estimates are done concurrently with the evolving conceptual design, allowing the facility team to have cost comparisons of alternative systems, configurations, and major materials and equipment choices.

Conceptual design and schematic design are not static phases of project evolution. Design undergoes adjustments, expansions, and contractions as alternatives are considered and budgets are resolved. Using the elemental approach, single composite rates per square foot or per gross floor area, or as a cost percentage of total cost, will be used. The ability to monitor and control costs is an overwhelmingly positive benefit of using parameters generated within elemental estimates. In-depth adjustments of systems and costs by the design-build team could not be well facilitated by trade-based product divisions within MasterFormat's Divisions 1 to 16.

One of Canada's great proponents of elemental classification systems, Robert P. Charette, P.E., cites multiple benefits of ASTM's UniFormat system, including improving the entire team's comprehension of the project at an early stage; allowing for brainstorming of alternatives without bogging down the process; providing for rapid calculation of the cost of alternative approaches; anticipating the layering of CAD drawings at the design development phase; tying the costs into life-cycle projections for the facility; and serving as an organizational tool for future operation, maintenance, and replacement activities for the property.

To improve design-to-cost efficiency, the conceptual estimator will normally work with a cost model that allows for a baseline reference for comparing alternative approaches, systems, equipment, and innovations. These cost models are organized from the top down (according to the UniFormat philosophy). Design-build conceptual estimating will serve as the foundation for value analysis when a design and construction cost model is combined with a project quality model. For each facility system, element, assembly, and subassembly, attributes of the quality model are ascertained by querying the owners and users of the facility. Common quality measures for a building facility are

1. Operational effectiveness
2. Site planning and environmental image and sustainability
3. Project schedule
4. Finance and capital cost issues
5. Energy and operation and maintenance
6. Image of the facility and architectural design
7. Flexibility, expandability, and adaptability to new uses
8. Environmental issues for the facility and improvements
9. User comfort and facility systems
10. Stakeholders and community input and concerns
11. Security and safety
12. Engineering performance

Parametric estimates transform the project scope, at the design criteria or conceptual stage, into total cost and system cost data. The Association for the Advancement of Cost Engineering (AACE) defines a parametric estimate as "semidetailed," with all project costs related to the parameter cost of a reference or similar project or projects.[2] Data for the reference projects are collected through four activities:

1. *General information* about the project is assembled, including the type of facility (e.g., hospital, library, garden apartment), location (city and state), project duration (start of design to occupancy), type of owner (private, public, public-private, or nonprofit), structural system (steel, concrete, other), type of exterior wall (masonry, metal, glass, other), and special site conditions (soils, wetlands, rock, water/dewatering).

2. *Standard parameters* to be measured are identified and agreed on, such as gross enclosed floor area, gross area supported, total basement floor area, roof area, net finished area, number of floors, area of masonry on elevations, area of other exterior wall, area of curtainwall and fenestration, interior partitions, HVAC systems, electrical power and low-voltage and control systems, and parking areas.

3. *Design characteristics* relating to the specific type of structure become additional parameters, including area of a typical floor, story height, lobby area, number of plumbing fixtures, number of elevators, number of rooms, num-

---

[2]*Skills and Knowledge of Cost Engineering*, 3d ed., AACE International, Morgantown, W.Va., 1992.

ber of occupants, and other measurements that may contribute to ratio indicators of cost for the specific project type.

4. *Design ratios* based on reference project data will contribute to adjusting the data based on performance and quality requirements. Design ratios may be architecturally oriented, such as the number of workstations per open office environment, or engineering-oriented, such as the HVAC tonnage per building square foot (mechanical) or parking allowance per building square foot (civil).

To complete the parametric estimating process, data are applied through a five-part process:

1. All cost elements must be listed, including UniFormat Divisions A through G, plus general conditions, overhead, markup, risk contingency, and other cost divisions.

2. Total cost of each element from the reference project must be calculated from the existing subelement data and project closeout data.

3. Each cost element must be related to a parameter considered to affect its cost. Each element will be assigned a code number (or parameter title) that corresponds to one of the parameters selected.

4. Cost element unit prices are determined relative to the selected parameters of the project. For example, for all kitchens and restrooms at the project site, tile floors are used. A unit price can be determined by dividing the total cost for installed kitchen and restroom floor tile by the total square footage of the kitchen and restroom floors to determine a unit price.

5. The percentage of each cost element in terms of the entire project cost must be determined. This percentage is obtained by dividing the total amount for each element by the overall project cost and multiplying the quotient by 100. The resulting percentages are ideal for analyzing cost trends among various cost elements in multiple projects.

Detailed parameter estimating provides for application of different cost parameters for project subelements and assemblies. These units can then be adjusted for building size and complexity through multipliers and indices obtained from internal and external sources. Design-builders going outside the geographic area from which their cost data have been collected will also employ a location multiplier.

In practice, the design-build conceptual estimator will use a detailed estimating approach for design items that have become fixed at the schematic phase; and will employ parametric estimating for systems that have not been well defined. The estimator's assumptions will be communicated explicitly to the design-build team for inclusion in the proposal submittal. As the success of the project is dependent on these assumptions, the role of the conceptual estimator is critical and indispensable during the design criteria and schematic design phases of the design-build project.

11.3.4   Detailed estimates

Detailed estimates for a true design-build project are the result of an evolutionary process, as opposed to a single flurry of activity as it was for "bid day" (for a 100 percent design–then bid traditional process). Today's use of CAD with links to estimating software allow for the generation of ongoing takeoffs of materials and systems costs from the inception of the project. These estimating techniques are fully adequate for financial and lending institutions. Nevertheless, the necessity of feeding into cost control systems for design and construction management will lead project managers to detailed estimating, especially for and with those trades who are still product-centric (such as the brick manufacturers who produce and the masons who install this time-proven material).

Experienced design-build companies will have created their own process for generation of detailed estimates; however, the detailed estimating steps that follow may be used by design or construction companies entering the design-build market, or by start-up design-build companies:

Structuring the detailed estimate

1. Set a detailed estimating schedule to provide timely information to the design-build team, and to ensure that the guaranteed maximum price or fixed lump-sum price is completed in time for contract milestones.

2. Review all project documents, including the request for qualifications (RFQ) and RFP, facility program, design criteria, performance requirements, instructions to proposers, and any conceptual design that may have been included with the original project manual. Also, review all documentation generated by the design-build team, such as schematic designs and outline specifications. Try to find and resolve any gaps between original design intent and design solution. Flag any special items that may have availability, fabrication, or delivery problems. Having a checklist of the usual items found in a design-build RFP will help prevent overlooking some needed project items.

3. Prepare a design-build project portfolio. The hard-copy or electronic data in the portfolio will consist of various files, including the design-build project proposal documents; designs and specifications generated by the team; costing information from the project evolution stages plus quantity takeoffs from schematic and design development drawings; clarifications and correspondence related to the project; price quotations from subconsultants, subcontractors, manufacturers, and materials suppliers; and any legal documentation that is tied to the project, such as contract forms, bonding and insurance coverages, and proposal forms.

4. Organize the estimate into direct project costs, indirect project costs, and other costs, such as escalation, contingency, fee, or profit. Naturally, the design-build process includes *design* direct costs and design office indirect costs along with traditional construction direct and indirect costs. Segregate information into categories pertaining to work that will be done by the prime design-build team from that which will be done by subconsultants and

subcontractors. Tasks that will be self-performed by the design-build team will have an overhead structure and productivity rating different from those that will be performed by parties contracted to the design-build entity. Inform the entire team, including designers, constructors, subcontractors, manufacturers, suppliers, insurers, and any other specialists, of the need for proposals and pricing by a firm date.

5. Have a system in place for accumulating services and products cost data for the entire design-build project. Both hard-copy and electronic estimating processes will consist of detail forms and summary data forms. After quantity takeoffs are completed, the labor and equipment pricing steps will be guided by geographic location, scarcity, wage rates, general and special conditions, and weather issues. Assembling the cost elements will depend on a design-build team's responsibility matrix. Having common data entry formats, whether paper- or electronic/Internet-based, and insisting on their use, will aid the design-build team in creating and tabulating its final estimate.

Developing a work breakdown structure.   A work breakdown structure not only will aid the estimating team but also is a tool for managing project activities from the proposal stage through design and construction. The structure directly feeds into project control and reporting. A work breakdown structure segments the project into various levels and units of activity, and tags each of these levels and activities with an alphanumeric prefix. The structure defines project deliverables and responsibilities at four levels: total project, major project elements, subproject or design elements, and disciplines/trades.

This project management tool enables the design-build estimator to (1) categorize portions of the project into identifiable and manageable units; (2) integrate costing and scheduling for planning and controlling the project; (3) provide a brief work description, task order number, and status for each identifiable segment of work; and (4) serve as a benchmark for keeping score on projected costs, contracted costs, and actual costs and associated schedules during the duration of the project.

Estimating the work items.   After classifying the work and describing the individual items through a work breakdown structure, the next stage of a detailed estimate is to determine the dimensions of the items of work. Detailed estimating relies on two basic methods of measurement: net measurement, in which the amount of construction work is determined by the length, area, volume, capacity, mass or number of units; and materials takeoff, where each element of cost is measured in units of commodity or time. The difficulty of using the materials takeoff method for design-build is that the estimator must know or infer the physical dimensions of the facility element or material itself to accurately develop the detailed estimate, often with less than 100 percent design documents.

Dimensional data are either taken directly from the project graphics; scaled from the plans, elevations, sections, profiles, and details; or derived by the estimator by assuming various aspects of the work that are not shown (such as caulking or firestopping at exterior walls). When units of work and material counts have been entered into the detailed estimating format, extensions are produced as interim gross measurements.

Costing the labor, material, equipment, general conditions, and overhead.   If the estimator's measurement of every aspect of the work is project analysis, then the process of pricing is project synthesis. When the assumptions are documented and takeoffs completed, and after the dimensions are extended, the estimator enters known costs of materials, labor, and equipment. Unless the materials are of a highly specialized nature, the manufacturer's product costs are readily available. Labor costs and equipment costs can be more problematic.

For architectural and structural products, materials prices are established by regional suppliers on the basis of market conditions, supply and demand, and corporate return on investment. In the mechanical and electrical fields, there is a strong tradition of list pricing with discounts for quantities and repeat customers. Occasional buyers pay the full list price that is set by the mechanical or electrical manufacturer or regional supplier.

The source of labor rates may be derived from government wage scales, local union rates, or the company's own hourly wage rates for permanent or project-specific employees. Labor costs normally consist of two parts: the gross paycheck amount and the payroll "burden" (social security and fringe benefits paid for by the employer). These rates are reduced to a per hour or per day rate to complete the estimating unit calculations.

Equipment costs for a design-build estimate are in the realm of project site equipment employed to complete the work (direct costs) plus other costs for tools and equipment that may be purchased, leased, or outsourced in the course of performing work under a design-build contract (indirect costs). These costs are distinct and separate from the installed equipment that becomes a permanent part of the project. Installed equipment costs are normally incorporated into the estimate under the labor and materials portion.

Civil infrastructure projects can be equipment-intensive, and the cost of specialized equipment can be the difference between obtaining or losing the contract. How the cost of equipment is calculated will impact the decision about its use. For example, if a design-build company obtains a railroad tunnel project, should an open-face boring machine be purchased or rented, or should the entire tunneling section of the contract be turned over to a specialty subcontractor? The design-build estimator will need to ascertain owning costs, operating costs, operator costs, and mobilization costs to make an informed decision. Equipment costs may be categorized as production equipment (unit costs are directly charged to the project), plant equipment (equipment used on multiple construction projects such as company-owned concrete pumps and buggies), and administrative equipment (often covered under overhead, but may be broken down as line items where the costs of such equipment, e.g., management control hardware and software or project Internet site with 24-hour cameras, add substantially to the project cost and value).

Subconsultant and subcontractor costs are crucial components of the design-builder's cost proposal. The value of the design-builder's solution in response to the owner's criteria is dependent on work that is shared or subcontracted to other members of the team. The challenge for the design-build estimator is to make certain that the subconsultants and subcontractors have taken responsibility for the UniFormat-based elements or systems of the project. One practical way to handle this problem is to ask the subcontractors and suppliers to cross-reference both UniFormat divisions and MasterFormat sections in their subcontracts. The design-build estimator must also make a determination as to whether the subconsultant and subcontractor cost proposals are realistic and competitive, and determine what markup is required to coordinate the supervision and coordination of the work.

Design-build project indirect costs are all costs that are not assigned to a particular element of work, individual work item, or equipment costing. Primary components of indirect costs are general conditions and overhead. The estimate format should carefully distinguish project overhead costs from overhead costs of the design-build company or consortia.

A listing of general conditions is derived from the project manual, specifications, and contract form, but will usually include

- Permits—environmental, building, and other
- Insurance—professional liability, comprehensive general liability, pollution, and other
- Bonds—proposal, performance and payment, operation
- Mobilization
- Project finance
- Safety program and equipment
- Supervision
- Professional services—quality assurance or project accounting, which may fall in this area
- Temporary facilities
- Miscellaneous costs—site cleanup, travel, and lodging

Firms participating in design-build have operating costs regardless of the number of contracts that may be ongoing in any given year. The general and administrative costs are usually calculated as a percentage of the total volume. Overhead cost components may include

- Office lease/facility costs
- Maintenance

- Office equipment and supplies
- Depreciation
- Salaries for home office personnel
- Employee benefits
- Advertising, marketing, and project procurement
- Business insurance
- Professional fees such as accounting services

Owners may require their design-builder to carry allowances for items that have not been fully defined, such as stone panels for a building entry facade. A specified sum of money will be included in the proposal that is expected to fund the allowance portion of the work. Cost adjustments during design and construction will reflect actual costs for the elements and materials covered under the allowance. When using a competitive or competitively negotiated design-build procurement process, an owner occasionally will request that alternates be included in the proposal. The alternates will be either additive or deductive, and will provide the owner with some amount of flexibility in obtaining project amenities that the owner would like to have, or eliminating some portion of the work that may be optional, given a tight budget.

Pricing design services. Professional services from architects, engineers, and other licensed design professionals are an important part of a design-build estimate. Compensation or fee may be based on an hourly rate, a stipulated sum, a unit cost (usually repetitive spaces or facilities), or a percentage of construction cost. The design-build estimator will include a price for design services that is often based on a percentage of construction cost. Costs are based on a menu or package of design services required by the project. Phases of the design process in design-build will be somewhat different from those in design-bid-build: beginning with feasibility analysis and progressing to programming/design criteria identification, conceptual and schematic design, design development, construction documents, and construction support services. With the advent of electronic CAD and three-dimensional imaging, further adjustments to the phases of professional design services are likely.

To develop the price of professional services, designers can assemble a step-by-step buildup of costs. Project tasks and roles at each phase are estimated and an hourly rate applied to each activity. The total design hours to be expended on the project will be transformed into direct salary expense. Add to the firm's direct labor costs the fringe benefits and other costs directly related to the project, and formulate indirect costs for the design firm/team. According to James R. Franklin,[3] the design estimate can be cross-checked by using rules

---

[3]*Current Practices in Small Firm Management.*

of thumb such as: (1) hours per sheet of drawings or specifications, (2) monthly allowances for direct office expenses based on facility type, (3) the typical number of hours expended per square foot of facility type, and (4) historical knowledge of fee as a percentage of construction cost.

After determining the prime design professional's costs, any subconsultants' costs are added to the design estimate. The prime professional should add an appropriate markup to cover project administration, coordination, and quality assessments of the subconsultants' work. Another cumulative line item in the design estimate is composed of reimbursable expenses. These expenses are incidental and out-of-pocket expenses incurred by the design firm during the course of the project. Reimbursables may include printing or electronic transfer of graphic information, travel and related expenses, and additional liability or other insurance. For the stipulated sum design-build project, a lump-sum amount for reimbursables must be tallied and included with the estimate.

Finally, the design estimate should include a contingency based on the degree of risk or uncertainty that is represented in the estimated design costs and a percentage of the subtotaled cost for profit. The amount of profit will often be a balance between the competitiveness of the marketplace (knowledge of what other designers are charging their clients) and the amount of value that professional design services bring to the project.

Capping the estimate with contingency and profit.   Contingency is directly related to the degree of certainty about the elements of work. For projects that are only slightly defined, the contingency amount will reflect the higher amount of uncertainty. For projects that have clear design criteria and performance specifications (meaning narrative and tabular information, not graphic design), the amount of contingency can be correspondingly low. Contingency amounts may be zero percent, and for work that is ill-defined or highly specialized may reach the low double digits. The most important variable in determining contingency is the knowledge of who has been assigned specific project risks and the ability to ascertain the magnitude of the risk. Under GMP contracts, creative contract clauses are being used that allocate ownership of portions of the contingency to the parties. Contingency allows the design-build entity to proceed with work with a reassurance that unforeseen cost elements will not damage their ability to function in a sound financial manner.

Entrepreneurial profit is closely tied to risk and uncertainty. There are many people in management who are adept at continuing the functions of an ongoing business. Distinguished from the operating managers, however, are the persons with vision, originality, and daring. Some economists think of profit as the reward for those who bring new approaches to business, who urge changes in strategic plan and market focus, and who innovate rather than maintain.

Profit in the design-build industry can be viewed as a reward for being efficient in the use of resources on behalf of a customer or client. Design consultants have long preferred commercial contract terms, such as cost-plus fee contracts, that insulate their firms from risk, but offer relatively low profit. General contracting companies have tended to prefer fixed-price, lump-sum contract terms that are not adjustable beyond agreed-on scope (except by change order or claim) but, with aggressive management, can provide relatively higher profit. The amount of profit is tempered by stiff competition in the construction industry, which tends to depress profit margins.

Some design-builders look at profit as the residue left to the business entity after all costs have been met. Profit, under this view, is a portion of the risk capital; that is, the amount of money required to operate the design-build firm as a business. Risk capital may range between 7 and 15 percent of the total annual volume of the company. Profit may be equated to an amount between 15 and 35 percent of risk capital. For example, a design-build firm with an annual volume of $10 million may have a risk capital of $1 million, with an annual profit of 30 percent of risk capital, or $300,000. In this example, profit on total company volume would equate to 3 percent. For many firms, this figure will appear too low to sustain the business; these firms will opt to use a 5 percent, 7 percent, or higher figure in their estimates. The actual profit margin that is used should be determined by corporate financial advisors and managers. A profit margin that is too low will lead the company to financial problems; a profit margin that is too high will result in diminished market share.

11.4   Productivity Considerations for Design and Construction

In manufacturing, output is measured in discrete physical units, and inputs include all labor, capital, and resources necessary to produce outputs. Most manufacturing inputs and outputs are relatively straightforward and easy to measure. In design and construction, however, there are both products and services involved, resulting in a variety of measurable units, such as lineal feet, hours, physical spaces, and pages of graphics. *Productivity* is defined by economists as the dollars of output per worker-hour of labor input. But productivity is far more than a labor issue. Better tools and equipment, use of innovative materials, improved education and training, and ideal weather conditions (or weather mitigation techniques) can all contribute to increased productivity.

For the efficient design-build project, improving materials management and labor productivity are primary ways of increasing output. In its *Construction Industry Cost Effectiveness* report in the 1980s, the Business Roundtable noted that the construction industry lagged far behind manufacturing in basic materials management for production. A materials management system consists of identifying, acquiring, distributing, and disposing of materials for a project. Because materials and equipment can represent as much as half of a new facility's overall cost, it is useful to separate materials management steps:

Responsibility and scope. A number of parties contribute to the materials management process, including manufacturers, suppliers, subcontractors, and direct signatories to the design-build contract. Responsibility is divided, and a problem can occur at any information transfer point. A wise materials management plan includes all participants, and the extent of their involvement is clearly stated at the outset. Delivery coordination information that is poorly communicated, or not communicated at all, can cause late deliveries or untimely deliveries of goods. A materials tracking system that begins at the estimating stage, and continues through requisition, buyout, handling and expediting, shipping, receiving, inventory, and facility installation can reduce the inefficiencies that are part of every construction project.

Procurement of materials. Most of the major materials procurement on an integrated design-build project can be handled by one organization (i.e., by the team, which may consist of a single firm or a consortia of companies). For smaller items, use of purchase orders as a short form of contract are used for commodities building supplies and some types of equipment. For materials requiring specialized labor to install the products on site, a subcontract will be used. The procurement cycle for materials on the design-build project may consist of the following steps in identifying the items during preparation of design criteria, conceptual estimating, conceptual design, or at a later phase of the project:

- Assigning design characteristics to the work item, so that the materials can perform the functions required
- Quantifying the products needed to complete the work and defining the materials in narrative and tabular form
- Writing and issuing a request for quotation for the materials or work items
- Receiving and evaluating proposals (RFPs) or bids [invitations for bid (IFB)]
- Issuing purchase orders or subcontracts for the materials or work items
- Reviewing and approving shop drawings and samples
- Off-site manufacturing or assembly of the materials
- Expediting of the materials
- Shipping, tracking, and delivery of the materials
- Receiving, handling, and use or storage
- Installation of the materials or equipment
- Testing and use/operation

Materials handling. A large component of on-site labor activity is in materials receiving and handling. Some studies report that materials handling consumes as much as 20 percent of every work-hour on a project.

Materials handling operations include shipping, unloading, sorting, storing, moving, and hoisting materials. Well-planned materials delivery will reduce idle time, increase site safety, and lessen product damage. Critical issues in site materials handling are space requirements and horizontal and vertical movement of products. The jobs of the design-build estimator and project manager are to find ways to plan and execute the project by moving the materials just once (or in the case of on-site borrow, just twice).

Materials control. The implementation of material control begins at the estimating stage, with calculations for quantities and pricing according to quality requirements. Long-lead items are ordered to meet aggressive design-build schedules. The quantities may be transferred (by hard copy or electronically) into field bills of material. Project managers and field engineers generally will use a just-in-time approach to delivery of time-sensitive building products, such as for cast-in-place concrete. However, some materials may be inventoried on or off site to assure continuous workflow. One line of product that seems to vary in its low cost availability and high cost shortages is gypsum wallboard. A high-volume design-builder may purchase and store such commodity items as they become available from manufacturers and suppliers to avoid schedule issues due to a materials shortage.

Labor productivity is a favorite topic among project managers. Some studies have indicated that only about half of the average worker-hour on a typical project site is spent on direct work. The other parts of the hour are consumed by idle time or waiting, receiving instructions, personnel breaks, late starts and early finishes to the workday, travel, retrieving or working on tools, and rework of work already begun. The point is that if only 25 to 75 percent of the time is used for direct work on the project, there is considerable room for productivity improvement. Labor productivity varies by trade, with carpenters achieving about 50 percent of the hour devoted to direct work and bricklayers as much as 70 percent, largely because of the variability of a carpenter's tasks and the specifics and organization of the mason's work.

Continued improvements in labor productivity are possible through efforts of management with the cooperation of labor. A design-build estimator should be aware of the productivity improvements being made by field management and ascertain whether the improvements can be replicated or continued on subsequent projects. Worker motivation and commitment directly affect the quality and speed of work at the site. To improve labor productivity, a number of steps can be taken:

- *Provide employment stability.* A stable workforce that is confident that their livelihoods are secure will produce better results than will a workforce that is unstable (undergoing constant changes in tasks and personnel).

- *Expand craft training.* Education of workers to perform their jobs with a higher degree of efficiency and with better quality will improve productivity. Craft training can lead to multiskilling, allowing the worker to perform different tasks throughout the schedule at the same project.

- *Maintain a safe project site.* Workers will have lower regard for management if there are unsafe conditions or a high accident rate at the site. Regular safety talks and inspections show regard for human safety and build confidence and morale.

- *Use planning and participatory management techniques.* Involve workers (of all trades, whether prime, subcontractor, or subsubcontractor) in the planning of pertinent work and in solving specific problems. Workers are closer to the task than managers, and can often devise the most cost-effective adjustments. Worker input can be obtained through good job-site interpersonal communication, quality circle (TQM) programs, surveys, and electronic or traditional suggestion-box programs.

- *Reduce downtime and disturbances.* The greatest enemies to productivity are the most obvious: interruptions and idle time (waiting for instructions, materials, or weather delays). When work is slowed or interrupted, it is often difficult to regain the momentum that was originally established. Downtime is an area where a good field supervisor can make a difference in project productivity.

- *Use findings from ongoing design and constructibility reviews.* Implementing the information that is learned in design reviews, submittal reviews, and site management meetings will prevent adverse consequences during facility construction. Constructibility reviews during design and construction should result in the most efficient means of performing the work.

- *Institute "best practices" management tools.* A project control system will integrate costs, schedules, materials delivery, and other aspects of the work to guide the project through to completion. Control of project scope, costs, and changes, which have a direct impact on project productivity, are dependent on good data and operating software. Modern management of human resources will lead to a culture of the workers being customers of office and field management. When the craft workers are aware of management's genuine interest in their well-being, commitment to the project and to the organization will result.

Productivity is a variable in the design-build cost estimator's calculations. Will the project have the most productive field and management team? Will the project be within historical productivity metrics or lower because of difficult site conditions? A technique that many firms are beginning to use is a worker incentive program directly tied to the goals of the project. Exceptional performance is recognized and rewarded through commendations, special luncheons, modest bonuses, and other positive reinforcement techniques. If these costs have been budgeted, management can leverage success from their implementation.

11.5   Integrating Capital Asset Costing for Operation and Maintenance with Estimating

The area of design-build estimating that is least understood is the determination of operation and maintenance costs as they are expected to be incurred over a set period of years. Recognition of the importance of operation and maintenance costs by asset portfolio managers in corporate and government settings has resulted in a renewed focus on annual operating expenses. For example, a decrease in $0.02/month on roof maintenance expenses for a 50,000-ft^2 building would result in an annual savings of $12,000.00. Similar savings can be projected for interior maintenance requirements for wallcovering or flooring of a building. In most instances, the owner's facility staff and their consultants have concentrated on the initial construction cost of a system, rather than looking into the longer-term operating cost efficiency of systems and materials. Offsetting the higher initial cost are lower-maintenance systems and materials, which contribute to the value of the facility in terms of depreciation and disposition valuations.

Costs for operation and maintenance can be derived from a facility management plan and operating budget. A management plan will outline the owner's objectives for the facility and enumerate programs that will contribute to the facility's performance and value. The plan will contain assumptions about demographic and economic growth conditions of the area, and will point out opportunities and risks for successful ongoing operations. The heart of the management plan will be the physical attributes of the facility, with particular attention paid to energy systems, security, power systems, and weather integrity. The plan will contain schedules for repairs and maintenance, as well as for major equipment replacement and modernization.

Rounding out the facility management plan will be sections devoted to administrative duties, financial procedures, and guidelines for decisionmaking and reporting about the facility's positive or negative performance. The operating budget will consist of year-by-year expense items for various operation and maintenance responsibilities:

- *Utilities*—operation of a central plant, zoned systems and distribution equipment, facility water supply, sanitary and storm sewer, chilled water, hot water, steam, and natural gas

- *Energy management*—operation of controls systems, mechanical systems (HVAC, plumbing), power, and lighting

- *Communications systems*—telephone, low-voltage and fiberoptic cabling, broadcast towers and dishes, and related communications equipment

- *Life safety systems*—sprinkler, fire alarms, detectors, emergency power and lighting, elevators and automated transportation, hazardous waste, and other safety issues

- *Facility maintenance*—interior cleaning, painting, caulking, upgrades, minor tenant fitup; custodial services; grounds maintenance, landscaping, snow or ice removal; and street cleaning

- *Materials handling*—receiving, storage, and distribution; loading dock; and inventory control
- *Equipment and vehicle maintenance*—power washers, landscape equipment, and personnel vans and company trucks
- *Insurance and taxes*
- *Fees and miscellaneous*

The design-build team can obtain current operating data from managers of similar facilities, or from comprehensive data compiled by owner organizations. Data compilations are available from the Urban Land Institute (ULI), International Facility Management Association (IFMA), Industrial Development Research Council (IDRC), Building Owners and Managers Association (BOMA), and other market-centered owners' associations.

Incorporating operation and maintenance costs into a design-build-operate-maintain estimate necessitates making an assumption on the economic life and design life (before major renovation or reconstruction) of the facility. A forecasted life of a local bridge may be 40 years, and the assumed life of an urban hotel, 10 years. School buildings may have an assumed life of 25 years, and a corporate office building 20 years. The contract may require the design-builder to design a school for a 30-year life cycle, and to operate the facility for 10 years before renegotiation or transfer of operation back to the owner. Regardless, it is important to base expected maintenance costs on a variety of factors, including normal wear and tear, equipment reliability, vandalism, and materials deterioration. Average expected operating and maintenance costs are converted into current dollars and added to the design-build-operate-maintain estimate.

11.6   Life-Cycle Costing

If the opinions and theory about life-cycle costing are stripped away, the basic life-cycle cost of a product or facility would be defined with five fundamental elements:

- The initial design and production cost of the product or facility
- The cost to finance the product or facility
- Maintenance and operating costs to keep the product or facility functioning
- Costs of downtime or failure of the product or facility that partially or totally eliminate owner or user benefit
- Costs (plus or minus) to dispose of the asset after its useful life (or economic life) has been fulfilled

Life-cycle costing is the process by which we analyze the longer-term economic consequences of our earlier decisions. Looking at the life-cycle costs of a private school building with a 30-year life at 7 percent interest may result in allocation of the total life costs to the following areas:

Finance	30%
Construction	40%
Design	2%
Operation and maintenance	15%
Replacement and major repair	12%
Other	1%

The advantages of life-cycle costing over most estimating systems is the ability to go beyond initial costs of facility design and construction, arriving at an evaluation of whole-life costs. A U.S. market sector that has begun to experiment with life-cycle costing models is the transportation sector. The federal and state governments continue to invest huge sums in highway and bridge construction and maintenance. Examination of the best ways to make optimum infrastructure investment decisions inevitably leads to life-cycle cost choices.

To investigate the life-cycle costs of highways, the project progenitors would construct a combined initial cost-finance-operations model. The planning, design, and construction costs would be dependent on the design guidelines of the road (interstate, primary, secondary), the amount of growth in the right-of-way corridor, the geographic location within the state/country, and the environmental assessment and mitigation steps that must be taken to gain permission to construct the roadway. Additional elements of the highway estimate will likely include geotechnical investigation, earthwork, subbase and pavement design, drainage, abutments, structures, landscaping, and signage.

A basic life-cycle question is immediately encountered with the design standards that accompany a highway project: Should the design standards be exceeded to provide a road that has better resistance to wear and tear? Or is there an optimum balance between initial construction costs and the cost of maintenance over time? Research conducted in the 1970s at the Massachusetts Institute of Technology began to incorporate external costs of vehicle operating expenses into life-cycle construction and maintenance costs. The study highlighted the importance of slight increases in design standards and initial costs or upgrading of periodic maintenance as a way of stimulating huge savings in user vehicle operating costs.

A simple table (Table 11.1) shows the impact of highway maintenance costs over a term of years for five competing schemes.

The application of life-cycle costing for design-build projects (see Fig. 11.1) provides the participants with predictions of mid- to long-term outcomes, instead of focusing only on short-term (largely initial cost) issues. When life-cycle costs are not taken into account, resource allocation is dependent on design standards and prescriptive specifications that have more to do with historical precedent than on current best practices. The challenge of integrating life-cycle costing into the design and construction process will continue to con-

TABLE 11.1   Five-Mile Highway Segment: Life-Cycle Comparison for Five Schemes
(All figures converted to net present values in millions)

Scheme	Initial cost	Maintenance cost	Lifespan, years	Value at reconstruction	Lifetime cost
A	$42	$8	20	$24	$178
B	$47	$7	20	$40	$147
C	$50	$6	20	$39	$131
D	$51	$5	20	$45	$106
E	$54	$6	20	$38	$136

front facility owners and industry providers. One major barrier to life-cycle costing is the lingering tendency for owners (and practitioners) to fragment each stage of the process. But whole-life costing, taking into consideration the assumed discount rate and project design life, will allow better decisionmaking. Even where forecasting of future facility demand, occupancy types, or highway traffic loads means a guesstimate by the estimator, the adage from the automobile transmission commercial of "You can pay now; or you can pay later" holds grains of truth for an industry that has largely ignored life-cycle costing.

11.7   Design-Build Planning and Scheduling

For the management team on a design-build project, it is not sufficient to plan and execute only the construction portion of the work. The team must be concerned with all activities that sustain a project from inception through design and construction. Planning a design-build venture will involve parties at each stage of the work: feasibility, design criteria, conceptual and schematic design, and design development and construction documents. Elemental packages of information enable design-build project estimators and schedulers to assemble cost data at each stage.

In practice, cost and time management tools are devised separately, then used in conjunction with one another. Cost management relies on labor, materials, and equipment metrics to generate costs per assembly and element. Time management is built on sequential and concurrent activities that take place over time. The greatest contribution of the critical-path method to the construction industry has been the need for designers and constructors to analyze projects in terms of individual activities and the expected productivity from those activities. Ideally, the assemblies and equipment cost breakdowns overlap with the project work activities, allowing the estimate and schedule to feed directly into project control systems.

To create computerized schedules for design-build projects, the scope of the project is segmented into activities with durations, many of which directly

Phase 4
Analysis/Judicial Phase: Analysis/Development

Key Question

- Select top ranked ideas (normally 8 and above)
- What is cost impact of ideas?
- Consider cost of ownership

Technique

- A Life Cycle Cost

Life Cycle Costing

Form WS-13, the LCC summary, summarizes all the life cycle cost. The present worth costs should be entered in the appropriate space for each design alternative.

The following general instructions for performing the LCC apply:

1. For initial costs, the present-worth factor always equals 1. Therefore, the estimated cost and the present worth are the same and should be so entered.

2. To obtain the present worth of each salvage & replacement cost, refer to Table 4, using the column with the corresponding discount rate. By knowing the building and item life, select the correct factor. Multiply this factor by the estimated replacement cost and enter the value in the present-worth column. For example, a replacement cost of $10,000 equivalent to today's buying power is estimated to occur at the 16th year. The PW factor is 0.2176. The PW of this cost is $2,180.

3. Before the estimated annual cost is converted to present worth, determine if a differential escalation rate is appropriate. Differential escalation rate is the difference one estimates the item will escalate in relation to the devaluation of the currency. For example, if the forecast is that energy will over a 20-year period escalate an average of 5%/year, and devaluation of the dollar ($) is estimated at 3% per year, the differential rate would be 2%. Refer to Table 1, "Present Worth of an Escalation Annual Amount," using a 10% interest factor as recommended by the U.S. Office of Management and Budget. The value PWA escalated for 2% differential over 20 years is 9.934 vs. 8.51 for the nonescalated value. If a differential escalation rate is not necessary, refer to Table 2, "Present Worth of Annuity," for conversion of an annual cost to present worth. By knowing the building life, select the correct factor. Multiply this factor by the estimated annual expenditure. Enter the computed amount in the Present-Worth column.

   For example, the table indicates that, assuming an annual cost of $1 per year at 10 percent interest, nonescalating (0%), the present worth of a 20-year cycle is $8.51.

4. Calculate the difference between the original design and the design alternative's life cycle costs. Enter this figure at the bottom of the form.

5. Attached is Example WS 13 illustrating a typical LCC.

Figure 11.1   Life-cycle template. (*From A. J. Dell'Isola, Value Engineering: Practical Applications for Design, Construction, Maintenance and Operations, R. S. Means Co., Kingston, Mass., 1997.*)

DEVELOPMENT PHASE — LIFE CYCLE COST (Present Worth Method)

Proposal No. _____ Date _____
PROJECT LIFE CYCLE (YEARS) _____
DISCOUNT RATE (PERCENT) _____

			Original		Alternative 1		Alternative 2		Alternative 3	
			Est.	PW	Est.	PW	Est.	PW	Est.	PW
Capital Cost										
A)										
B)										
C)										
D)										
Other Initial Costs										
A)										
B)										
Total Initial Cost Impact (IC)										
Initial Cost PW Savings										
Replacement/Salvage Costs	Year	Factor								
A)										
B)										
C)										
D)										
E)										
F) Salvage (neg. cash flow)										
Total Replacement/Salvage PW Costs										
Operation/Maintenance Cost	Escl. %	Factor								
A)										
B)										
C)										
D)										
E)										
Total Operation/Maintenance (PW) Costs										
Total Present Worth Life Cycle Costs										
Life Cycle (PW) Savings										

PW - Present Worth   PWA - Present Worth of Annuity

WS-13

Figure 11.1 (Continued)

impact other activities. Best practices for design-build project scheduling include

- Encompass feasibility, environmental permitting, finance, and other front-end activities in a comprehensive schedule for the entire project.
- Prepare the schedule in phases that match the estimating phases. As more information is known, add more detail to the schedule activity descriptions.
- Build in information from all stakeholders, prime and subcontractors, and consultants and suppliers. The schedule will have better information, will become a priority, and will enjoy project-wide buyin with this emphasis.
- Have activity durations that make up the schedule reflect the correct quantities, productivity, and project constraint information.
- Ensure that resource availability (e.g., labor, materials) matches the technical requirements of the activities.
- Update schedule periodically to incorporate project progress, changes in the cost estimate, and floats in order to make good management decisions.
- Integrate estimating, scheduling, and project control tools to provide periodic reporting for work items, including budgeted quantities, budgeted work-hours/dollars, actual quantities to date, percent of quantity in place, actual labor-hours/dollars to date, and percentage of hours/dollars to date, leading up to a forecast of variations by hours and by dollars. Variances can be explained and managed with an up-to-date computerized job control system.

A strong benefit of the integrated design-build is the freedom of design-builders to employ systems or proprietary processes that produce efficiencies and market advantage. For example, a firm that specializes in design and fabrication of large-scale refrigeration systems will have an advantage in the food processing and storage markets over firms that do not have this capability. The ability to immediately begin work on this long-lead item would produce ripple benefits in other areas of the project management, including scheduling advantages. But a caveat is necessary here; it is possible to determine the schedule for construction activities within tight tolerances—but not possible to drive environmental assessment or permitting, land acquisition or right-of-way, or feasibility/programming with the same degree of certainty. It is prudent to start the clock running (for strict contractual reasons such as incentive/disincentive or liquidated damages) after those aspects of the project are well within the control of the design-builder.

11.8 Risks and Challenges of Conceptual Estimating

11.8.1 Guaranteeing a price

Project owners tend to gravitate toward price certainty on their projects. Whether the legal arrangement is for firm fixed price, guaranteed maximum price (GMP), or merely confirmation of a budget price, the commercial terms of the legal agreement are usually dictated by the parties' degree of trust and tolerance for uncertainty. Many sophisticated design-build firms prefer cost-plus fee arrangements when the project program is not well defined or when new technologies are being implemented. In these situations, the owner does assume the overall financial risk but can also benefit from cost underruns. (These underruns would not otherwise be known by the owner under a lump-sum agreement.)

On the other hand, the owner will not be responsible for cost overruns under a lump-sum contract. With a lump-sum arrangement, the design-builder will seek to place boundaries on the scope of work to carefully identify what is included and what is excluded. Work that is beyond the scope (as spelled out in the owner's solicitation and design-builder's proposal) would require a contract change order. Changes to the basic work can have far-reaching cost and schedule implications and may result in heated negotiations. Clarity of project documents, records, and communication can help during these occasional management challenges.

The growing popularity of GMP contracts for design-build is partially the result of the flexibility that it gives the parties in accepting minor adjustments to scope but at the same time capping overall cost exposure. If the actual costs of a project are higher than the level of the GMP, the design-builder is not entitled to additional compensation. If the project costs are less than the GMP then the difference may revert back to the owner or be shared under some formula between the owner and the design-builder.

The point at which the design-builder sets the target price of the project has been a subject of much debate. Some design-build entities will guarantee a firm fixed price at 10 percent design. Other design-builders, particularly in the traditional civil infrastructure markets, will not offer firm project costs until design is 25 to 30 percent complete (approaching early design development). The basic design-build estimate will have been prepared without adding extra margins or allowances to individual line items. Prior to fixing and submitting the price, however, the design-build team must consider the need for escalation, contingency, and fee based on the amount of expected risk.

11.8.2 Contingencies

The amount of contingency that should be included with the overall cost of a design-build project is directly related to the degree to which the program or design criteria have been defined. If the owner or user needs have been fully articulated in the programming narrative, and the design-build team can initiate its conceptual design from the program, then the amount of contingency can be lowered. If the overall project goals are not verbally defined, or if the facility program document is sketchy or poorly prepared, contingency amounts will soar.

To better understand degrees of uncertainty, design-build teams can prepare a risk analysis matrix. The analyzers will look at both project elements and proposal issues. The team will place a percentage of confidence in whether the

TABLE 11.2   Design-Builder's Contingency Matrix

Project/proposal component	Component $	Contingency %		
		Low	Medium	High
High risk				
Demolition (F)	200,000	4	7	10
Sitework (G)	1,500,000	3	5	8
Medium risk				
Substructure (A)	1,000,000	2	4	6
Services (D)	2,000,000	2	3	4
General Condition (Z)	400,000	2	3	4
Low risk				
Design	800,000	1	2	3
Shell (B)	4,500,000	0	1	2
Interiors (C)	1,500,000	1	2	3
Equipment and furniture (E)	600,000	0	1	2

owner's project information and/or understanding of the work is complete enough to ward off unforeseen or unexpected costs. The amount of contingency is directly related to the costs that would be incurred if the estimator(s) or scheduler(s) on the design-build team, in the absence of complete project information, assumed one or more project design solutions that do not reasonably fulfill the ultimate goals of the project. A critical aspect of developing a contingency matrix is an understanding of who assumes what portion of the risk: Is the responsibility for an overrun totally the design-builder's? The owner's? Or is the responsibility shared? If so, is it shared evenly or on a different scale? Table 11.2 is a design-builder's contingency matrix done at the design criteria stage (amounts shown are for illustration purposes only) for a small office building that incorporates low-, medium-, and high-probability scenarios.

Project subtotals. Another way to conduct a contingency analysis is to extract only those items from the estimate for which there is a degree of uncertainty. This form of contingency matrix may include schedule; major equipment; environmental permitting, mitigation, and construction; and labor for specialized tasks. As a final caveat, contingency is an excellent tool that should be used only when all the other tasks of generating costs and schedules have been completed. If contingency amounts are distributed within the base estimate, the design-build team will be working with a distorted view of project costs.

The final proposal amount will contain contingency, overhead, and profit as determined by management of the design-build entity. Among the considerations for final proposal fee or project cost is the lost-opportunity costs of the team's capacity being tied up for a specific project when other equally enticing or more enticing projects may become available.

Performance Specifying for Integrated Services Projects

12.1 Fundamental Differences between Prescriptive Specifications and Performance Specifications

Prescriptive specifying (as used in traditional design-bid-build) describes individual products and the methods of construction used to put the products in place. By contrast, *performance specifications* concentrate on end results that are expected for entire facilities or the systems that make up those facilities.

Think of traditional prescriptive specifications as grains of sand that are compiled to make up a shoreline; performance specifications as stretches of beach. An environmental restoration project could include descriptions of the grains of sand and prescriptive details of how to place the grains in lifts to satisfy the inspection of the environmental engineer. The same environmental restoration project could be accomplished by describing the results wanted: a stretch of beach that will sustain landside grasses 200 ft from the mean low waterline, and provide seasonal recreation for local citizens. Obviously, the grains-of-sand approach is products/methods-based; the stretch-of-beach approach is end results–based.

Building codes have been steadily shifting to a performance-based orientation, allowing designers and builders to devise alternative solutions to structural, fire protection, or sanitary requirements. The building code organizations recognized that traditional building requirements discouraged innovation and sometimes prevented the use of products or systems that added new safety or efficiency features to facilities. With the new performance orientation in building codes and standards, the specifications community has an incentive to update its publications and products, thereby redressing the current bias toward prescriptive specifications.

12.2 "Traditional" Specifying in the Nineteenth and Twentieth Centuries

For centuries, owners and designer-builders have used written specifications in conjunction with graphical drawings to record their agreed-on requirements for facilities. Generally, the drawings have conveyed how the project's components would be arranged and connected, and how the structure's visible surfaces would appear. The specifications describe (through narrative language) quality and materials issues that are not explainable through graphical representations. For most of recorded history, although individual practitioners developed organized systems for explaining a building or a bridge with nongraphical communication, there were no widely accepted formats for how the descriptive material should be written, organized, numbered, or named for use as a communicating device among the building team.

Nineteenth-century specifications were organized sequentially; in parallel with what trades were appearing on the building site. With such an ordered scheme, earthmoving descriptions would precede masonry and foundation work, which would precede exterior wall construction, which would precede glazing, and so on. The specifications systems were evolving into a trades-based system. Many of the construction trades have relied on an apprenticeship training and closed guild process that has perpetuated a communications approach that relies on trades-based and materials-based language.

An early twentieth-century specification comprised as many chapters or divisions as there were trades on the project. By the 1950s, a de facto standard was emerging, wherein many designers were writing prescriptive specifications with 18 divisions. Still, many architects and engineers retained their own systems, using anywhere from a dozen to two dozen divisions to format their specifications. Industry frustration over the lack of uniformity led the Construction Specifications Institute (CSI) (founded in 1949) to finally move to a standardized 16-division master format in 1963.

The 16-division format for ordering and numbering product specifications was later named MasterFormat, and this classification system has been endorsed by the major professional associations of the design and construction industry, including the American Institute of Architects (AIA), the American Society of Landscape Architects (ASLA), the Associated General Contractors of America (AGC), the Associated Specialty Contractors, and the National Society of Professional Engineers (NSPE).

Within MasterFormat, general requirements for a construction project are in Division 1, while the products and trade methods portion of the work are organized into 15 subsequent sections:

Division 2. Site construction

Division 3. Concrete

Division 4. Masonry

Division 5. Metals

Division 6. Wood and plastics

Division 7. Thermal and moisture protection

Division 8. Doors and windows

Division 9. Finishes

Division 10. Specialties

Division 11. Equipment

Division 12. Furnishings

Division 13. Special construction

Division 14. Conveying systems

Division 15. Mechanical

Division 16. Electrical

For the traditional design-bid-build project, specifications are written during design development or after products have been chosen by the design project manager. The purpose of prescriptive specifications drafted at this phase is to fix product quality as part of the legal agreement between owner and general contractor. Some architects submit short form specifications to owners at key points within the design process as a way of keeping the owner advised of the progress and conformance with the original design intent. These preliminary prescriptive specifications may also be transmitted to estimators at the conclusion of each design phase, to verify whether the design is staying within the owner's budget. But the predominant use of prescriptive specifications is to establish methods of construction and record product quality. The majority of specifications currently used, whether commercially available guide specifications or house master specifications, are detailed descriptions for use of individual products and materials.

12.3 Genesis and Adoption of Performance Specifications and UNIFORMAT

In Europe following World War II, the use of performance specifications emerged within countries that did not want to take the time to fully articulate material requirements of a completed design before embarking on a building project. Instead, British and mainland Europe specifiers turned to performance-oriented solicitations containing, in part, technical performance requirements for whole structures and for systems within the structures. Using an adaptation of the design-build method, cities and towns that were devastated by war were able to design and construct vital infrastructure and housing in roughly half of the time that it would have taken under the traditional design–then bid–then build method.

In the United States, a group of school construction officials in California formed the School Construction Systems Development Project in the 1960s to manage the tremendous demand for educational space in the state. The objective was to stimulate flexible designs for 13 school districts by organizing them into one large owner/client. While the projects would still be issued in groups, the need for several million square feet of buildings within a confined time period made the projects attractive to building material manufacturers.

The Systems Development group wanted classroom layouts that could be altered overnight to accommodate changes in student use or age groups. After some analysis, the team focused on description and design of four building subsystems considered critical to the need for adaptable space:

- Structural and roof system
- Air-conditioning system
- Ceiling and lighting system
- Movable interior partitions

Essentially, the California schools program wanted a "kit of parts" that would be both flexible in themselves and capable of being combined in flexible ways. The chief goals of the proponents were high-quality design, swift fabrication of components, and reduced price. The program became an early exercise in innovative design-build delivery. The school facilities team recognized that they were experts in what they needed, but not in how to meet their requirements. On a system-by-system basis, they employed performance specifications to define what they needed, rather than resorting to the prevailing convention of prescribing how to construct the projects on a product-by-product basis.

Twenty-six manufacturers responded to the California schools' solicitation for proposals, and four teams were chosen to design and build particular subsystems. The four winners were required to work together as an overall team to ensure that the ensemble was totally integrated and all systems were directly compatible with one another. At about the same time, as CSI was beginning to standardize and popularize an organizational format for product-based specifications, the California schools project was advancing a specifications organizational system for performance specifications.

12.3.1 Early GSA and NIST sponsorship of performance specifications

During the early 1970s, the General Services Administration (GSA) became interested in the possibility of using performance specifications as a way to raise the quality of government office buildings above the level generally provided in the private sector. The impetus came from the section chief of the Building Systems Section at the National Bureau of Standards, Mr. Robert Blake. In addition to the project goals of design innovation and increased quality, GSA wanted cost reductions and a compressed design and construction schedule.

As was the case with the earlier California schools initiative, GSA used economies of scale to make it worthwhile for manufacturers to innovate. The concept for the program was to define systems, subsystems, and some products in performance terms, and to invite construction industry firms to participate in their design and construction. The experiment began with a group of three virtually identical modular buildings: Social Security Administration payment centers in Philadelphia, Chicago, and Richmond, California.

The *Peach Book,* named for the color of its cover, contained these specially assembled performance requirements. GSA awarded contracts to design firms for developing modular designs and for producing construction documents containing the information from the *Peach Book.* The projects also incorporated new products born of the experiment, including carpet tiles and an unusual light fixture that was integrated with both the sprinkler system and the heating system. The three Social Security Administration pilot projects were designed and constructed at quality levels that met expectations, and the projects came in on time and at or less than budgeted. In light of these successes, the *Peach Book* was used on other government buildings, including office facilities in Norfolk, Virginia and Baltimore, Maryland.

During the 1970s and 1980s, two significant organizing formats were developed for the design and construction industry. Hanscomb Associates created an elemental format for costing conceptual design documents entitled *MasterCost* beginning in 1973. At about the same time, the General Services Administration was developing an elemental format called UNIFORMAT. Both systems, which were later merged, were intended for use in preliminary project estimates.

12.3.2   ASTM and CSI/CSC custodianship

The American Society for Testing and Materials (ASTM) Subcommittee on Building Economics embarked on a project to standardize a classification for building elements in 1989. The Subcommittee used the original UNIFORMAT as a starting point. The ASTM Working Group, consisting of representatives from AACE, NSPE, CSI, GSA, NAVFAC, U.S. Army Corps of Engineers, and others, used these guiding principles:

A formally created classification format for building systems and assemblies is needed by facility owners and industry.

The format should be based on the UNIFORMAT system originally devised in the 1970s, since it has already been adopted by cost database providers.

The format should be expanded to all market sectors and all types of structures, beyond the original use for buildings alone.

The format should have some interface with MasterFormat to provide entry from one system to the other.

Three years later, the National Institute for Standards and Technology published *UNIFORMAT II: A Recommended Classification for Building Elements and Related Sitework* for review and discussion by the user community. In 1993, after discussion and balloting within the national standards process, ASTM issued ASTM Standard E1557, *Standard Classification for Building*

Elements and Related Sitework—UNIFORMAT II. The standard was revised somewhat in 1997 as UNIFORMAT II, E1557-97 and includes three levels of detail. (There is active discussion about adding a fourth level of detail as described in NIST publication NISTIR 6389 dated Oct. 1999.)

The Construction Specifications Institute and Construction Specifications Canada began revising UNIFORMAT in 1995 to ensure that the system was coordinated with MasterFormat. As of this book's publication, the CSI/CSC UniFormat and ASTM E1557 have the same higher-level numbering format; however, ASTM does not include CSI's Category Z—General, a section for project description, or additional levels of detail. The CSI/CSC UniFormat levels 1 through 3 are listed in the appendix of the *CSI Manual of Practice* (1998 edition of UniFormat) (the core of this system was developed by governmental agencies; however, the system *as shown* is copyrighted by CSI/CSC and is included for educational purposes only):

PROJECT DESCRIPTION (Level 1)

10	Project Description (Level 2)	
	1010	Project Summary (Level 3)
	1020	Project Program
	1030	Existing Conditions
	1040	Owner's Work
	1050	Funding
20	Proposal, Bidding and Contracting	
	2010	Design Delivery
	2020	Qualifications Requirements
	2030	Proposal Requirements
	2040	Bid Requirements
	2050	Contracting Requirements
30	Cost Summary	
	3010	Elemental Cost Estimate
	3020	Assumptions and Qualifications
	3030	Allowances
	3040	Alternates
	3050	Unit Prices

Element A: SUBSTRUCTURE

A10	Foundations	
	A1010	Standard Foundations
	A1020	Special Other Foundations
	A1030	Slabs on Grade
A20	Basement Construction	
	A2010	Basement Excavation
	A2020	Basement Walls

Element B: SHELL

B10   Superstructure
      B1010   Floor Construction
      B1020   Roof Construction

B20   Exterior Closure
      B2010   Exterior Walls
      B2020   Exterior Windows
      B2030   Exterior Doors

B30   Roofing
      B3010   Roof Covering
      B3020   Roof Openings

Element C: INTERIORS

C10   Interior Construction
      C1010   Partitions
      C1020   Interior Doors
      C1030   Fittings Specialties

C20   Stairs
      C2010   Stair Construction
      C2020   Stair Finishes

C30   Interior Finishes
      C3010   Wall Finishes
      C3020   Floor Finishes
      C3030   Ceiling Finishes

Element D: SERVICES

D10   Conveying Systems
      D1010   Elevators and Lifts
      D1020   Escalators and Moving Walks
      D1030   Materials Handling
      D1090   Other Conveying Systems

D20   Plumbing
      D2010   Plumbing Systems
      D2020   Domestic Water Distribution
      D2030   Sanitary Waste
      D2040   Rain Water Drainage
      D2090   Other Plumbing Systems

D30   Heating, Ventilating and Air Conditioning
      D3010   Fuel Energy Supply Systems
      D3020   Heat Generation Systems
      D3030   Heat Rejection Systems Refrigeration

D3040   Heat HVAC Distribution Systems
D3050   Heat Transfer Terminal and Packaged Units
D3060   HVAC Instrumentation and Controls
D3070   HVAC Systems Testing, Adjusting and Balancing
D3090   Other Special HVAC Systems and Equipment

D40   Fire Protection Systems
D4010   Fire Protection Sprinklers Systems
D4020   Standpipes and Hose Systems
D4030   Fire Protection Specialties
D4090   Other Fire Protection Systems

D50   Electrical Systems
D5010   Electrical Service and Distribution
D5020   Lighting and Branch Wiring
D5030   Communications and Security Systems
D5040   Special Electrical Systems
D5050   Electrical Controls and Instrumentation
D5060   Electrical Testing
D5090   Other Electrical Systems

Element E: EQUIPMENT AND FURNISHINGS

E10   Equipment
E1010   Commercial Equipment
E1020   Institutional Equipment
E1030   Vehicular Equipment
E1090   Other Equipment

E20   Furnishings
E2010   Fixed Furnishings
E2020   Movable Furnishings

Element F: SPECIAL CONSTRUCTION AND DEMOLITION

F10   Special Construction
F1010   Special Structures
F1020   Integrated Construction
F1030   Special Construction Information
F1040   Special Facilities
F1050   Special Controls and Instrumentation

F20   Selective Demolition
F2010   Building Elements Demolition
F2020   Hazardous Components Abatement

Element G: BUILDING SITEWORK

G10   Site Preparation
G1010   Site Clearing

G1020   Site Demolition
G1030   Site Earthwork
G1040   Hazardous Waste Remediation

G20   Site Improvements
G2010   Roadways
G2020   Parking Lots
G2030   Pedestrian Paving
G2040   Site Development
G2050   Landscaping

G30   Site Civil/Mechanical Utilities
G3010   Water Supply
G3020   Sanitary Sewer
G3030   Storm Sewer
G3040   Heating Distribution
G3050   Cooling Distribution
G3060   Fuel Distribution
G3090   Other Site Mechanical Utilities

G40   Site Electrical Utilities
G4010   Electrical Distribution
G4020   Site Lighting
G4030   Site Communications and Security
G4090   Other Site Electrical Utilities

G90   Other Site Construction
G9010   Service Tunnels
G9090   Other Site Systems

Element Z: GENERAL

Z10   General Requirements
Z1010   Administration
Z1020   Procedural General Requirements Quality Requirements
Z1030   Temporary Facilities and Temporary Controls
Z1040   Project Closeout
Z1050   Permits, Insurance and Bonds
Z1060   Fee

Z20   Bidding Requirements, Contract Forms and Conditions Contingencies
Z2010   Bidding Requirements Design Contingency
Z2020   Contract Forms Escalation Contingency
Z2030   Conditions Construction Contingency

Z90   Project Cost Estimate
Z9010   Lump Sum
Z9020   Unit Prices
Z9030   Alternates/Alternatives

Users of ASTM's UNIFORMAT II have expressed a need for a fourth level of detail in the ASTM standard that would allow more complete descriptions of some subelements, more complete cost data, and better transitions to MasterFormat. As an example, the suggested level 4 for D3040 Distribution Systems as listed in the October 1999 NIST report would include

Element D: SERVICES

D30	Heating, Ventilating and Air Conditioning	
	D3040	Distribution Systems
	D3041	Air Distribution Systems
	D3042	Exhaust Ventilation Systems
	D3043	Steam Distribution Systems
	D3044	Hot Water Distribution
	D3045	Chilled Water Distribution
	D3046	Change-Over Distribution System
	D3047	Glycol Distribution System

12.3.3   Elemental classification systems for cost estimating

Estimates based on the Elemental Classification Format, whether ASTM's or CSI's, aid in the cost analysis and cost monitoring of a project from its feasibility and conceptual stages through all phases of design. The same attributes are not available from the traditional 16-division trades estimates because of their reliance on specific design and products decisions.

As elemental descriptors for facilities have evolved, they have generated a common language for generating preliminary project descriptions and technical program information. With this commonality, the direct relationship of project scope and project cost is immediately apparent. Parametric rates indicate quality levels for each element and allow the project proponents to obtain a realistic distribution of costs among the "competing" elements. Another benefit of elemental formatting is the ability to use cost-risk analyses that identify the probability and magnitude of cost overruns using ASTM Standard E1496, *Standard Practice of Measuring Cost Risk of Buildings and Building Systems,* which employs UNIFORMAT II elements.

12.3.4   UniFormat as a basis for developing a performance specification system

At the feasibility and design criteria stages of a project, the UniFormat system can be used to generate a narrative description of a facility. The technical performance portion of that description will set forth the results to be achieved, but not the means for achieving those results. Application of performance specifications to design-build projects allows the design-build team to have the freedom of choice between systems, equipment, products, and methods of design and construction.

Performance specifications organized by UniFormat elements allow owners or their consultants to define the entire facility in performance terms at level 1 or to intermix a number of elements at levels of definition that satisfy the facility owner. The electronic master guide performance specification system from the Construction Specifications Institute (CSI) and the Design-Build Institute of America (DBIA) uses the CSI/CSC UniFormat system as the basis for organization. This relatively new product, called *PerSpective,* is described in more detail later in this chapter.

12.4 Why Performance Specifications Are Appropriate for Integrated Services Delivery

12.4.1 Placement of responsibility for design and construction on a single entity

The major motivations for use of performance specifications in design-build have been the need for flexibility and innovation and also the desire for an integrated design and construction process that has the potential to deliver projects faster. The California schools projects invented their new process because needed performance requirements were not being met by existing procurement methods or product suppliers. The General Services Administration (GSA) developed its famous *Peach Book* because it wanted innovation, speed of delivery, and higher-quality buildings.

Requests for proposals (RFPs) for design-build projects place responsibility for both design and fabrication of the facility in the hands of a single entity. Innovation requires very close collaboration among designers, materials and equipment experts, and those skilled in the means and methods of construction. The more complex the building component, or the more components to be integrated into a single complex facility, the more likely it is that an innovating owner will concentrate responsibility for the tasks in the hands of a single entity with the primary responsibility of making all the elements work together.

12.4.2 Performance specifications define end results that are desired; not means and methods of how to achieve the results

If an owner makes specific design decisions (through the use of prescriptive specifications) prior to contracting with a design-builder, control over significant design and construction approaches for the project are lost to the design-build team. Specifications that are prescriptive rather than performance in nature diminish the range of creative solutions. It is likely that some of the creative options may have been faster to implement or less expensive to construct, and all equally acceptable to the owner.

Suppose that performance specifications for the exterior wall of a building require the following basic results. The building interior must be kept dry, heat transmission through the wall must be limited to a low value, and a set amount of sunlight should penetrate to the interior. With only these set performance characteristics to guide the design-builder, the team may propose a wide variety of solutions, such as (1) aluminum and insulated glass curtainwall with integral horizontal shading; (2) brick and block wall with rigid insulation on the interior, plus fixed composite frame windows with integral blinds; or (3) stuccoed hay-bale walls with carefully placed glass block windows.

Each of these creative schemes might meet the stated requirements, depending on the climate of the project site. While it is unlikely that all three would offer equal satisfaction or cost to the owner, the advantage of performance specifying is clear; there is an unusual amplitude for creativity when project requirements are stated in performance terms. Part of the range of solutions in the previous example stems from the fact that so few performance requirements were used to define the needs. Should the owner ever want to open the windows, for instance, none of the suggested schemes would be acceptable, since all glazing is fixed. It is imperative that owners state their "must have" performance needs in their entirety at the outset. At the same time, in an effort to ensure that a list of performance requirements is complete, specifiers may tend to overspecify, either stating their needs in prescriptive terms or becoming more specific than necessary in their performance requirements.

12.4.3   Overspecifying deprives design-builder of flexibility of choosing among systems and components to meet owner's desired end results

Owners contemplating the delivery of new facilities using design-build and performance specifying may lose the benefits of these tools by overspecifying. Overspecifying can result from any or all of the following:

- Owners tell consultants what they need, but err on the side of too much detail, too many demands, and too much control over the process.

- Consultants divert their energies into solving the problem, either narratively or graphically, rather than articulating the problem for the architect-engineer (A/E) of record (design-build team) to solve.

- Consultants are unused to performance specifications at an elemental level, and because of their zeal for extreme detail rather than clarity of performance expectations, begin to strangle the creative process rather than provoke outside-the-box proposals.

Although the hay-bale walls scheme may seem outlandish, this unusual solution could be an ideal approach for single-family housing in a treeless, hot, and arid country with few sources of glass, metals, or cement. However, if the owner had used prescriptive specifications or defined the program too narrowly, this scheme may never have been considered.

Architects are trained to conceptualize a problem in abstract terms. Some architects can elicit unrealized client needs. But the chasm that is interposed between designer and constructor in the traditional design-bid-build process denies most designers the up-to-the-minute information about materials and methods of construction that would allow the project to be as innovative and efficient as humanly possible. Without convenient access to the availability of new materials and construction approaches, designers tend to fall back on past experience and preconceived notions. Unless there is in-house expertise, it is generally too expensive to acquire the knowledge on a project-by-project basis unless fees are increased.

In sum, it benefits the owner to facilitate regular consultation between the designer and builder through the use of a process that requires such interaction. When designer and builder are on the same team, both types of expertise are effectively "in-house" and the owner gains the benefits of the resulting synergies.

12.5 The Performance Specifying Process

A performance specification is a statement of required results that includes criteria for verifying compliance. Four essential elements of a performance specification are

1. *Attributes*—characteristics of a facility element, such as water tightness, heat transmission, or stain resistance
2. *Requirements*—statements of desired results, usually framed in qualitative terms
3. *Criteria*—measurable or observable definitions of performance, expressed in either quantitative or qualitative terms
4. *Tests or Substantiation*—objective means of confirming conformance with performance criteria, usually involving industry-recognized test methods, but sometimes using calculation, analysis or professional judgment

Used together, the four elements of a performance specification allow an owner and a design-builder to agree on what performance is required for components of a building, how it will be measured, and what test will be considered definitive. A performance requirement for an exterior window, for example, may include a *condensation resistance requirement* (CRF) for windows and frames to be a minimum of 35 when measured in accordance with AAMA 1503.1-1988. Deconstructed into its constituent elements, the specification would include

- *Attribute*—condensation resistance of window glazing and frames
- *Performance requirement*—window glazing and frames that resist condensation
- *Measurable criterion*—achievement of a CRF rating of 35
- *Agreed-on test*—measure in accordance with AAMA 1503.1-1988

A traditional prescriptive specification for the exterior window would have included the means and methods to be used to achieve the results, but a performance specification focuses on the results themselves coupled with a means of observing or testing that the results provide the quality that is being sought.

Here is an illustration of a performance specification as applied to a real life situation. A person consults a physician and proposes a performance specification: "By the time I leave for vacation next week, I want to be free of this back pain and play at least nine holes of golf." Deconstructed, this performance specification would include *attribute*—back health or function; *performance requirement*—absence of back pain by next week; *measurable criterion*—direct observation by the patient; *test*—ability to play at least nine holes of golf. The physician may respond to the owner of the body with a more detailed performance or even a prescriptive specification: "Consume this medication, in this dosage 3 times per day with food, and perform these exercises every morning until your vacation."

12.6   Elemental Attributes

Describing the expected performance of whole facilities, facilities elements, systems, or assemblies requires the listing of attributes. Attribute classes include various aspects of amenity and comfort, health and safety, structure, durability, and operation and maintenance. By contrast, a traditional prescriptive specification would begin with products and materials, using terms such as "mix," "fabricate," "place," and "attach" to describe the methods that shall be used by the contractor.

Specifiers can follow a systematic approach to developing performance specifications by listing each fundamental facility element by function (using UniFormat), and then identifying the attributes that may have an effect on the listed elements. Attribute classes and discrete attributes that can be attached to facility elements are the building blocks of a performance specification system. Briefly, the organizational structure for the specifier's elemental analysis would appear as

Element	Basic function
Attribute class	Attribute (detriment)
	Requirement
	Criteria
	Test (substantiation)

Although some of the individual "attributes" identified for a facility are really "detriments" (e.g., combustibility), the system uses the generic term attribute to describe all potential issues that could attach to an element. Individual *attributes* could be construed as being either positive or negative, and the requirement and criteria portions are used to outline the expected performance of the element in terms of the discrete attribute.

12.6.1 Amenity and comfort attributes

These concerns may be derived directly from the facility program or from a pre-design facility audit that explains (through narrative and compilations of facility demographics) the goals of the facility. Amenity and comfort factors are the needs of the facility users translated into measurable attributes. As an example, a roof system must keep out rain and snow (water intrusion, leakage) and retain heat for occupants in winter and allow for exterior ventilation (heat/cold; humidity/condensation; odor); and the roof system may be seen from a distance (appearance; texture). For a building, one might argue that amenity factors are not absolutely required, but are important to the inhabitants or users of the facility. Comfort factors, on the other hand, which may include wind and drafts, light and glare, sight and privacy, and sound and noise, are the very impetus for designing and constructing the facility in the first place.

12.6.2 Health and safety attributes

Professional designers are required under their codes of ethics to hold paramount the health, safety, and welfare of the public. Concern for safety is reinforced by state and local building and fire codes, which guide the design and use of assemblies for various types of facilities. Issues related to safety and well-being of the public are generated by natural or nonnatural dangers, accidents, or disasters that could affect the facility. Some of these attributes would include combustibility, smoke, explosion, electrical shock, exposure to chemicals or diseases, animal infestation, and violent storms (hurricanes and tornadoes). The requirements and criteria for health and safety attributes usually begin with "protect," "control," "limit," or a similar declarative expression.

12.6.3 Structural attributes

Performance issues related to the strength of the structure are usually based on minimum requirements of local codes, augmented by facility design goals that may add increased structural performance to support, for example, higher wind loads or overhead crane loads. These structural loads, including horizontal and wind loads, live and vertical loads, impact and concentrated loads, static and dead loads, and seismic loads, could affect any of the facility elements. The performance specifier will want to look at each element and its constituent parts to determine which structural attributes apply to the facility. The requirements and criteria for structural attributes are stated in terms of stiffness; deflection; compression, tension, and torsion capability; resistance to movement; ability to move; and other structural descriptors.

12.6.4 Durability attributes

As the facilities design and development industry moves more toward life-cycle thinking, the importance of durability factors continues to rise. Issues of durability are usually of prime importance to the owners or operators of a facility who will be owning or operating the structure for 10 years or more. A discussion of these topics during the facility programming or performance specifying phase will help identify how the owner envisions the longer-term performance (and longer-term care and attention) that will be expected for each of the various elements. Individual attributes within this attribute class will include (among others) resistance to dirt, grease, or stains; resistance to corrosion; resistance to abrasion or scratching; ability to withstand impact; and ability to withstand temperature changes. At the heart of durability attributes is the concern over expected life of the elements and assemblies that make up the facility. When the design life has been determined, the individual elements can be selected according to how they will contribute to the physical life and economic life of the structure.

12.6.5 Operating and maintenance attributes

The systems, assemblies, materials, and equipment incorporated into a facility are essential to operation and maintenance costs. As equipment and materials manufacturers and suppliers become more attuned to performance-based contracting, long-term performance data will be provided along with technical properties and first costs to designers and design-builders. Individual attributes within the operating and maintaining class include cleaning and serving; repairing and replacing; power, water, and fuel use; and relocating/adapting. Interestingly, operating and maintenance data for entire facilities has been recorded and published by associations that represent major owners; however, comparable performance data for facility elements or for equipment and material is currently lacking. With the growth of integrated facilities delivery and increased interest in life-cycle costing, it is likely that operation and maintenance metrics will soon become commonplace.

12.7   Criteria, Testing, and Substantiation

To implement a facility performance specification system, the specifier must convert every attribute that is attached to a facility element into a needs or goals statement. These statements are the performance requirements and criteria that embody the functional and technical goals of the facility. Appended to each requirement and criterion are ways for the design-build entity to substantiate (prove) that the team's design and construction solution will meet the goals of the facility.

Design-build contracts are signed well before facility design is completed and sometimes before any graphical representation has been created. Owners and their consultants are aware that there is a risk in having the facility performance requirements correctly translated into a completed facility. Therefore, owners use substantiation measures to mitigate the project risk, and to confirm, as the design develops, that the design-builder is providing the systems, assemblies, and materials that will prove acceptable once they are in place.

Owners may ask for substantiation updates at progressive phases of the project: design criteria, conceptual design, schematic design, design development, construction documentation, or commissioning. Certainly, an owner will want the design-builder's assurance of substantiation to be embodied in the proposal and at least at one milestone during project execution.

12.7.1 Systems and assemblies standards

There is a growing body of systems and assemblies standards that identify performance of facility elements and subelements. Examples are found in the UL (Underwriters Laboratories) assemblies guide and in national standards sanctioned or produced under the auspices of ANSI, ASTM, and other recognized consensus standards bodies. Confirmation by the design-builder that proposed design and construction will meet these standards is a basic form of substantiation.

12.7.2 Tests and testing

Laboratory testing of standard assemblies or of in-place construction is another form of substantiation, such as traditional concrete slump testing, test cylinder compression testing, and core sample testing. To be cost-effective, testing must be by accepted and available methods; special testing could be extremely expensive and deemed onerous by the design-build proposers

12.7.3 Mockups and in-place observations

Field mockups are a popular approach for some types of systems or assemblies, such as an exterior masonry wall sample. The mockup allows the owner and his advisors to observe whether aesthetic and/or performance goals will be met by the facsimile. Use of mockups usually occurs very late in the design process (or early in the construction process).

12.7.4 Field testing and verification

During construction, field testing of elements and materials can verify whether the systems are meeting code or specification requirements. For example, compaction testing of soils or fill will reassure the owner and design-build entity that quality-work procedures have been followed to ensure the structural integrity of the facility. In other cases, simple observation and recordation of findings can verify that performance goals are being met.

12.7.5 Certification and calculations from registered professionals

Some elements or assemblies are comprised of a number of different products or pieces of equipment. For structural systems or for HVAC systems, it is reasonable to expect that calculations and sealed drawings provided by professionals would serve as evidence of compliance with facility requirements.

12.7.6 Manufacturer's warranties

Owners may require the design-build team to submit literature that verifies compliance with performance goals. A manufacturer's warranty carries more weight because it spells out terms and is signed by the manufacturer or its representative. Warranties are often requested for systems that are expected to provide facility comfort or occupant protection, such as roof systems and heating systems.

12.7.7 Long-term guarantees and operations bonds

Occasionally, owners include requirements for extended guarantees for designated facility elements. Especially in the transportation sector, there are public agencies seeking reassurance of performance through the use of multiyear guarantee or warranty clauses. A few owners have begun asking for operations bonds that cover operating the facility for a term of years after the standard performance and payment bond expires one year after project completion.

12.8 Recommended Organization of a Design-Build Performance Specification

Organizing a performance specification system according to a predetermined format provides the originators and the users with a disciplined checklist. The format should follow a logical systems-elements scheme (such as UniFormat) to examine all the performance factors that accompany a facility, from whole-facility performance on down to assemblies and subassemblies performance.

Incorporating the performance specifications contents described in this chapter, the organizing format would include the following:

Part 1: Performance

Element
  Basic function
  Attribute class
    Attribute (or detriment)
    Requirement
    Criteria
    Test or substantiation
  Assembly
    Basic function
    Attribute class
      Attribute (or detriment)

Requirement

Criteria

Test or substantiation

Part 2: systems, assemblies, and products

Systems or types of systems that meet or exceed the performance requirements. Design-build entity may use X system or assembly—owner originated communication; or design-builder will furnish X system in compliance with performance requirement—design-builder originated communication.

Systems or types of systems that do not meet performance requirements or are otherwise unacceptable. Design-build entity may not use X system or assembly—owner originated communication; or design-builder will not use X system or assembly—design-builder communication.

Part 3: requirements of design and construction

Facility programmatic goals for element or assembly (used when program is not available or as a summary).

Expected outcomes of design and construction.

Specific instructions for design and construction. This is not a design-build "best practice" and is used only when owner insists on a specific approach, technique, or assembly, or for the design-build prime to communicate with specialty contractors and manufacturers or suppliers.

The organizational outline contains the critical performance aspects of a project and its constituent parts first (Part 1), guiding the spec(ifications) developer and user through a series of requirements for "layers" of the facility. The format begins with the overall project, then addresses elements (for a building: UniFormat elements would include substructure, shell, interiors, services, etc.), then assemblies and subassemblies. At each layer or level, the specifier will identify a basic function of the element or assembly, and begin to attach attributes, requirements and criteria, and tests and substantiation to ensure that the performance expectations are clearly communicated.

Attributes are individual issues found within attribute classes that relate to a particular element or assembly. As outlined earlier, attribute classes include amenity and comfort, health and safety, structure, durability, and operation and maintenance. Attributes found within the *health and safety* class, for example, may include combustibility, smoke generation, and flame spread.

To verify that the element or assembly will meet performance goals, the specifier will include criteria and testing or other verification measures that confirm that the designer-builder's facility solution is consistent with those goals. To substantiate performance, the design-build entity can be expected to provide calculations or results of tests; however, the owner and its consultants should refrain from adding unreasonable substantiation requirements. Unreasonable requirements (such as unusually difficult or expensive testing) will drive away worthy offerors or greatly increase the cost of the work.

Owners procuring design-build services should use Part 2 (systems, assemblies, and products) sparingly, if at all. Instead, Part 2 should be used by design-build teams to propose systems and assemblies to owners in response to Part 1 performance requirements. Similarly, Part 3 is primarily intended for use by design-build entities for communicating within the team of subconsultants and subcontractors, or for informing the owner (if necessary or required) about the kinds of processes that will be used to carry out the design and construction work.

12.9 Importance of Reference Standards

The earlier example of the hypothetical performance specification was that of someone stating a "free of back pain in one week" goal to a physician. There is subjectivity in the "measurable criterion" of direct observation by the patient. If payment to the physician were dependent on the success of the cure, and the only measurement of success was the patient's own statement of pain or lack of it (which has no widely accepted measure), then subjectivity would be a problem. There may be stages of misery between complete freedom from pain, and barely finishing the ninth hole before collapse.

Fortunately, the design and construction industry has better means of measuring degrees of performance achievement than does the health industry. The facilities design and development industry relies on independent and objective reference standards, such as the Architectural Aluminum Manufacturer's condensation resistance requirement. These consensus industry standards are essential to clear definition of performance requirements and enforcement of the terms of design-build contracts.

12.9.1 Whole-facility standards and performance building codes

The popularity of performance contracting has stimulated ongoing interest in new systems-based standards. Performance specifications rely directly on performance standards to demarcate levels of quality. The integration of the facilities delivery process will create significant demand for performance standards that treat whole facilities, systems, elements, and assemblies, going beyond the current proclivity toward material and equipment standards.

In 1996, The American Society for Testing and Materials Subcommittee E0.25 issued *ASTM Standards on Whole Building Functionality and Serviceability*. The standard was intended to focus a facility auditor away from traditional architectural programming (where the end result is a design solution) to a more universal focus on how the facility serves its owners, managers, and occupants. There are a series of standards numbered E1660 (*Serviceability of an Office Facility for Support of Office Work*) to E1671 (*Serviceability of an Office Facility for Cleanliness*) in which the user may identify and record the requirements of the owner-users on a sliding scale and then to rate the existing or proposed facility (again on a scale) to determine the serviceability and functional capability of that element.

A typical section in the standard uses a layout that is conducive to recording the observations and thinking of the evaluator (see Fig. 12.1).

Figure 12.1   ASTM E1662 (A.3. Sound and Visual Environment; Scale A.3.4. Lighting and Glare) scale layout for determination of occupant requirements and facility rating.

Occupant Requirement Scale	Facility Rating Scale
9. Operations require that illumination levels suit very different types of work. Require freedom from lighting defects including glare, e.g., all staff have VDUs [visual display units] and many work at them for long periods of time. Very low tolerance for any defects in illumination.	9. **Illumination Level:** The level of illumination is varied to provide lower levels needed for VDU screens and higher levels needed for paperwork. **Visual Defects:** There are no apparent or reported lighting defects. **Glare:** There is no glare on VDU screens from windows and lights.
7. Operations require different illumination levels in parts of the office. Nearly all staff have VDUs. Very low tolerance for any illumination defects, or glare from windows or lights.	7. **Illumination Level:** The level of illumination is sufficient to read fine print without strain in any workplace. Lower levels of illumination are provided in areas with many VDUs, or could be provided with partial relamping. **Visual Defects:** Any lighting defects do not affect staff at their workstations, and are not reported as a problem. **Glare:** VDU screen from windows and/or lights is barely visible, and can easily be decreased.
5. Operations require predominance of lighting levels appropriate to reading fine print. Many VDUs are used. Can tolerate some lighting defects, e.g., some glare from windows or ceiling lights.	5. **Illumination Level:** The level of illumination is sufficient to read fine print without strain at any workplace. No lower levels of illumination are provided for VDUs, and it is not practicable to make the necessary changes. **Visual Defects:** There is one visual defect, e.g., gloomy appearance of the ceiling, flicker, extreme contrasts, or different color fluorescent lamps.

(Continued)

Figure 12.1 (*Continued*).

Glare: VDU screen glare from windows and/or lights is clearly visible, but could be decreased at moderate cost, e.g., by installing parabolic reflectors on ceiling luminaries.

3. Most work is visually undemanding, so can tolerate a wide range of levels and quality of illumination without affecting health or productivity. Few VDUs are used.

3. **Illumination Level:** The level of illumination is excessively high, or too low, in a few of the areas used for general office activity.

Visual Defects: There are one or two visual defects, e.g., gloomy appearance of the ceiling, flicker, extreme contrasts at workstations.

Glare: At many workstations, e.g., 40% to 60%, there is unavoidable VDU screen glare from windows and/or lights, with no effective glare control.

1. No requirement at this level

1. **Illumination Level:** The level of illumination is excessively high, or too low, in most areas used for general office activity.

Visual Defects: There are three or more visual defects, e.g., gloomy appearance of the ceiling, flicker, extreme contrasts at workstations.

Glare: At all workstations there is or would be unavoidable glare on screen of a VDU, from windows or lights, with no effective glare control.

Ranking of importance of the selected requirement: Exceptionally important ———; Important ———; Minor Importance ———.

Space for Notes on Requirements or Ratings:

When used as intended, the *ASTM Whole Building Functionality and Serviceability Standards* can serve as the basis for translating owner or user requirements into technical performance requirements for each facility element; and can provide the groundwork for facility programming.

The application of systems standards is driven by both user/user preferences and building code requirements. Many governments are currently supporting a change toward performance-based building regulations to minimize the negative effect of prescriptive requirements while still upholding the priority of protecting the public. Performance-based building codes provide greater clarity of the base intent of the law; add consistency to the scope of the codes; reduce the need for frequent change in the guidance; respond more readily to innovative materials, assemblies, and methods; and are more adaptable to international trade in the global economy.

The International Code Council's draft for a performance code provides a structure that is similar to the recommended performance specification format shown previously:

- *Objective*—topic-specific expectation

- *Functional statement*—explanation of purpose

- *Performance requirements*—more detailed criteria to achieve the goals of the functional statement

- *Solutions*—options for meeting the technical requirements and acceptable verification measures

The ICC Performance Committee has been challenged by the lack of national performance standards in some areas. However, the majority rejected the recitation of specific prescriptive standards in the requirements and solutions sections in favor of a global approach, that is, using performance language that simply provides criteria as to what would be considered acceptable to code officials.

12.9.2 Major sources of reference standards

Of the 94,000 national standards available from U.S. standards-writing organizations and government agencies, approximately 11,000 (12 percent) pertain directly to the design and construction industry. Many of the building and construction standards tend to be overlapping or redundant. To overcome the product-centric nature of prescriptive-oriented standards, many of the standards-writing groups are trying to recast their standards in performance terms. Performance-based language and data, according to the National Institute of Standards and Technology, are more flexible and adjustable to changes in technology, whereas older prescriptive-oriented standards tend to document obsolescent technology.

The most significant standards-writing or developing organizations (please note that this is only a partial list) in the building and civil infrastructure sectors include the following:

Buildings standards	Civil infrastructure standards
American Architectural Manufacturers Association	American Association of State Highway and Transportation Officials
American Lighting Institute	American Public Works Association
American Institute of Steel Construction	American Railway Engineering Association

Buildings standards
(*Continued*)

American Plywood Association

Asphalt Roofing Manufacturers
Association

American Society of Heating,
Refrigeration and Air Conditioning
Engineers

American Society of Mechanical Engineers

American Welding Society

Builders Hardware Manufacturers
Association

Brick Institute

Concrete Reinforcing Steel Institute

Door and Hardware Institute

Gypsum Association

Illuminating Engineering Society

Metal Building Manufacturers
Association

National Concrete Masonry Association

National Electrical Manufacturers
Association

National Elevator Industry Association

National Forest Products Association

National Fire Protection Association

National Electrical Manufacturers
Association

National Roofing Contractors Association

Painting and Decorating Contractors
of America

Safety Glazing Certification Council

Resilient Floor Covering Institute

Steel Deck Institute

Thermal Insulation Manufacturers
Association

Sheet Metal and Air Conditioning
Contractors Association

Tile Council of America

Vinyl Window and Door Institute

Civil infrastructure standards
(*Continued*)

American Society of Civil Engineers

American Water Works Association

Asphalt Institute

Association of Engineering Geologists

American Welding Society

Concrete Plant Manufacturers Bureau

American Concrete Institute

Construction Industry Manufacturer's
Association

Contractors Pump Bureau

National Clay Pipe Association

National Corrugated Steel Pipe
Association

National Lime Association

National Stone Association

Plastics Pipe Institute

Portland Cement Association

Steel Tank Institute

Water Environment Federation

Other major private nonprofit standards organizations include

Institute of Electrical and Electronics Engineers

American Society of Mechanical Engineers

Major government standards organizations include

General Services Administration

Department of Defense

U.S. Army Corps of Engineers

Naval Facilities Engineering Command

United States Postal Service

The large number of individual standards organizations is representative of an American system that is decentralized and complex. A selected list of standards-related organizations is of particular importance to performance specifiers:

American National Standards Institute (ANSI). ANSI does not develop standards but coordinates development of voluntary standards in the United States and approves standards in the ANSI system. ANSI also coordinates participation in the International Organization for Standardization (ISO). There are approximately 10,000 approved American National Standards. ANSI is headquartered in New York City.

American Society for Testing and Materials (ASTM). ASTM is a nonprofit organization formed to develop standards on characteristics and performance of materials, products, systems, and services. The standards include classification systems, guidelines, practices, specifications, terminology, and test methods. ASTM, a technical society, has published approximately 9000 standards. It is headquartered in Conshohoken, Pennsylvania.

National Fire Protection Association (NFPA). NFPA is a national nonprofit organization dedicated to gathering and analyzing fire statistics, investigating significant fires, working on antiarson projects, and operating service divisions in areas such as electrical systems, flammable liquids and gases, marine environment, and safety to life. NFPA published the National Fire Codes, including standards, fire protection, prevention and suppression codes, guidelines, and practice manuals. NFPA, a custodian of approximately 300 standards, is headquartered in Quincy, Massachusetts.

Underwriters Laboratories (UL). UL tests and evaluates products and systems to determine their safety and performance under various conditions, including hazardous conditions. The organization provides a listing service, which allows use of the UL mark for products that have been investigated with respect to risks to life or property (i.e., reduction of foreseeable risks to an acceptable degree). UL also provides classifications and guidances about products and assemblies with respect to (1) specific risks, (2) performance under specified conditions, (3) regulatory codes, and (4) standards of domestic and international organizations. UL presides over approximately 700 standards. The organization is headquartered in Northbrook, Illinois.

National Institute of Building Sciences (NIBS). NIBS is a nonprofit technical organization that works in the area of building sciences and regulations. The organization seeks to develop and maintain performance criteria and technical provisions, evaluate existing and new building technology, conduct investigations in support of building regulations, and assemble and disseminate technical data relating to building sciences. NIBS has focused on areas such as building seismic safety, lead paint abatement, energy efficiency of facilities, and asbestos risk reduction. NIBS also produces the Construction Criteria Base (CCB), a CD-ROM product that contains many of the guide specifications, design and construction criteria, and standards produced by the aforementioned organizations. NIBS is headquartered in Washington, DC.

12.10   Available Performance Specifying Tools

12.10.1   CSI-DBIA PerSpective

In 1995, the Construction Specifications Institute and the Design-Build Institute of America formed an alliance to produce a performance specifications tool to assist owners and practitioners engaging in design-build. The joint venture reasoned that the existing master specifications products were designed to document completed designs, whereas a product was needed to define facility performance in the absence of graphic design. Several commercial master systems were available using the CSI 16-division format, but none were available using the systems outline of UniFormat.

First and foremost, PerSpective can serve as a vehicle for communications between owners and design-build proposers. The owner (or the owner's consultant) creates a new project file or overlay atop the underlying master performance specification. The design-build entity can use the same project file to build a response to the call for proposals and, in turn, communicate with subcontractors, manufacturers, and suppliers with the same data and information.

Using this CSI-DBIA Microsoft Windows-based computer application, owners or their criteria professionals may select, edit, or add to preformulated performance requirements in the database and assemble them into a *requirements statement.* For sole-source contracts, the document may be given directly to the design-builder. For competitive proposal contracts, the document can be embedded within a *request for proposals* (RFPs). Design-build proposers, whether responding to an RFP or involved in a negotiated selection process, add the information that is requested within the electronic file and submit the response back to the owner. Throughout the process, design-build team members may continue to exchange information electronically (or on paper output) using PerSpective as a common or universal language.

Features of the product.   The CSI-DBIA PerSpective product contains performance descriptions for the commercial buildings sector. Included in this sector are "office buildings," "low-technology commercial facilities," "hospitality and multitenant residential buildings," and "low-technology institutional buildings." Users may be able to produce performance requirements for more sophisticated building types, such as hospitals, prisons, or industrial facilities with the database, but the developers admit that the information is not comprehensive and lacks many special requirements needed for specialized types of facilities. In addition, the product does not currently address civil infrastructure or heavy-process projects.

Benefits of the product include its great flexibility, allowing specifiers to move up or down levels of specificity for each project requirement. For example, an owner may specify the substructure of a building at the highest level (using Chapter A—Substructure), but deaggregate the building shell (Chapter B) into

- System groups are represented in Chapter B—Shell.

- Assembly groups are represented by Chapters B1—Superstructure, B2—Exterior Closure, and B3—Roofing.

- Elements are represented by Chapter B31—Roof Coverings, B32—Roof Openings, and B33—Roof Fixtures.

For this project, the owner wanted or needed to go beyond the first two levels (level 1 contains whole facility requirements; level 2 comprises system groups) to specify roofing assemblies and elements in performance terms.

The composition and software platform of PerSpective make it an extremely flexible tool for a wide variety of users and design-build variations. The ability to drill through multiple levels of hierarchy to specify a single component in greater detail is a plus, although owners are strongly cautioned by the product's producers about overspecifying. In reality, the owner and its consultants would stay at a systems level of performance specifying, diving lower only when there were "must have" issues brought forth by the owner or users of the facility. Users of performance specifications have different notions about performance requirements and fall along an entire continuum according to the project type and their knowledge of the design-build process. Most owners should be able to find a comfort level that balances (1) their concern about detailed control over the finished project and (2) the desire to use the tool of design-build as an incentive for the marketplace to apply its creativity in solving the owner's design-construction problem.

Many benefits of the PerSpective product are tied to automation. The distribution of the information within the product on a database platform permits one cell to link with another cell via a mathematical formula. Traditional 16-division specification products are not configured in this manner and do not have this capability. Computerized databases are currently used for cost estimating and scheduling, but until the introduction of PerSpective, there were no companion specification products available to the design-construction industry.

Parcels of data within PerSpective are connected to other data by connecting paragraphs of text information. When a user selects a paragraph for inclusion in the performance specification, links to and from the paragraph may automatically select related text, eliminate conflicting paragraphs, or highlight others for the user's consideration. Helpful links like these occur with the chapters and across them throughout all the PerSpective hierarchies.

Another helpful feature is the product's ability to compare two documents generated from PerSpective and to highlight the differences between them. The owner can compare the proposal received from the design-builder with the original RFP; or the design-builder can compare its proposal file with the PerSpective-generated proposal received from a subcontractor. Owners may also compare several proposals (and design-build RFPs sometimes generate apples and oranges proposals) to ascertain how the various proposers have met or *exceeded* the performance requirements contained in the RFP.

As the CSI-DBIA PerSpective continues to improve with new software versions, the utility of the product for a wide range of project applications will mainstream this type of approach for design-build.

Potential linkages with other data.   Computer software products, such as the CSI-DBIA PerSpective, are being configured for future linking with other products, via central proprietary and open-system linking capabilities. Linking management software will be able to receive, interpret, and route requests for information about related databases that are live-linked to one another, allowing performance specifying systems to exchange information with cost estimating and computer-assisted drawing (CAD) applications. The potential for adding even more efficiency to the design-build process through such exchanges is enormous; with misunderstandings, transposition errors, or outright omissions virtually eliminated from the integrated services process. Information technology is helping to blur the distinctions between all the important disciplines in design and construction, and the new synergies mirror the potential of the design-build project delivery method.

12.10.2   Future tools

Systems-based thinking is slowly permeating into the consciousness of the design-construction industry. Nonetheless, *widespread* understanding of an elemental-based system for facilities is nonexistent. At a national conference of the American Institute of Architects, a group of 20 architects were asked whether they were familiar with UniFormat. Only 3 of the 20 (15 percent of this sampling of AIA state and national leaders) were aware of the standard classification system that has been an ASTM standard since the late 1980s.

Through the continued adoption of information technology and CAD systems that integrate value analysis, pricing, and product performance, greater numbers of architects, engineers, and other industry professionals will be exposed to systems-based classifications systems. As soon as a significant number of design professionals are knowledgeable about systems approaches, these industry stalwarts will start demanding systems-based data from code organizations, facility product manufacturers, and materials suppliers. The collective needs of facility owners and design-construction industry professionals will drive the adoption of performance-based building codes. Current publishers of specification data in 16-division formats will convert their online and hard-copy publications to systems-based organizational formats within the decade. All of these changes will have been spurred by the realization that owners make decisions about an organization's capital-asset portfolio well in advance of design development. The ability to align design practice and construction practice with customer desires will require management-based rather than materials-based thinking and tools.

When systems-based classification formats become widespread, materials and equipment manufacturers will fulfill the potential of performance-based specifications by writing, at the "brick" level as well as at the exterior wall level, performance-based specifications at a variety of quality levels. When this change occurs, it will be possible to write an entire specification system, from the whole building level to the individual brick level, within the organizing discipline of overall facility quality (e.g., "class A office building"), facility systems quality, assemblies quality, or product quality. Whether the facility is a 10-year economic life R&D building for a technology start-up or a 100-year design life bridge for two boroughs in New York City, the software products will be flexible in handling the requirements.

The design-construction industry does not currently embrace an organizing system that is systems-based, but electronic products that are currently in development will revolutionize the business of delivering facilities. Powerful tools such as electronic recordation of user requirements, the ability to tie feasibility studies to facility demographics, and the marriage of conceptual costing and schematic design will be available to the master builders of tomorrow.

Design-Build Contract Structures and Payment Methodologies

13.1 Introduction

The relationship between an owner and a design-builder is defined largely by the type of contract used on the project and how the design-builder will be paid for its services. The four most common types of contract structures are (1) lump sum, (2) cost-plus, (3) cost-plus with a guaranteed maximum price (GMP), and (4) unit prices. Each structure has unique characteristics—particularly relative to when the price for the work will be determined—that can affect the rights, responsibilities, and roles of the parties. This chapter addresses some of the most significant characteristics of these contracting methods.[1]

13.2 Lump-Sum Contracts

Under a lump-sum contract, the design-builder agrees that for a specific price it will provide the owner with the services called for in the contract. The design-builder will receive that price regardless of the actual costs it incurs in performing the work. Lump-sum contracts are frequently used when the owner does one or more of the following:

- Selects the design-builder through cost competition
- Provides design development documents to prospective design-builders under the bridging concept[2]
- Selects the design-builder directly, without competition, and uses the design-builder's programming and design skills to achieve a specific budget

---

[1]Many of the concepts set forth in this chapter are derived from the *Design-Build Contracting Guide,* Design-Build Manual of Practice Document 510 (Nov. 1997), published by the Design-Build Institute of America, Washington, DC. This is also referred to as the *DBIA Contracting Guide.*

[2]For a more comprehensive discussion of bridging, see Secs. 6.3 and 8.4 herein.

- Specifies a fixed price for the design and construction of the project and selects the design-builder through a design competition

While these procurement scenarios can also be used under a cost-plus or GMP type of contract arrangement, they are particularly well suited for lump-sum contracts. The owner who selects a lump-sum contract typically does so because of its simplicity and the ease with which the contract is administered.

When an owner's goal is to obtain a competitively bid lump-sum price for design-build services, some suggest that the owner should provide detailed design specifications to the bidder. This ensures that the owner is comparing bids on an "apples to apples" basis and actually getting what it wants.[3] Many owners, however, reject this idea, contending that it ties the hands of the design-builder and hinders design creativity, which is one of the major benefits of design-build. In fact, as noted in the procurement scenarios set forth above, many competitively bid lump-sum design-build projects have been successfully executed when the owner provided only performance specifications and gave the design-builder flexibility regarding the overall design. However, while lump-sum pricing can be established well before significant design work is performed, the project will proceed with greater mutual understanding and cooperation between the parties if the price is established after the design-builder reaches a clear understanding of the owner's program.

The construction industry has long used lump-sum contracting under the design-bid-build and construction management delivery systems. As a result, most owners, design professionals, and contractors will easily understand how to administer a lump-sum contract on a design-build project. Some of the strategic contracting issues associated with this process are noted in the sections that follow.

13.2.1 Establishing the lump-sum price

Certain design-build procurement methods require that the lump-sum price be established during the contracting process (Fig. 13.1). If the owner, for example, uses a cost competition selection process, the design-builder will include a lump-sum price in its proposal. The price will be incorporated into the contract when the contract is executed. Similarly, if the owner has a fixed budget and intends for the design-builder to provide design alternatives that fit within the budget, the lump-sum price will usually be finalized when the contract is executed.

It is important to remember, however, that many owners will sign a contract with a design-builder before agreeing on a lump-sum price. In these cases, the parties typically establish a price (lump-sum or cost-plus) for programming and design development services. After the design-builder and owner agree on an appropriate design, the design-builder provides a firm fixed price for executing

---

[3]C. B. Thomsen, FAIA, "Bridging," in *The Architect's Handbook of Professional Practice*, The American Institute of Architects, Washington, DC, 1994, pp. 415–417.

	FACTORS SUPPORTING USE OF LUMP-SUM PAYMENT METHODOLOGY
1.	Project scope is capable of being clearly defined.
2.	Project is competitively bid and fixed project price is one of the selection factors.
3.	Owner does not need to have access to design-builder's cost records.
4.	Owner does not have the resources to evaluate design-builder's cost records.
5.	Owner is comfortable with design-builder's selection of subcontractors and does not expect to have significant control or input.
6.	Owner believes that the design-builder's lump-sum price is appropriate for the scope of work being performed.

Figure 13.1

the final design and construction. In fact, this concept is specifically anticipated by the two-part system used in the AIA contracting approach. Under Part 1 of AIA Document A191, *Standard Form of Agreement between Owner and Design-Builder,* the owner receives preliminary design documents and a price and time proposal for executing the work pursuant to the preliminary design. If the owner agrees with the proposal, the parties execute Part 2 of AIA 191, which requires the design-builder to finalize the design and complete the construction.

Regardless of whether the lump-sum price is determined before or after contract execution, it is critical that the owner and design-builder clearly understand the scope of work and the services that the design-builder will provide. If the price will be developed when the contract is executed, the parties should have clear and complete performance specifications with specific and objective acceptance criteria. If the price will be provided after contract execution, the design-builder should advance the design to the point where subcontractors can provide relatively firm estimates of the cost to complete the work.

13.2.2 Payment process

The progress payment process under a lump-sum contract typically calls for the following:

- A schedule of values to be used in determining percentages of completion of various portions of the work[4]

---

[4]Engineers Joint Control Documents Committee, EJCDC 1910-40, Paragraph 13.01 (1995 ed.); Associated General Contractors of America, AGC Document 415, Paragraph 9.2 (1993 ed.).

- Progress payments to be made at specific intervals (typically monthly) on or before specific dates established by the contract[5]
- Retainage in an agreed-upon percentage[6]
- Appropriate documentation to be submitted as a condition to both progress payments and final payment[7]

This standard method of addressing the payment issue can be modified if the parties determine that another approach is more advantageous to them.

There are several creative ways to administer lump-sum contracts which may both provide financial benefits to the design-builder and translate into cost savings to the owner. For example, standard language calls for retention to be withheld at a fixed rate through substantial or final completion. Many parties modify this on lump-sum contracts by discontinuing the withholding of retention when 50 percent of the work is complete, provided the work is proceeding satisfactorily. They may also choose to reduce retention at substantial completion to a multiple of the outstanding punchlist work. These accommodations can financially benefit the contractors working on the project (and thus lead to a reduction of their prices) while still providing the owner with reasonable protection that its work will be satisfactorily completed.

Parties may also modify the progress payment process by associating specific payment milestones with the completion of discrete portions of work, such as foundations and superstructure, rather than general percentages of completion. Because this approach is a more predictable measure of performance completion and is easier for lenders to evaluate, it is commonly used on industrial and power generation design-build projects. Even if the construction work is to be measured from percentages of completion, it may be appropriate for progress payments on the design portion of the work to be based on milestones such as the completion of design development documents and construction documents.

Finally, it is important for the parties under a lump-sum design-build contract to identify who will determine whether progress payments are due and what the recourse will be in the event the parties are in dispute. Under a traditional design-bid-build approach, this role is fulfilled by the project design professional working for the owner. In a design-build project, the owner or its designee will typically determine whether the design-builder has achieved the necessary level of completion to support a progress payment. However, some owners and design-builders rely on payment certifications provided by the project's design professional of record, even if it is a subconsultant to the design-builder.

---

[5]EJCDC 1910-40, Paragraph 13.02; AGC Document 415, Paragraph 9.2.1.

[6]EJCDC 1910-40A, Paragraph 5.01.

[7]EJCDC 1910-40, Paragraph 13.02.

13.2.3   Extra work

It is well established that a lump-sum contract price will be increased or decreased as work is added or subtracted from the design-builder's scope. The contractual vehicle for accomplishing this is the "changes" clause of the contract, which specifies the types of changes that are permissible as well as how such changes are to be priced.[8]

At the time of contracting, it is critical for both the owner and design-builder to consider the markup associated with changes. Many contracts address this issue by simply stating that the design-builder will be entitled to a "reasonable markup for overhead and profit." Some owners, however, believe that this point should be resolved as part of the precontract execution process. Identifying a specific markup will not only enable the competitive process to work to the owner's advantage by establishing a figure for overhead and profit, but also provide certainty as to what the owner's exposure will be for changes. Attention should also be paid to the markups allowed for subcontractors as many owners require the design-builder to flow down to its subcontractors specific overhead and profit limitations on subcontractor change order markups.

Another important issue related to the pricing of changes on a lump-sum design-build contract involves determining what costs associated with the change will be considered reimbursable. As noted in Sec. 13.3.2, cost-plus contracts contain a specific list of reimbursable and nonreimbursable cost items which will be used for determining pricing for change orders. With the exception of the EJCDC design-build contract family, which contains a specific delineation of reimbursable costs for changes,[9] standard form design-build contracts do not identify which costs can be recovered if a change occurs. Instead, the contracts have more general language that leaves the answer to the ultimate discretion of the parties. For example, Paragraph 8.2 of AGC 415 states as follows:

> *Determination of cost*   An increase or decrease in the Contract Price resulting from a change in the Work shall be determined by one or more of the following methods:
>
> 1. unit prices set forth in this Agreement or as subsequently agreed;
> 2. a mutually accepted, itemized lump sum; or
> 3. if an increase or decrease cannot be agreed to as set forth in Subparagraphs 8.2.1 or 8.2.2 and the Owner issues a written order for the Contractor to proceed with the change, the adjustment in the Contract Price shall be determined by the reasonable expense and savings of the performance of the Work resulting from the change. If there is a net increase in the Contract Price, a reasonable adjustment shall be made in the Contractor's overhead and profit.

---

[8]See Sec. 14.8 herein. EJCDC 1910-40, Article 9; AGC 415, Article 8; AIA Document A191, Part 2, Article 8 (American Institute of Architects, Washington, DC, 1997).

[9]See Article 10, EJCDC 1910-40.

This process requires that the design-builder maintain accounting records which document these expenses. Given this clause, the parties are well advised to meet early in the project and determine an appropriate means of dealing with changes.

Finally, the parties under a lump-sum contract should specifically address what happens if the parties cannot agree on either the design-builder's entitlement to or cost of a change. As noted in Sec. 14.8, an effective way to balance this risk is the approach recommended by the *DBIA Contracting Guide,* which suggests the following language:

> If the parties are unable to agree [upon a change], and Owner expects Design-Builder to perform the work in accordance with Owner's interpretations, Design-Builder shall proceed to perform the disputed work, conditioned upon Owner issuing a written order to Design-Builder directing Design-Builder to proceed and specifying Owner's interpretation of the work that is to be performed. In such event, Design-Builder shall be entitled to submit in its payment applications an amount equal to fifty percent (50%) of its estimated cost to perform the work, and Owner agrees to pay such amounts, with the express understanding that such payment by Owner does not prejudice Owner's right to argue that it has no responsibility to pay for such work and that receipt of such payment by Design-Builder does not prejudice Design-Builder's right to seek full payment of the disputed work under the applicable terms of the Agreement in the event the Owner's order is deemed to be a change to the Work.[10]

This serves as an incentive for both parties to resolve the issue. It also prevents the design-builder from having to finance the work.

13.2.4 Access to financial records

One benefit of using a lump-sum contract is that the owner is not required to analyze the cost records of the design-builder to determine what is or is not due under the contract. This not only saves the owner the expense of employing accounting personnel to administer the process but also saves the design-builder the cost associated with complying with the owner's accounting reviews. If the owner needs access to specific cost information for financing, regulatory, or other reasons, it should specifically address this in the contract and precisely define the scope of its audit and review rights.

Even though the owner will not be given access to the design-builder's financial records for the base contract work, it must be able to review records of change order work performed on a cost-plus or similar basis. It is in the design-builder's best interest to provide this information because the owner will not pay unless it receives appropriate support for the change. If, however, a claim is seriously disputed, such as one dealing with delays and disruptions, an access-to-records provision ensures that the owner will have the right to obtain information if the design-builder is not reasonably forthcoming with such information.

---

[10]*Design-Build Contracting Guide,* p. 22.

13.2.5   Allowances and alternates

Many owners are hesitant to use lump-sum contracts if a portion of the work is not well defined at the time the contract price is being set. In such cases, the owner may still be considering some design alternatives for portions of the work, such as level of finishes or specific brands of equipment. Because lump-sum contracts afford the owner limited access to financial records, the owner may have a difficult time determining the true cost to the design-builder for deletions or additions to the work. To overcome this problem, owners may want to consider the liberal use of allowances and alternates in their lump-sum contracts.

An allowance has been defined as

> A sum of money stated in contract documents to cover the cost of materials or items in those documents, the full description of which is not known at the time of bidding. All contractors bid the allowance as part of their proposals. The actual costs of the items are determined by the contractor (not including installation) at the time of their selection by architect or owner and the total contractual amount is adjusted accordingly. Examples: brick, carpet, appliances.[11]

Thus, for example, if the owner is not certain of the type of landscaping that will be used, it can establish an allowance within the lump sum of a designated amount (such as $200,000) for the expected cost of the work. This enables those bidding the project to rely on a "plug number" for this undefined work. It also provides the owner with a fixed price for the work. Once the work is defined, the design-builder will be paid for the actual cost of performance regardless of the amount established in the contract for the allowance.

13.3   Cost-Plus Contracts

Under a cost-plus contract, the design-builder is paid by the owner for the actual costs of performing the design and construction services called for in the contract. These contracts are frequently used when

- The owner selects the design-builder on a negotiated basis.
- The owner and the design-builder have a strategic partnering or alliance agreement that will govern the work.
- The project is technically difficult or unique and the owner wants to avoid any conflict with the design-builder over cost issues.
- The owner selects the design-builder directly, without competition, and uses the design-builder's programming and design skills to meet a specific budget.

As noted in the *DBIA Contracting Guide*:

> Cost-plus arrangements are particularly appropriate on projects which cannot be adequately defined by the parties in advance of executing the contract, such as

---

[11]R. W. Dorsey, *Project Delivery Systems for Building Construction,* Associated General Contractors of America, Alexandria, VA, 1997, p. 250.

projects that (a) will be advancing the state-of-the-art and therefore do not have an effective design benchmark or (b) contain unusually high contingencies because of site conditions, locations or other unknowns—e.g., hazardous waste remediation or unclear technical interfaces. While lump sums can always be provided, the ultimate price to the owner for the contingencies associated with the uncertainty may not justify the certainty of up-front pricing.[12]

Because of the situations where cost-plus is used, there is a strong perception that cost-plus contracting promotes strong and cordial working relationships between the owner and the design-builder. This differs from the widely held view that lump-sum contracts lead to disagreements over scope and thus strain relationships. Although each of these perceptions has merits, one should remember that cost-plus contracting is disadvantageous to the design-builder for several reasons.

1. The owner will be privy to how money is being spent on the project and to the amount of the fee received by the design-builder. This can cause an owner to decide that there is little financial risk to the design-builder and thus prompt it to reduce the design-builder's fee. A lump-sum arrangement, on the other hand, enables the design-builder to benefit from wise decisions by, for example, keeping the cost savings from beneficial subcontractor buyouts and productivity increases.

2. Risks are associated with cost-plus contracting, particularly if budgets— and hence the owner's expectations—are not achieved. In these situations, the owner may attempt to hold the design-builder responsible for some or all of the difference between the budgeted and actual costs.

The sections that follow review these, and other, issues unique to cost-plus contracting.

13.3.1 Current standard forms

The lump-sum design-build approach is handled fairly consistently by each of the industry standard forms. There are, however, structural differences in how the various design-build organizational sponsors address cost-plus contracting. The AGC and EJCDC families of design-build contracts include relatively well-developed provisions governing cost-plus/GMP agreements between the owner and the design-builder. EJCDC defines both reimbursable and nonreimbursable costs.[13] AGC 410 defines reimbursable costs, but does not identify non-reimbursable costs.[14] AIA, on the other hand, uses general rather than specific language regarding payment and is not user-friendly for cost-reimbursable relationships. Thus, if the AIA Document A191 is used on a cost-plus arrangement, it will need to be modified.[15]

---

[12]*DBIA Contracting Guide,* Section 5.0.

[13]EJCDC 1910-40, Article 10.

[14]AGC 410, Article 8.

[15]It should be noted that AIA Document A491 (the *Design-Builder/Contractor Agreement*) specifically enables the design-builder to contract with a contractor on a cost-plus basis, using a format similar to AIA Document A111, the cost-plus contract applicable to a traditionally delivered project.

13.3.2 Defining reimbursable and nonreimbursable costs

The primary challenge in drafting and administering a cost-plus contract is to have a clear definition of those costs that are reimbursable by the owner and those that are to be assumed by the design-builder (as part of its fee or at its risk). Section 6.3 of DBIA Contract Document 530 identifies the following as reimbursable expenses:

6.3.1. Wages of direct employees of Design-Builder performing the Work at the Site or, with Owner's agreement, at locations off the Site, provided, however, that the costs for those employees of Design-Builder performing design services shall be calculated on the basis of prevailing market rates for design professionals performing such services or, if applicable, those rates set forth in an exhibit to this Agreement.

6.3.2. Wages or salaries of Design-Builder's supervisory and administrative personnel engaged in the performance of the Work and who are located at the Site or working off-Site to assist in the production or transportation of material and equipment necessary for the Work.

6.3.3. Wages or salaries of Design-Builder's personnel stationed at Design-Builder's principal or branch offices and performing the following functions. The reimbursable costs of personnel stationed at Design-Builder's principal or branch offices shall include a _____ percent (_____%) markup to compensate Design-Builder for the Project-related overhead associated with such personnel [Insert the names, job description or job title of personnel.]

6.3.4. Costs incurred by Design-Builder for employee benefits, premiums, taxes, insurance, contributions and assessments required by law, collective bargaining agreements, or which are customarily paid by Design-Builder, to the extent such costs are based on wages and salaries paid to employees of Design-Builder covered under Sections 6.3.1 through 6.3.3 hereof.

6.3.5. The reasonable portion of the cost of travel, accommodations and meals for Design-Builder's personnel necessarily and directly incurred in connection with the performance of the Work.

6.3.6. Payments properly made by Design-Builder to Subcontractors and Design Consultants for performance of portions of the Work, including any insurance and bond premiums incurred by Subcontractors and Design Consultants.

6.3.7. Costs incurred by Design-Builder in repairing or correcting defective, damaged or nonconforming Work, provided that such defective, damaged or nonconforming Work was beyond the reasonable control of Design-Builder, or caused by the ordinary mistakes or inadvertence, and not the negligence, of Design-Builder or those working by or through Design-Builder. If the costs associated with such defective, damaged or nonconforming Work are recoverable from insurance, Subcontractors or Design Consultants, Design-Builder shall exercise best efforts to obtain recovery from the appropriate source and credit Owner if recovery is obtained.

6.3.8. Costs, including transportation, inspection, testing, storage and handling, of materials, equipment and supplies incorporated or reasonably used in completing the Work.

6.3.9. Costs less salvage value of materials, supplies, temporary facilities, machinery, equipment and hand tools not customarily owned by the workers that are not fully consumed in the performance of the Work and which remain the property of Design-Builder, including the costs of transporting, inspecting, testing, handling, installing, maintaining, dismantling and removing such items.

6.3.10. Costs of removal of debris and waste from the Site.

6.3.11. The reasonable costs and expenses incurred in establishing, operating and demobilizing the Site office, including the cost of facsimile transmissions, long-distance telephone calls, postage and express delivery charges, telephone service, photocopying and reasonable petty cash expenses.

6.3.12. Rental charges and the costs of transportation, installation, minor repairs and replacements, dismantling and removal of temporary facilities, machinery, equipment and hand tools not customarily owned by the workers, which are provided by Design-Builder at the Site, whether rented from Design-Builder or others, and incurred in the performance of the Work.

6.3.13. Premiums for insurance and bonds required by this Agreement for the performance of the Work.

6.3.14. All fuel and utility costs incurred in the performance of the Work.

6.3.15. Sales, use or similar taxes, tariffs or duties incurred in the performance of the Work.

6.3.16. Legal costs, court costs and costs of meditation and arbitration reasonably arising from Design-Builder's performance of the Work, provided such costs do not arise from disputes between Owner and Design-Builder.

6.3.17. Costs for permits, royalties, licenses, tests and inspections incurred by Design-Builder as a requirement of the Contract Documents.

6.3.18. The cost of defending suits or claims for infringement of patent rights arising from the use of a particular design, process, or product required by Owner, paying legal judgments against Design-Builder resulting from such suits or claims, and paying settlements made with Owner's consent.

6.3.19. Deposits which are lost, except to the extent caused by Design-Builder's negligence.

6.3.20. Costs incurred in preventing damage, injury or loss in case of an emergency affecting the safety of persons and property.

6.3.21. Other costs reasonably and properly incurred in the performance of the Work to the extent approved in writing by Owner.[16]

This list is similar, but not identical, to the list of reimbursable costs set forth in AGC 410[17] and EJCDC 1910-40.[18]

---

[16]DBIA Document 530, *Standard Form of Agreement between Owner and Design-Builder—Cost-Plus Fee with an Option for a Guaranteed Maximum Price*, Section 6.3.

[17]AGC 410, Article 8.

[18]EJCDC 1910-40, Paragraph 10.02.

Nonreimbursable expenses are also typically identified in the contract. These include the following:

- Salaries and other compensation of the design-builder's personnel stationed at the design-builder' principal office
- Expenses of the design-builder's principal office and offices other than the site office
- General overhead expenses
- Capital costs
- Costs due to the negligence of the design-builder

Many costs, such as those direct costs incurred for labor, materials, subcontractors, and equipment, are not contentious. Experience has shown, however, that determining the responsibility for other costs can cause disagreement between the parties and should be considered more carefully. Some cost issues of particular note are addressed below. See also Fig. 13.2.

Payment for design services.   The method by which the design-builder is paid for design services can be confusing on a cost-plus arrangement. If the design-builder is using an architect-engineer (A/E) as a subconsultant, the costs for such work are generally based on the contract between the design-builder and the A/E. This is neither conceptually challenging nor contentious if the A/E's fees are reasonable and customary. If, however, the design services are performed by an employee of the design-builder, there can be confusion as to whether they should be paid in accordance with the design-builder's actual employee cost or on a predetermined rate schedule that is consistent with prevailing rates for such services.

Some design-builders use their in-house design capabilities as a marketing tool. They offer design services on a "loss leader" basis, charging either nominal rates or actual hourly rates plus a multiplier. In these situations, the amount of the multiplier and how it is to be applied must be given some serious attention by both the owner and the design-builder. Several factors should be considered in developing and evaluating design services multipliers. The multiplier can be applied to (1) a "bare" labor rate (actual salary of the employee); (2) total payroll cost of the employee, including payroll taxes and insurance; or (3) a predetermined wage rate which has nothing to do with the employee's actual salary but is used for all employees of a particular classification ("junior engineer," etc.). Such multipliers should reflect the competitive environment for the project and the conditions of the job.

Most owners and design-builders prefer to pay for design services by using fixed hourly rates, lump sums, or percentages of construction cost. These methods enable the design-builder to bill its design services in the same manner as an independent A/E. Additionally, they are easier to administer than the process of using actual hours and multipliers.

SHOULD THESE COSTS BE REIMBURSABLE?		
Cost	**Arguments Against Reimbursement**	**Arguments in Favor of Reimbursement**
Home office expense of personnel supporting the project but stationed at the home office	1. Home office expenses are not directly incurred by a single project, making it difficult to determine whether cost is properly incurred on a given project. 2. Most standard forms consider home office overhead as nonreimbursable and covered by the fee.	1. An increasing number of contractor organizations have their personnel work on multiple projects and not simply stationed at the jobsite. This reduces overall project costs by avoiding full-time salaries being charged to project. 2. Parties can agree at the time of contracting to a markup to cover home office expense, avoiding the need to evaluate actual costs.
Contractor-owned equipment	1. Some owners believe they should only pay for the actual, depreciated cost, of the equipment. 2. Design-builders can use their owned equipment as a separate profit center.	1. Design-builders should not be penalized for having made an investment in equipment, which has a large capital expense. 2. Owners should pay for the fair rental value of equipment, whether or not it is owned by the contractor. 3. Design-builders would be incentivized to rent equipment rather than use their own equipment.
Costs incurred to correct defective work	1. Owners should not have to pay for work done improperly. 2. Costs of corrective work should be absorbed by design-builder's fee.	1. Design-builders will not perform the work perfectly the first time (e.g., bent nails, misaligned formwork, etc.). The question should be whether the design-builder's mistakes exceeded industry standards. 2. If design-builders are not paid for any corrective work (whether or not they were negligent), it could force them to become too conservative, which may jeopardize the speed of the project.
Legal fees incurred on the project (other than those incurred against the owner)	Owner should not have to pay for a design-builder to litigate with its subcontractors.	This will encourage design-builders to spend the time to recover monies that will ultimately benefit the owner.

Figure 13.2

Costs due to fault or negligence. One of the most challenging questions in cost-plus contracting is who will pay for those costs that result from rework or corrective action. As noted in the *DBIA Contracting Guide,* it is unreasonable to expect that any construction project will be executed perfectly.[19] In fact, given the schedule pressures inherent in fast-track projects (whether delivered by design-build or otherwise), it may be more beneficial to the owner and the project to accelerate the work even though it may result in more corrective work.

The EJCDC approach to this issue is to exclude from reimbursement any "cost due to the negligence of DESIGN/BUILDER, and Subcontractor, or anyone directly or indirectly employed by any of them."[20] AGC Document 410 does not address this issue directly, leading to the conclusion that these costs are reimbursable regardless of the severity of the negligence. Although AIA Document A191 does not specifically address cost-plus contracting, AIA Document A491 (the *Design-Builder/Contractor Agreement*) contains language similar to EJCDC's, except that it discusses payment for costs arising from "fault or negligence."[21] This increases the design-builder's liability because, while "negligence" gives it freedom to make errors within the industry standard, the term "fault" can include any error of any magnitude.

As noted in the *DBIA Contracting Guide,* the design-builder will have to absorb through its fee the costs of performing rework, unless a reasonable accommodation is made. This places more risk on the design-builder than would be the case under a lump-sum contract, where it can have some contingency, subject to market competition, for such rework.[22] Responsibility for the rework costs may also force the design-builder to take an ultraconservative approach and avoid intelligent risk taking. Such would be the case if the design-builder were to defer placing a concrete slab because of a 20 percent chance of a thunderstorm rather than risk the possibility of replacing the slab in the event that rain prevented it from curing properly. This approach could adversely affect the owner by slowing down the fast-track pace of the project while, at the same time, leading to an increase in certain reimbursable costs, such as on-site project management.

As a result of these considerations, the DBIA contract form eliminates the "fault" standard and bases the rework cost standard on the design-builder's "ordinary mistakes or inadvertence, and not the negligence of Design-Builder."[23] Under this approach, if the design-builder were prudent, based on the overall project requirements and industry standards, in placing concrete despite the threat of a thunderstorm, the costs of remediating the slab should

---

[19]*DBIA Contracting Guide,* p. 10.

[20]Subparagraph 10.02(B)4, EJCDC 1910-40.

[21]*DBIA Contracting Guide,* p. 10.

[22]Idem.

[23]Section 6.3.7, DBIA Document 530, *Standard Form of Agreement between Owner and Design-Builder—Cost-Plus Fee with an Option for a Guaranteed Maximum Price,* 1998.

be reimbursable. If, on the other hand, the decision was a poor one (e.g., dark clouds overhead with audible thunder, visible lightning and rain starting to fall), the costs would likely not be reimbursable.

Personnel stationed at the home office. Neither EJCDC nor AIA standard forms allow reimbursement for the costs associated with project personnel stationed at offices other than the field office.[24] This restriction does not reflect the realities of modern contracting, since it is often much more efficient and cost-effective to use home office personnel—such as accountants, schedulers, and safety managers—on an as-needed basis for a project rather than stationing such personnel at the job site. DBIA and AGC take a different approach, allowing reimbursement for salaries and associated benefits of specific employees from the principal or branch office performing specific functions. This allows the owner and design-builder to specifically discuss the issue and agree on what personnel is appropriately required and reimbursable for the project, without establishing a preconceived notion that these expenses should be absorbed by the design-builder's fee.

DBIA's standard contract also includes a clause which reimburses the design-builder for the reasonable costs of supporting those project personnel who are directly reimbursable and stationed at the home office. DBIA's clause states as follows:

> The reimbursable costs of personnel stationed at Design-Builder's principal or branch offices shall include a _____ percent (_____%) markup to compensate Design-Builder for the Project-related overhead associated with such personnel.

DBIA's clause is intended to cover such employee-related costs as rent, utilities, telephone, secretarial services, and other miscellaneous costs which cannot be readily segregated on a per-employee basis.

Wage markups. Burden and fringe benefit rates to be charged on the base labor-hours can be a problem on some projects. Unless the design-builder is unionized, virtually every design-builder will take a slightly different approach to defining its burden markups on labor. Some will include vacation time in this general burden, while others may charge it directly to the job on which the individual is working. Some design-builders may have very generous fringe benefits, while others may not. Some may be self-insured. All of these issues affect this markup. At the outset, the owner should determine what burden rate the design-builder proposes and the basis for its calculation. Ideally, the parties should do this in advance of contracting and directly specify the burden rate in the contract.

Costs necessarily incurred in proper performance of work. The entire category of reimbursable costs is subject to the caveat that the costs be "necessarily

---

[24]AIA Document A491, Subparagraphs 7.1.1–7.1.3; EJCDC 1910-40, Subparagraphs 10.02 (B)1 and (B)2.

incurred" in the "proper performance" of the work. Although there is little case law that addresses these terms, it is clear that they are intended to place some limits on the extent to which a design-builder can recover its costs on a cost-plus contract. For example, if costs are inflated as a result of mismanagement by the design-builder, the owner may be able to argue that they are the financial responsibility of the design-builder. Likewise, the owner could claim that any significant overruns in costs were caused by the design-builder's failure to "properly perform" the work and thus use this as a basis for not paying the design-builder.

13.3.3   Project budgets

Although cost-plus contracts do not place a ceiling on the actual costs that the design-builder will be paid for performing the work, the owner will undoubtedly expect the design-builder to perform its work to meet the owner's budget expectations. The design-builder should be candid and honest with the owner about the budget, and avoid simply telling the owner what it wants to hear, as this can lead to unfulfilled expectations and problems later. The parties should also agree on a process for notifying the owner when a budget or target benchmark is about to be reached. This will enable the owner to, among other things, make decisions about allocating additional funding or restructuring the project. The budget should be updated at regular intervals as the design is advanced and firmer prices from the trades are obtained.

Some owners may attempt to base the design-builder's fee on how successful the design-builder is in achieving the project's budgetary goals. This "carrot and stick" approach can be beneficial to both parties as it provides an incentive to the design-builder to perform well. In fact, it could enable the design-builder to substantially increase its potential fee. If the parties use this approach, the design-builder should be careful to (1) establish a clear record as to the assumptions for the baseline budget, (2) document the reasons for any overruns, and (3) ensure that the owner is aware of and agrees with both the baseline budget assumptions and overrun causation.

Finally, the design-builder should remember that if its cost estimating on the project does not accurately predict the actual costs to be incurred by the owner, the owner may look for recourse against the design-builder regardless of what is stated in the contract. The owner's argument will be based on its reliance on the design-builder's expertise. This could result in claims for an adjustment to the design-builder's fee or even damages over and above the fee. To mitigate against this type of risk, the design-builder may want to consider including a limitation of liability clause in the contract.

13.3.4   Staffing the project

Because the owner will pay the design-builder for all its reimbursable costs to perform the work, it is important for the owner to understand how its money will be spent. This is particularly true with regard to the staffing of the project, which can both dictate the success of and be one of the major costs generated on the project. Owners may be concerned that the design-builder will use the cost-plus contract to cover the cost of project management personnel who are not needed on the current project but are idle because of the lack of other projects to which they could be assigned. By the same token, design-builders have a legitimate fear that the owner will use the cost-plus vehicle to determine who the design-builder uses on the project, which may impact the ability of the design-builder to successfully execute it. To address these legitimate, and often competing, concerns, the owner and design-builder on a cost-plus arrangement should identify together the best way to staff the project from a cost and personnel qualification standpoint.

13.3.5   Control over subcontractors

The cost-plus arrangement enables the owner to be more actively involved in the selection of trade subcontractors. This involvement may be as little as having "veto power" over which subcontractors are used. It can also be as extensive as working with the design-builder to obtain a shortlist of qualified subcontractors, participating in the bid evaluation meetings with the design-builder's team, and conducting final negotiations with the apparently successful subcontractor.

Regardless of the owner's role, it is essential that the design-builder be given control over the selection and management of the project's subcontractors. In particular, the owner should not require the design-builder to use a subcontractor with whom the design-builder does not feel comfortable. These "forced marriages" not only can potentially impact the design-builder's ability to meet its budget, time, and quality commitments to the owner but may also lead the subcontractor to assume that it has direct access to the owner. This may create arguments that the subcontractor is in a de facto contractual relationship with the owner, which is seldom a desirable situation for an owner, because it increases the financial exposure and risk to the owner.

13.3.6   Payment processes

Unlike a lump-sum contract, where payments are based on percentages of completion, a cost-plus contract payment process is more complex. The design-builder's internal costs, such as general conditions expenses, will be paid as the costs are incurred and billed by the design-builder. Payments due to subcontractors—presuming that the subcontractor is working on a lump-sum contract—will be made on a percentage of completion basis. As a result of this hybrid process, the owner typically receives substantial documentation from the design-builder with each progress payment request. The owner and the design-builder must reach an agreement early in the project regarding what will be required for payment approval to ensure that the design-builder structures its billing program accordingly.

13.4   Cost-Plus with a Guaranteed Maximum
Price Contract

Although some design-build projects are performed on a straight cost-plus basis, most owners have finite resources to commit to a project. Thus, it is more common for the design-builder to ultimately provide the owner with a GMP for the work. The GMP contract is essentially a hybrid contract combining the cost reimbursement features of a cost-plus contract with the cost certainty of a lump-sum contract. The owner benefits by paying only the actual reimbursable costs of the work for the design-builder's performance and by knowing that its project will not exceed a preestablished price. This preestablished price will, however, be adjusted for changes made by the owner or for the effect of risks that the owner has contractually assumed. The GMP also offers both the owner and the design-builder the opportunity to realize savings on the project in the event the actual costs are less than the GMP.

The issues addressed in Sec. 13.3, particularly relative to reimbursable costs, are applicable to the GMP contract. In addition, several unique issues should be considered by both the owner and the design-builder. These issues are discussed below and summarized in Fig. 13.3.

13.4.1   Establishing the GMP

As noted in the *DBIA Contracting Guide,* the GMP should be established when the owner's program is sufficiently defined to make the GMP number realistic and meaningful.[25] Setting it too early does not give either the owner or the design-builder a reasonable opportunity to define the scope or evaluate the project risk and may result in the owner receiving an unrealistic price. On the other hand, setting it too late may not achieve the owner's objective of having an early price guarantee on the project to enable it to make a go/no-go decision.

Establishing a GMP is similar to other areas of the design-build process in that its success depends on the willingness of the owner to balance its needs and restraints (i.e., owner financing requiring an early GMP) with those of the design-builder and vice versa and to address them in the contracting process. For example, the parties may be able to agree on a "target price" in accordance with very preliminary programmatic requirements of the owner. As the design evolves, this price could be refined to result in a final GMP.

13.4.2   Savings

One benefit to the owner in using a GMP is the possibility that, with good management by the design-builder and timely support from the owner, the actual costs of and fee for the work may be less than the GMP. This creates a savings

---

[25]*DBIA Contracting Guide,* p. 11.

SPECIAL ISSUES ASSOCIATED WITH GUARANTEED MAXIMUM PRICE CONTRACTS	
Establishing the GMP	1. Should be established when owner's program is well defined. 2. Setting too early may result in an unrealistic price. 3. Setting too late may eliminate owner's goal to have an early price guarantee.
Savings (difference between actual cost-plus fee and GMP)	1. Owner has the right to all savings unless contract says otherwise. 2. Need to define what costs will not be subject to savings (e.g., design-builder's contingency). 3. Contract should creatively evaluate how to share savings given nature of project. For example, perhaps owner should receive first portion of savings if GMP is developed early and contains substantial contingency pending refinement of design. 4. Need to determine any ceilings on savings.
Design-Builder's Contingency	1. Need to determine how much contingency will be added into the GMP and where it will be shown (e.g., separate line item or interspersed throughout line items) 2. Decide on owner's right, if any, to approve draws on contingency.
Changes in Scope	1. Determine effect of changes on fee and general conditions allowance. 2. Consider having specific markups in the design-builder's contract with owner that will flow-down to subcontractors.

Figure 13.3

pool that should benefit both the design-builder and the owner.[26] If the contract does not address this situation, the owner will retain any and all project savings. However, most GMP contracts specify that the savings be shared, thus providing the design-builder with an incentive to save costs. In fact, creativity in the savings clauses can provide the parties with flexibility in addressing the fees paid for services.

Although there are numerous ways to share savings, some of the most common include

- *Single percentage split between the owner and design-builder.* Under this approach, the amount of the percentage (e.g., 60% owner, 40% design-builder) should be determined by how realistic the GMP was and the manner in which the fee was developed.

---

[26]See J. D. Hollyday and K. I. Levin, "The Design-Build Agreement," in *Design-Build Contracting Formbook,* Wiley Law Publications, New York, 1997, Chapter 5.

- *Owner receives first portion of savings.* If the GMP is inflated because it was developed prior to the refining of plans and specifications, it may be appropriate for the owner to have the first right to the savings—taking $500,000, $1,000,000, or any other amount of the savings before the remainder of the savings is shared on a percentage basis.

- *Design-builder receives first portion of savings.* If the fee is small, it may be appropriate for the design-builder, rather than the owner, to receive the first portion of the savings.

- *Variable percentages based on levels of savings.* The parties establish allocation percentages for several potential levels of savings. For example, if the savings is $500,000, the owner will receive 50 percent and the design-builder, 50 percent. If the savings exceed $500,000, the owner will receive 75 percent and the design-builder 25 percent.

The parties should also consider when the savings will be shared and how to address reimbursable costs (such as warranty expenses) that arise after the savings have been distributed.

13.4.3 Contingency

As noted earlier, the limitations in a cost-plus design-build contract may make it prudent for the parties to agree on a contingency line item within the GMP. This will be used by the design-builder in the event that a cost is incurred that is not addressed in a specific line item and does not qualify for an increase in the contract price. The *DBIA Contracting Guide* provides the following as examples of items that fall into the contingency category[27]:

- Trade buyouts which are not as favorable as projected
- Rework
- Trade subcontractor default
- Scope of work which falls into the cracks between trade contractors
- Labor disputes
- Overtime or acceleration costs (including impact costs) that the design-builder may choose to implement that would not be reimbursable through a change order increasing the contract price
- Nonnegligent design errors or omissions

Note, however, that the design-builder's contingency should not be used by the owner for risks that the owner is contractually assuming, such as changes in the scope of work and differing site conditions.

---

[27]*DBIA Contracting Guide,* p. 13.

13.4.4   Adjustments to the GMP

As with lump-sum contracts, the typical GMP agreement provides for adjustments to the GMP to reflect conditions that occur after further refinement of the design of the project or that unexpectedly arise as a result of the actual construction of the work. In deciding when adjustments are appropriate to a GMP on a design-build project, it must be determined whether the disputed work is extra or simply a refinement of the original scope of work. To help avoid disputes over this determination, the parties should identify at the time the GMP is provided the design, scheduling, and other assumptions that have been made, as well as the areas where the work is yet to be meaningfully developed. This approach is specifically taken in AGC 410, which calls for a GMP proposal that includes "a list of the assumptions and clarifications made by the Contractor in the preparation of the GMP proposal to supplement the information contained in the drawings and specifications."[28]

13.4.5   Line-item guarantees

The purpose of a GMP contract is to give the owner the benefit of an overall cap to the project cost. Some owners attempt to convert this into a contract that places ceilings on individual line items. This is typically done through contract language that precludes the design-builder from shifting underruns in one cost category to fund overruns in another—notwithstanding that the overall GMP remains the same. This approach is inappropriate for several reasons. Except in unusual situations, few owners care what happens to the costs within the GMP and whether there are savings or overruns in an individual line item. Additionally, bidding and estimating practices are not so precise that the design-builder can be absolutely certain at the time the GMP is prepared what the actual costs will be for each line item.

Although few owners persist in attempting to maintain line-item guarantees within trade categories, some remain concerned about the shifting of costs between the general conditions (the design-builder's costs to manage the process) and trade categories. They believe that the general conditions costs can be used by the design-builder to defray the costs of excess personnel and otherwise unabsorbed overhead—which serves to reduce savings. Likewise, if the design-builder was selected competitively on the basis of a fee and general conditions proposal before the GMP was developed, the owner will want to preserve this bargain in the contract. If an owner insists on having a general conditions cap within the GMP, the the design-builder's cost to manage the process should attempt to make the contingency funds available to pay for legitimate general conditions expenses that exceed the budgeted amount.

---

[28]AGC 410, Subparagraph 3.2.4.3.

13.4.6   Extra services

Two issues are associated with additional work on a GMP contract. First, the parties should ensure that they have a meeting of the minds on whether the design-builder will perform services that relate to the development of the project. These services may include (1) preparation of a project feasibility study, (2) building site investigations, (3) subsurface investigations required by differing site conditions, (4) consultation with the owner regarding damaged or destroyed work performed by the owner's other contractors, and (5) services related to tenant space design and buildout. Design-builders who do not have the expertise to perform these services, or are concerned about the liability associated with them, should make it clear in the contract that these services are not their responsibility and, if performed at all, will be undertaken by the owner. If the design-builder is amenable to performing these services, the contract should make it clear as to whether the costs are within the GMP and, if so, whether they are part of the fee or separately reimbursable.

Second, the GMP contract should identify how the parties will be compensated for the cost or time impacts of additional services. As discussed in Sec. 13.2.3 on lump-sum contracts, many GMP design-build contracts limit the profit and overhead to a predetermined rate for extra work, often stated as a percentage of the cost of the work. Some owners will look for a "free zone" of changes, whereby there will be no additional general conditions or fee associated with any changes on the work—or changes up to a certain monetary level.

13.4.7   Retainage

The treatment of *retainage* under a GMP contract is similar to that under a lump-sum contract. A percentage is established for each progress payment and the owner has the discretion to modify this by having (1) no further retention withheld after a designated percentage (e.g., 50 percent) of the work is complete, provided the work is proceeding satisfactorily; and (2) retention reduced at substantial completion to a multiple of the outstanding punchlist work. The nature of a GMP relationship also permits the parties to agree to a release of retention to early trades, such as sitework and foundations contractors, since the amount and terms of these subcontracts will be known by the owner.

13.5   Unit Price Contracts

Unit prices have a long history of use in the domestic and international construction industry. Under a unit price method of payment, the parties agree on a specific rate for units of materials, time, or completed construction items. The total compensation is dependent on the number of units used for the project. Unit prices are often employed on design-build projects when

- The owner wants to use the design-builder's expertise to develop the design and then pay for construction on a unit basis.

- The design-build project is in a sector of the construction industry accustomed to dealing with unit prices, such as environmental remediation, tenant improvements, and road projects.

- The actual quantities of materials or construction cannot be accurately determined at the time of contracting.

Unit prices are typically used as a complement to lump-sum and cost-plus contracts, with compensation for specific portions of the work being defined pursuant to the unit price.

Except for EJCDC, the current sponsors of standard form design-build contracts do not address unit price contracts. EJCDC forms are extensively used on public-works projects subject to competitive bidding, where unit prices have great practical application—particularly in the transportation and wastewater industry sectors. The EJCDC provision establishes mechanisms for

- Identifying the areas of work subject to unit prices

- Providing estimated quantities for such unit prices

- Reminding the design-builder that the estimated quantities are not guaranteed[29]

It also provides a means for adjusting the unit price if quantities are exceeded.[30]

Regardless of whether the entire contract or just portions of it are performed on a unit price basis, owners and design-builders should consider several issues that are unique to this pricing method. See Fig. 13.4.

13.5.1 Specifying the units

Most unit price contracts are based on units of materials (such as number of light fixtures or feet of pipe), time (such as hours of design services), or completed construction items (such as linear feet of lined tunnel). Two problems commonly arise relating to the specification of the units.

First, there can be confusion as to precisely what is covered by the unit price. This can be particularly troublesome on items that are components of or accessories to the completed work—such as pipe supports and hangers, excavation, and rebar. These components, accessories, and work activities are sometimes identified within the contract with their own unit prices to provide for a convenient means of pricing changes. However, this may create confusion as to what the trades expect to be paid.

A contract, for example, may specify a unit price for each linear foot of pipe placed, as well as unit prices for excavation, backfill, and bedding materials,

---

[29]EJCDC 1910-40-A, Subparagraph 4.01.B.

[30]EJCDC 1910-40, Subparagraph 10.04.

SPECIAL ISSUES ASSOCIATED WITH UNIT PRICE CONTRACTS		
Topic	**Problem**	**Proposed Solution**
What is covered by the unit price?	There can be confusion over the scope of a unit price (e.g., does unit price for underground pipe include excavation, bedding, backfilling, etc.?).	Clearly define all that is included and excluded from a unit price in the line item definitions.
What happens if there is a major change in estimated quantities?	Unit prices are developed based, in part, on the estimated quantities of the unit. Major changes in the quantities can disadvantage the owner or design-builder, as the case may be.	Include a Variations in Estimated Quantities clause in the contract, which will allow adjustments to be made to the contract unit price if quantities are changed by certain percentage (e.g., 25% over or under).
Are the unit prices automatically to be applied for changes in the work?	If the assumptions underlying the unit prices change, the prices established in the Contract may no longer be appropriate.	Owners must be open to possibility of a renegotiation of unit prices based on changed situations.
How are the quantities claimed by the design-builder to be monitored?	The concept of unit prices anticipates that the design-builder actually provided the quantities claimed. Conflicts can develop over whether the accuracy of quantities (e.g., how many cubic yards of rock were actually removed?).	Both parties should agree on a mechanism for verifying actual quantities. It is appropriate for both parties to sign off on these quantities in the field.

Figure 13.4

activities which are necessary for proper pipe placement. The trade contractor performing this scope may calculate each unit price, expecting it to apply to any pipe placed. The owner, on the other hand, may expect that the base unit price for pipe would take into account all of these required activities. The safest way to avoid problems is to precisely define what is included in each unit price and ensure that the definitions are not ambiguous. This is particularly true relative to overhead and profit on the work.

Second, problems can arise when unit prices are used as a substitute for payment milestones. Unit prices are intended to price variable quantities. Activities such as design, mobilization, permit acquisition, and procurement of specialized equipment are not appropriate for unit prices, since they are not measured in quantities. Therefore, it may be wise for the parties on a predominately unit price contract to have the balance of work paid for on a lump-sum or cost-plus basis.

13.5.2   Variations in quantities

The contract should forecast what happens if there is a variation between the estimated and actual quantities of work. If the contract is silent on this point, the risk of overruns and underruns falls to both the owner and design-builder. Because the price per unit is fixed, the design-builder bears the risk of an increase in the costs of performing the unit and receives the benefit of decreases in such costs. As a result, if the actual cost of the unit is far less than the contract unit price, the design-builder will receive a windfall if more units are required than planned. On the other hand, if the design-builder performs far fewer quantities of an item than expected, it may not be able to recover its expected overhead and profit attributable to such item.

To avoid this problem, contracts have long used "variation in estimated quantities" clauses. These clauses essentially call for an adjustment of the contract unit price if there is a variation of plus or minus 20 to 25 percent between actual and estimated quantities. This provides some assurance to both parties that if they are incorrect on either the estimated quantities or the price of the unit, there will be an equitable adjustment in the unit price.

13.5.3   Applicable unit prices

Even though unit prices may be established in the contract, it may not always be appropriate to use them as the work proceeds if the assumptions behind how they were determined change. For example, if unit prices for drywall installation were calculated on the assumption that the work would be performed on straight time, but problems with the project require overtime, then the contract unit prices would not adequately compensate the design-builder. Likewise, if excavation were to be done according to the assumption that shoring would not be needed, but differing site conditions did not allow this to take place, unit prices for excavation would have to be adjusted.

Design-builders should take appropriate steps to notify the owner of precisely what assumptions are being made in establishing unit prices and to document their positions if a problem arises. Likewise, owners should be sensitive to the possibility that changed situations may necessitate a renegotiation of unit prices.

13.6   Combining Payment Methodologies

Although each system described above can be used as the sole method of payment on a project, it is far more common for the systems to be combined to meet the needs of the owner and the unique aspects of the project. For example, on a high-tech project that will be advancing the state of the art, the owner who has a strong working relationship with a design-builder may directly select the design-builder without competition and then (*a*) use a cost-plus arrangement until the program is better developed and the preliminary design more certain, (*b*) shift to a GMP once the design is advanced and pricing more definite, and (*c*) shift to a lump sum when the design is final and the design-builder feels comfortable with providing more certainty to its price for the work.

The move to a lump-sum arrangement should be financially beneficial to the owner, since the price for a lump sum will likely be lower than the price for a GMP, given the accounting work associated with a GMP. Unit prices can be used at each step in the above-described process for discrete portions of the work.

13.7 General Terms and Conditions of Contract

The parties should remember that the business terms used in the contracting approach—whether lump-sum, cost-plus, GMP, or unit price—must be consistent with the general terms and conditions of the contract that will establish the overall rights and responsibilities of the parties. Particular attention should be paid to the payment and changes clauses of the general conditions, since these will establish much of the working mechanisms for implementing each of these payment methodologies.

Contracting Issues in Design-Build

14.1 Introduction

One key challenge to successfully executing any commercial venture is for the parties to give proper attention to their contracting arrangements. This is particularly true on design-build projects, since the participants may not be familiar with the process and there is little legal precedent defining the relationship among the parties.

It is well understood that the primary purpose of a contract is to define the rights and responsibilities of the parties to the contract. However, contracts in the construction industry also serve the purposes of

- Providing a uniform basis for competitive bids and/or proposals from those performing services on the project

- Assigning and managing risk among the parties to the contract

- Identifying the goals and expectations of the parties relative to the project and each other

- Establishing a project planning and management tool for the project

This last point is particularly important, since a well-drafted contract can become a handbook for performance on the project, addressing such important procedures as inspections, payments, and interpretation of the contract documents.

This chapter focuses on some of the most common issues associated with the contracts between the owner and design-builder, including (1) principles of risk allocation, (2) standard form of contracts available for design-build relationships, and (3) unique issues in the design-build contract.[1] The reader should also refer to Chaps. 13, 15, and 16, each of which addresses areas that are important to the contracting process.

14.2   Contractual Risk Allocation

The construction industry has long been recognized as risk-laden. Some of the risks associated with the construction process are fairly predictable or readily identifiable, while others may be totally unforeseen. Some risks (such as the nonperformance of subcontractors) are assumed by the contractor as part of its business. The assumption of other risks (such as unforeseen site conditions) can be the subject of extensive negotiation and debate between the owner and contractor as to who is in the best position to deal with the risk.

14.2.1   Risk allocation principles

Much attention has been devoted to the issue of risk in the construction process and how to most appropriately allocate such risk among the members of the project team. At one time, many public and private owners followed the philosophy that virtually all risks should be shifted to the contractor, on the theory that contractor acceptance of risk was the price of doing business and that the profit potentials justified the contractor assuming considerable risk.[2] This philosophy manifested itself in one-sided contracts that forced contractors into taking unrealistic gambles, improvident corner cutting, or making commitments that were not realistic. Coupled with the highly competitive nature of construction contracting, these unbalanced contracts tended to create disputes when an unexpected risk occurred.

Beginning in the late 1970s, the philosophy of shifting all risk to the contractor was challenged by many thoughtful and compelling industry studies. These studies concluded that (1) by shifting all risk unreasonably to contractors, owners were paying significantly more for the constructed project through increased bid prices, change order disputes and litigation costs[3]; and (2) contractual misallocation of risk was the leading cause of construction disputes in the United States, with parties spending more time positioning for the dispute than in getting the job done right, on time, and within the budget.[4]

---

[1]Many of the concepts set forth in this chapter are derived from the *Design-Build Contracting Guide,* DBIA Manual of Practice Document 510 (Nov. 1997), published by the Design-Build Institute of America, Washington, DC. This is sometimes referred to in this chapter as the *DBIA Contracting Guide.*

[2]See R. J. Smith, "Risk Identification & Allocation—Saving Money by Improving Contracts and Contracting Practices," *International Construction Law Review,* p. 40 (1995).

[3]See, e.g., Levitt, Ashley, and Logcher, "Allocating Risk and Incentive in Construction," *Journal of the Construction Division,* ASCE, **106**(CO3) (Sept. 1980); *Preventing and Resolving Construction Disputes,* Center for Public Resources, Inc., New York, 1991.

[4]*Preventing and Resolving Construction Disputes,* pp. 2–3.

Two studies were particularly instructive. The 1979 Conference on Construction Risks and Liability Sharing, sponsored by the American Society of Civil Engineers (ASCE), was the forum for several papers, which collectively concluded that

- Risks belong with those parties who are best able to evaluate, control, bear the cost, and benefit from the assumption of risks.
- Many risks and liabilities are best shared.
- Every risk has an associated and unavoidable cost which must be assumed somewhere in the process.[5]

In addition, the Construction Industry Institute (CII) published a report which found that

- The ideal contract—the one that will be most cost effective—is one that assigns each risk to the party that is best equipped to manage and minimize that risk, recognizing the unique circumstances of the project.[6]
- The typical situation of owners shifting risk to contractors, who, in turn, shift risk to subcontractors, will not result in an efficient allocation of risk. An equitable allocation of risk would assign each risk to the party best able to control it. Such an allocation minimizes the indirect costs associated with misaligned incentives, mistrust, and disputes. Because of the advantages and disadvantages associated with efficient and equitable allocations of risk, each project should be assessed individually to determine for each risk what allocation will reduce the project's total cost of risk.[7]
- Proper allocation of risk between parties involved in a construction project will reduce the overall cost to the project and promote a much more positive working relationship between the parties. This can be accomplished by assigning each risk addressed in the contract to the party that (1) has a comparative advantage in regard to risk-bearing ability and (2) has control over the risk.[8]

As the result of these studies, numerous public and private owners changed their contracting practices to reflect risk allocation philosophies consistent with these conclusions. The results have been positive, with many owners experiencing reduced claims, fewer adversarial relationships and reduced construction costs.[9]

---

[5]American Society of Civil Engineers, *Construction Risks and Liability Sharing,* Reston, VA, 1980, Volume II, p. 2.

[6]CII Impact Report, *Assessment of Construction Industry Project Management Practices and Performance,* April 1980, note 1, at 6.

[7]Idem, at 6.

[8] Idem, at 7.

[9]See Smith, "Risk Identification" (see fn. 2, above), endnote 1.

14.2.2   Risk allocation on design-build projects

While some owners believe that reasonable risk allocation may have application in a competitively bid, design-bid-build environment, this does not apply in design-build and that substantial risk should be shifted to the design-builder. This is an inappropriate way to handle the design-build contractual arrangement. Although a design-builder by virtue of the single point of responsibility feature of design-build, assumes more risk than does a design professional or general contractor, the design-builder is not in the best position to control or manage every conceivable risk that could occur on a project.

Risk allocation principles need to be considered in design-build as carefully as under other delivery structures to deal with such items as differing site conditions, hazardous materials, governmental interference, and force majeure events. Owners who attempt to avoid responsibility for all project risk are likely to experience not only higher project costs but also conflicts and disputes, eliminating one of the major advantages of the design-build delivery approach.[10]

To balance how risks should be allocated on a design-build project, the owner should carefully assess its goals and the unique attributes of the project. It should focus particularly on whether some of its goals conflict with others and will create contractual tensions. For example, if schedule is an absolute project driver, an owner could shift the risks associated with project delays to the design-builder in exchange for agreeing to pay the design-builder the costs associated with accelerating its performance to overcome the delays. Likewise, if the owner's primary goal is to develop a cutting-edge design that advances the state of the art, it should consider assuming some of the risk in the event that the design-builder, despite its good-faith efforts, fails to achieve the expected results.

An approach commonly used by owners in advance of establishing the contract is to develop a matrix that identifies potential risks and the party most capable of assuming the particular risk (Fig. 14.1). Once project risks are identified, the project team can use any of several approaches to allocating or transferring the risks.[11] The design-builder, for example, can assume the risk in the contract and adjust its price accordingly. The design-builder can attempt to transfer the risk to a third party who is best able to handle the risk, such as a subcontractor—although the design-builder will remain contractually obligated to the owner. Insurance or third-party guarantees can be used in dealing with the contingency created by the risk. The risk can also be retained by the owner, who would pay for the costs associated with the risk if it becomes a reality.

---

[10]M. C. Loulakis and O. J. Shean, *Risk Transference in Design-Build Contracting,* Construction Briefings, 2d series, 96-5, Federal Publications, April 1996, endnote 5, p. 2. See also R. J. Smith, "Allocation of Risk—The Case for Manageability," *DBIA Contracting Guide,* p. 2.

[11]See also *Owner's Guide to Saving Money by Risk Allocation,* Task Force of the Associated General Contractors of America and American Consulting Engineers Council, Washington, DC, 1990.

DEVELOPING A RISK ALLOCATION PROFILE	
1.	Identification of project risks
2.	Determination of which risks will be retained by owner
3.	Determination of which risks will be transferred to design-builder and the pricing ramification of this risk transfer
4.	Determination of which risks assumed by design-builder will be transferred to subcontractors and other third parties
5.	Development of insurance program to handle project risks

Figure 14.1

14.3   Industry Standard Form Contracts

Standard form contracts have long served an important function in the United States and international construction markets. The common purpose of these forms is to provide an economical and convenient way for parties to contract without having to resort to expensive lawyering to create a contract for each new project. However, the goals of the organizations sponsoring standard forms go far beyond convenience. In fact, many believe that as the terms of standard form contracts are used and accepted, they actually become a model for how the industry expects parties to relate to each other contractually.[12]

As design-build has grown in popularity, several organizations drafted families of design-build contracts. These include the Design-Build Institute of America (DBIA), American Institute of Architects (AIA), Associated General Contractors of America (AGC), and the Engineers Joint Contract Document Committee (EJCDC). The forms developed by each of these organizations are briefly reviewed below.[13] See also summary in Fig. 14.2.

14.3.1   DBIA documents

In 1997 DBIA published the *DBIA Contracting Guide* as a first step in the development of its own standard forms. The intended purpose of the *DBIA Contracting Guide* was to provide owners, design-builders, and other participants in the design-build process with guidance on how to most effectively

---

[12]This is readily apparent simply by examining how the role of the architect and engineer has changed since the late 1950s, as the standard forms attempted to minimize liability for job-site safety by eliminating the A/E's role in supervision and the A/E's right to stop the work.

[13]For a more comprehensive review of the standard form contracts in design-build, see the following publications: M. C. Loulakis and W. B. Fisher, "Comparison of the New Design-Build Contracts," *Construction Briefings,* 2d series, 95-5, Federal Publications, April 1995; M. H. McCallum, D. E. Ellickson, and H. G. Goldberg, "The 1996 Editions of AIA Design/Build Standard Form Agreements," *The Construction Lawyer* **16**(4) (Oct. 1996).

STANDARD FORM DESIGN-BUILD CONTRACT CHARACTERISTICS				
TOPIC	**AIA**	**EJCDC**	**AGC**	**DBIA**
Most recent edition	1997	2000	1999	1998
Lead design professional	Architect	Engineer	Either architect or engineer	Either architect or engineer
Procurement approach	Anticipates negotiated procurement process	Anticipates competitive procurement process	Anticipates both negotiated and competitive procurement	Anticipates both negotiated and competitive procurement
Pricing approach	Pricing not addressed in agreement between owner and design-builder	Both lump-sum and cost-plus with GMP option	Both lump-sum and cost-plus with GMP option	Both lump-sum and cost-plus with GMP option

Figure 14.2

address the critical contract issues between owners and design-builders.[14] In addition to providing specific commentary and suggested language for critical contracting issues, the *DBIA Contracting Guide* also compared how such issues were treated in the latest model design-build contracts published by AIA, AGC, and EJCDC.

14.3.2   AIA documents

AIA released its second generation of design-build contract forms in late 1996, publishing updated versions of the following three documents:

- AIA Document A191, *Standard Form of Agreements between Owner and Design / Builder*
- AIA Document B901, *Standard Form of Agreements between Design / Builder and Architect*
- AIA Document A491, *Standard Form of Agreements between Design / Builder and Contractor*

The AIA documents are distinguished by their use of a mandatory two-part contracting system. Part 1 of each form listed above deals with services for preliminary design and budgeting. The deliverable under Part 1 of AIA

---

[14]*DBIA Contracting Guide,* p. 1.

Document A191 is a proposal that provides the owner with price, time, and scope commitments, including a preliminary design and outline specifications. If the owner accepts the proposal, the parties execute Part 2 of AIA Document A191, which covers final design and construction. If the parties do not reach agreement, there is no further involvement by the design-builder on the project. Moreover, the owner is not permitted to use the work product developed during Part 1 if Part 2 is not executed.

It is common for AIA's contract forms (such as AIA Documents A101, A111, and A201) to be used on commercial projects that are designed by architects and constructed through an independent general contractor or at-risk construction manager. Because the AIA design-build family of contracts uses much of the same terminology as AIA's other contracts, those familiar with AIA principles may tend to automatically resort to using AIA Document A191 as their design-build contract of choice. Although AIA Document A191 *could* be the most appropriate form for the project, potential users should clearly understand the substantive content of this form and the situations for which its use is intended.

14.3.3 EJCDC documents

The EJCDC is a consortium of the National Society of Professional Engineers (NSPE), the American Consulting Engineers Council (ACEC), and the American Society of Civil Engineers (ASCE). EJCDC has prepared a variety of standard form contracts for projects where the engineer is the lead design professional. These contracts are extensively used by a variety of public and private owners on projects such as sewage treatment plants, roads, bridges, prisons, and cogeneration facilities. EJCDC's design-build family of documents, released in February 1995, is as follows:

- EJCDC 1910-40, *Standard General Conditions of the Contract between Owner and Design-Builder*
- EJCDC 1910-40-A, *Standard Form of Agreement between Owner and Design-Builder on the Basis of a Stipulated Price*
- EJCDC 1910-40-B, *Standard Form of Agreement between Owner and Design-Builder on the Basis of Cost-Plus*
- EJCDC 1910-41, *Standard Form of Subagreement between Design-Builder and Engineer for Design Professional Services*
- EJCDC 1910-48, *Standard Form of General Conditions of the Construction Subagreement between Design-Builder and Contractor*
- EJCDC 1910-48-A, *Standard Form of Construction Subagreement between Design-Builder and Contractor on the Basis of a Stipulated Price*
- EJCDC 1910-48-B, *Standard Form of Construction Subagreement between Design-Builder and Contractor on the Basis of Cost-Plus*

- EJCDC 1910-42, *Guide to Use of EJCDC Design/Build Documents*
- EJCDC 1910-43, *Standard Form of Agreement between Owner and Owner's Consultant*

The content of EJCDC's design-build contracts is quite similar to its standard forms for design-bid-build projects. The EJCDC drafters did this specifically to ensure that applicable concepts successfully used with EJCDC's design-bid-build documents would be retained. However, the language may be more detailed and comprehensive than required for some projects, creating some potential obstacles in fast-tracking and administering the project.

EJCDC's *Guide to Use of EJCDC's Design-Build Documents* states that the EJCDC documents can be used regardless of the level of completion of the design. The EJCDC Guide also notes that the owner should perform (on its own or through a consultant) certain tasks during a "study and report" phase, which is typically the first step taken by an engineer in a traditional design-bid-build project. The product from this phase is inserted into conceptual documents and provided to prospective design-builders in a request for proposals (RFP). The typically comprehensive nature of an EJCDC study-report phase effort suggests that the design furnished by the owner to the design-builder in the RFP may be fairly well developed. If this is the case, significant design discretion and cost/benefit optimization opportunities will be taken away from the design-builder. While the level of detail in the conceptual documents is always the owner's decision to make, DBIA believes that minimizing the design-builder's design discretion is generally not in the best interests of the owner, given that innovation and value engineering are strong benefits of the design-build process.

14.3.4 AGC documents

AGC published in late 1993, and released in the first quarter of 1994, a comprehensive family of design-build contract documents which includes the following documents:

- AGC 400, *Preliminary Design/Build Agreement between Owner and Contractor*
- AGC 410, *Standard Form of Design/Build Agreement and General Conditions between Owner and Contractor (Where the Basis of Payment Is the Actual Cost Plus a Fee, with a Guaranteed Maximum Price Option)*
- AGC 415, *Standard Form of Design/Build Agreement and General Conditions between Owner and Contractor (Where the Basis of Payment Is a Lump Sum)*
- AGC 420, *Standard Form of Agreement between Contractor and Architect/Engineer for Design/Build Projects*
- AGC 450, *Standard Form of Agreement between Design/Build Contractor and Subcontractor*

- AGC 460, *Standard Form of Agreement between Design/Build Contractor and Design/Build Subcontractor (Where the Subcontractor Provides a Guaranteed Maximum Price)*

The second edition of these forms was issued in early 2000. Unlike the AIA and EJCDC documents, AGC was not constrained to use similar contract language from its previous contracting forms in developing its design-build family of documents. This has both advantages and disadvantages. On the positive side, the forms contain fresh, relatively easy to understand language, rather than "legalese." The negative, however, is that the absence of familiar language means that one must actually read and carefully consider the content of each clause and determine how it might differ from expectations.

A strength of the AGC approach is its use of AGC 400 as an optional preliminary agreement to assist the owner in defining its program and determining budgets for price and time. To use this service, the owner can execute AGC 400 and then later contract through either AGC 410 (cost plus) or AGC 415 (lump sum). If the owner's program is well developed, or there is a competitive selection process, AGC 400 is not needed.

14.3.5 International standard form contracts

In addition to U.S. standard design-build contract forms, a variety of international standard form design-build contracts are available. Three such forms are briefly described here:

- *FIDIC Orange Book.* The full title of the *Orange Book* is *Conditions of Contract for Design-Build and Turnkey* (1996). FIDIC, an international federation of independent consulting engineers headquartered in Lausanne, Switzerland, is the leader in international standard form contracts, primarily because its forms are widely used in developing countries and are recognized by the World Bank. FIDIC's design-build form appears to be most appropriate for industrial, power, and process facilities.

- *JCT Standard Form of Building Contract with Contractor's Design, 1981 ed.* The JCT (Joint Contracts Tribunal for the Standard Form of Building Contract) has published documents which are in widespread use throughout the United Kingdom. This design-build form, which has been updated through a series of amendments since 1985, contains unique terminology and concepts.

- *Engineering Advancement Association of Japan (ENAA).* ENAA published in 1996 a new model form of contract intended for design-build power-plant construction. This ENAA contract form, entitled *ENAA Model Form International Contract for Power Plant Construction,* is often used on World Bank power projects constructed on a design-build basis.

Although there are standard form contracts in international design-build, it should be noted that the complexity of most projects, particularly in the infrastructure area, mandate that project specific contracts be drafted and negotiated.

14.3.6 Standard forms versus project-specific contracts

In traditional design-bid-build project delivery, standard form contracts have long served an important function—they are economical and can be used with little modification. They are viewed as promoting an equitable allocation of risks and responsibilities. Furthermore, because there are so many court decisions interpreting clauses in standard form contracts, participants on a project typically know in advance how a court is likely to view the meaning of certain clauses.

This environment is rarely present in design-build. Because each design-build project is unique—impacted by prior relationships among the parties, financing arrangements, and specific design requirements—there is often more than one means of appropriately achieving contracting goals. Parties should be careful about attempting to "cookie-cut" their design-build contracts or blindly following the model design-build contracts. Effective contracting is accomplished when the parties have given specific thought to their contracting goals and then tailored the contract to meet the unique needs of the project and the design-build team.

Notwithstanding the need to think through the contracting issues, current industry standard form design-build contracts can provide an excellent starting point for contract discussions. They identify many of the issues that need to be addressed in the design-construction process that owners and design-builders may not have considered. Additionally, as design-build becomes more widely used, these forms promote some important concepts about the most appropriate way to handle certain risks. See Fig. 14.3.

14.4 Correlating Design-Build Selection with the Contract

As noted in Chap. 8, numerous methods are available to an owner for selecting a design-builder.[15] These choices range from purely qualifications-based selection methods to purely price-based selection methods; the four most commonly used methods are

- *Direct selection.* The design-builder is chosen directly by the owner on the basis of qualifications, prior relationships, and other factors.
- *Competitive negotiation.* Several prospective design-builders compete with respect to a combination of factors, including qualifications and general cost criteria (fees, prospective general conditions, etc.).

---

[15]*DBIA Manual of Practice,* Document 101, Design-Build Institute of America, Washington, DC, 1996 (1st ed.).

FACTORS TO CONSIDER IN USING STANDARD FORMS	
Benefits	**Disadvantages**
Economical and efficient to use as a baseline	Each design-build project is unique and the contract relationships should not be "cookie-cut"
Provide excellent starting point for establishing the relationships	Standard design-build forms do not address many of the issues associated with certain construction industry sectors (e.g., transportation and power generation)
Promote equitable allocation of risks	Difficult risk issues, such as allocation of responsibility for design errors, are not addressed in most standard forms
Participants generally know how courts interpret standard form clauses	Design-build is a relatively new area and there is little case law as of yet addressing standard form interpretation

Figure 14.3

- *Cost / design competition.* The proposers are shortlisted according to qualifications and selected through an evaluation system after responding to an RFP.

- *Cost competition.* Proposers are shortlisted as above, but proposals are evaluated on a pass/fail basis after responding to an RFP. The low bidder that meets the technical requirements is awarded.

Because each of these choices entails a different level of relationship at the start of the design-build relationship, each envisions a different approach of contracting.

Owners who use the direct-selection method often rely on the design-builder's expertise to develop the overall program and scope of work. Under this arrangement, it is not unusual for the initial program definition services to be performed on the basis of an oral understanding or simple letter agreement, with a more complete contract developed once the project is better defined. On the other hand, when an owner plans to use a competitive selection process and fully develops its program and scope documents before involvement of the design-builder, the contracting arrangements will be more formal and complete from the beginning of the relationship.

One should carefully evaluate any proposed standard form to determine whether they are appropriate for the selection process chosen by the owner. All major standard forms currently in the industry are not universally applicable for all types of procurement techniques.

- AIA's two-part design-build contracting format works well when the design-builder will be retained on a qualifications basis, either directly or through negotiations. It requires major modifications if the owner is interested in selecting the design-builder on the basis of a competitive RFP process.

- EJCDC's contracting approach is well suited for a competitive design-build selection, but not as accommodating to the owner who prefers to engage a design-builder directly or through competitive negotiation. This can be attributed to the fact that EJCDC documents are widely used in the public sector, where some competition is generally needed.

- Both DBIA's and AGC's contracting approaches offer more flexibility than do those of either AIA or EJCDC. Each has a preliminary form that is similar in purpose to the AIA Part 1 Agreement and enables the parties to engage in direct negotiations. However, unlike AIA, both DBIA and AGC forms allow the parties to also contract on a competitive basis on either a lump-sum or cost-plus approach. The primary difference between DBIA and AGC, relative to the contracting process, is the manner in which they treat ownership of documents, which is discussed in Sec. 14.6.

Given the wide range of standard forms available, an owner should be cautious about simply using a form without careful consideration of whether it matches its overall procurement plan for the project. Both owners and design-builders must also be sensitive as to how issues such as ownership of documents and termination rights established in the contract documents affect the selection method used on the project.

14.5   Scope of Work

Establishing the scope of work on a design-bid-build project is fairly simple. The plans and specifications are represented by the owner to be 100 percent complete, and the contractor's price is based on such documents. The only issue that typically arises as to the scope of work in this situation is whether the contract documents "reasonably inferred" that something was necessary, even though it was not specifically called out in the documents.

Resolving this issue in the context of design-build is not as simple. Unlike the simple reference to the completed plans and specifications, the scope of work on a design-build project may be defined by the RFP, the design-builder's proposal, and several iterations of plans and specifications delivered after contract execution. As a result, there is potential for confusion as to what the parties have actually agreed on relative to scope.

14.5.1   Differences between the RFP and proposal

Scope issues will often arise in the context of differences between what the owner asked for in its RFP and the proposal received from the design-builder. The resolution of this issue is partly dependent on the manner in which the design-builder is selected.

It is less likely to be a problem when a direct design-build selection or a competitive negotiation process is used. In these cases, the owner and the design-builder will have had some dialogue about the owner's needs and the proposed solutions, which minimize the likelihood of anything significant having been misunderstood. Moreover, the parties are free to agree on a "conformed" set of project technical requirements that integrates the owner's program and the design-builder's proposal (if any) into a single scope document. This will enable the parties to have an appropriate benchmark as to performance obligations and should result in the elimination of ambiguities between these two documents.[16]

These practical remedies are not readily available in a competitive selection environment. In these cases, there will be little informal discussions between the owner and the design-builder, and the design-builder will be required to pay careful attention to precisely and literally what is being identified in the RFP. Scope disputes are particularly prone to arise when the owner uses detailed design specifications to define its program in the RFP and seeks lump-sum or GMP pricing from the shortlisted design-builders. During this process, ambiguities can develop between the owner's program and proposals which may not otherwise be discovered until postcontract award or execution. This might also occur during the iterative process of design development in a GMP context, where pricing assumptions are being based on updated plans and specifications. The problem is exacerbated when the owner's program references a laundry list of standards or guidelines that may or may not have real application to the project.

14.5.2 Differences among the contract documents

Because much of the design-builder's work will be defined through design documents developed after contract execution, there is an issue as to what happens if there is a difference between such documents and those documents that were developed prior to execution—such as the RFP or proposal. This can arise if the design development documents and construction documents change a level of detail or other item specified within the RFP and this change is not noticed by the owner until after construction.

14.5.3 Interpretation of permit requirements

Some scope of work will be specified by simply requiring the design-builder to obtain and comply with all permits and "laws and regulations" applicable to the project. While this language does not seem objectionable, it should be recognized that many permit requirements, such as those relating to fire marshal, state hospital codes, and historic codes, are "interpretive" and can be enforced

---

[16]Note, however, that even with a conformed set of contract documents, there is still the chance that something was missed that will create a problem as to intended scope.

subjectively and capriciously. Given this, it may be appropriate for the parties to specify the assumptions being used for these "interpretive" codes and permits in the RFP or proposal so as to establish an appropriate baseline for pricing and time purposes.

14.5.4 Contract treatment

Several contract clauses affect the definition of the scope of work. These typically include those clauses which (1) define the contract documents, (2) specify the "order of precedence" for the contract documents, and (3) establish the GMP.

The standard forms each treat the issue differently. Part 1 of AIA Document A191 states that if a design-builder deviates from the owner's program, the deviation should be disclosed in the proposal:

> The Design/Builder shall submit to the Owner a Proposal, including the completed Preliminary Design Documents, a statement of the proposed contract sum, and a proposed schedule for completion of the Project....Deviations from the Owner's program shall be disclosed in the Proposal. If the Proposal is accepted by the Owner, the parties shall then execute the Part 2 Agreement.[17]

There is no guidance as to what is meant by the expression "disclosed in the proposal." How far does the design-builder have to go to "disclose" its deviation? There is also nothing which expressly states what happens if there is an ambiguity between the owner's program and the proposal. Moreover, because Part 2 does not make the owner's program a contract document, it is questionable as to what happens if the final design or construction is different from what the owner specified in its program. It appears that the owner would be without recourse, given that the contractual significance of the program has disappeared during Part 2.

AGC handles this issue by developing a hierarchy of contract documents which establishes the following order of precedence:

- Change orders and written amendments signed by both owner and design-builder

- The agreement

- The most current design documents (such as schematic, design developments and construction documents) approved by the owner

- Information provided by the owner (permits, approvals, etc.)

- Contract documents in existence at the time the Agreement was executed

- The owner's program

This philosophy essentially establishes the principle that the latest contract document will govern over earlier contract documents.[18]

---

[17]Subparagraph 1.3.5, Part 1 of AIA Document A191 (1996 ed.).

[18]Paragraph 2.4, AGC 410.

EJCDC's approach is a mixture of both AIA and AGC philosophies. It specifies both the owner's RFP documents and the design-builder's proposal to be contract documents. However, there is no order of precedence clause establishing a hierarchy among the documents. In fact, because of the following language, all documents are given equal weight:

> It is the intent of the Contract Documents to describe a functionally complete Project (or part thereof) to be designed and constructed in accordance with the Contract Documents. Any Work, materials or equipment that may reasonably be inferred from the Contract Documents or from prevailing customary trade usage as being required to produce the intended result will be furnished and performed whether or not specifically called for.[19]

As a result, there is some difficulty in assessing what to do in the event of a conflict between the proposal and owner's program operating under the EJCDC format.

The *DBIA Contracting Guide,* after evaluating the language shown above and the policy behind establishment of the scope of work, recommends that (1) an order of precedence clause be used to establish a hierarchy of documents and (2) the owner's program take precedence over the design-builder's proposal unless the proposal calls attention to deviations from the program. DBIA's suggested language relative to this latter issue is as follows[20]:

> Design-Builder's Proposal shall specifically identify any deviations from the Owner's Program, which identification shall be set forth on a separate exhibit in the Proposal identified as "Deviation List." In case of any inconsistency, conflict or ambiguity between the Owner's Program or the Design-Builder's Proposal, the inconsistency, conflict or ambiguity shall be resolved in accordance with the following order of precedence:
>
> Deviation List
>
> Owner's Program
>
> Design-Builder's Proposal (excluding the Deviation List)

DBIA also recommends that the parties include a clause that specifically identifies which documents and assumptions form the basis for any GMP that might be developed if the work is proceeding on a cost-plus basis.

14.6   Ownership of Documents

Another issue that is unique to the design-build relationship is who, as between the owner and design-builder, owns the contract documents and design work product. This issue can present itself in at least five different contexts:

---

[19]Subparagraph 3.01.B, EJCDC Document 1910-40 (1996 ed.).

[20]*DBIA Contracting Guide,* pp. 13–15.

- What happens if the owner and the design-builder are unable to consummate a contract after the preliminary design has been performed and the owner terminates the relationship?

- What happens if the owner decides to terminate the services of the design-builder for convenience after the final design has been completed under the contract?

- What happens if the owner terminates the agreement because of the default of the design-builder?

- What happens if the design-builder terminates the agreement because of the default of the owner?

- What happens if the owner decides to use the design-builder's work product on another project after each party performed the underlying contract in full?

Each question creates a unique set of concerns on the part of parties that should be addressed in the contract.

14.6.1   Early termination of the relationship

The question of what happens if the parties are unable to agree on commercial terms after the design-builder has provided the preliminary design is of vital importance to both parties. If the owner is confronted with commercial terms that exceed the owner's expectations, the owner's typical choices would be to reduce the scope of work, increase the budget, or cancel the project. As noted in the *DBIA Contracting Guide,* if an owner proceeds in good faith down the path of developing the project through design-build yet is ultimately unable to reach agreement with the prospective design-builder on commercial terms, it should have the option to shift to another proposed design-builder for the project. If the owner cannot use some or all of the original work product in going forward with another course of action, it will have its schedule impacted, which is one of the key drivers leading owners to the design-build approach.

By the same token, the *DBIA Contracting Guide* notes that it is not appropriate for a design-builder to be manipulated by an owner who takes advantage of the design-builder's expertise in early design development and preconstruction services and then "shops" the work product to other bidders. The *Guide* notes that the critical decisions affecting the success of the project and greatest intellectual effort are typically developed during the early-phase activities. Furthermore, many design-builders do not seek market value for project development and conceptual design activities, since they are proceeding with the understanding that they will be executing the project should the owner decide to go forward.

The current standard forms do not provide significant help to the owner on this issue. Both Part 1 of AIA Document A191 and AGC 400, which are the preliminary agreements geared to enabling owners to get started with design on a direct design-build selection basis, state that ownership of the work product is not vested in the owner. As a result, under these approaches, the owner would have to start the process over if it were unable to reach accommodation with the original design-builder over the proposal for going forward with the work.

DBIA approaches this matter differently. It recommends a twofold approach if an owner and design-builder are unable to reach agreement on contracting terms and the owner wishes to use the design-builder's preliminary work product for the project. In exchange for receiving the right to use the design work product, the owner is to (1) compensate the design-builder with a premium over its initial compensation if the owner elects to go forward on the project with another design-builder and (2) provide appropriate indemnifications to both the design-builder and its A/E that protect them from liability relating to the project and design. It recommends the following language relative to "ownership of documents"[21]:

> All drawings, specifications and other documents and electronic data furnished by Design-Builder to Owner under this Agreement ("Work Product") are deemed to be instruments of service and Design-Builder shall retain an ownership and property interest therein. In the event Owner fails to enter into a design-build contract on the Project with Design-Builder and proceeds to design and construct the Project through its employees, agents or third parties, Design-Builder shall grant Owner a limited license to use the Work Product to complete the Project, conditioned on the following:
>
> (a) Use by Owner of the Work Product is at Owner's sole risk and without liability or legal exposure to Design-Builder or anyone working by or through Design-Builder, including design professionals of any tier (collectively referred to as the "Indemnified Parties"). Owner shall defend, indemnify and hold harmless the Indemnified Parties from any and all claims, damages, losses and expenses, including attorneys' fees arising out of or resulting from the Work Product.
>
> (b) Owner agrees to pay Design Builder the sum of $_____ as compensation for the right to use the Work Product in accordance with this Article.

The key to this issue is in adequately determining how much value the owner is truly obtaining from the design-builder through this early design work. Some design-builders do not believe that their work product at this early stage justifies a premium fee for use by others. Others are more concerned about the liability issue than the money to be paid for a premium. Still others would rather not affect the relationship with a good client by insisting on this language. Regardless, it is important for both owner and design-builder to consider all the circumstances in assessing how to address this particular issue.

14.6.2 Termination for convenience after final design

A more difficult question arises if the owner waits until the design-builder has completed its design and then elects to terminate for convenience. As discussed

---

[21]*DBIA Contracting Guide*, p. 8.

more fully in Sec. 14.15.5, the owner's right to issue a termination for convenience allows the owner to end the contractual relationship for any reason in exchange for payment to the terminated design-builder of its costs for performance plus a fee on the performed work. These clauses have long been used in the construction industry to enable an owner who is confronted with an unviable project to mitigate its costs without being in breach of contract.

Although an owner should have a process for terminating the design-build contract for convenience, the process should consider the interests of the design-builder as well. It should not enable an owner to obtain the design-builder's valuable design concepts and management approaches, terminate for convenience, and then bid the construction—since compensating the design-builder for its costs in the preconstruction phase of the work will likely not be adequate compensation.

Even though the owner may have the ability to terminate for convenience, it still needs to obtain the right to use the design-builder's work product if the project is to go forward. This requires an integration of the "ownership of documents" clause with the "termination for convenience" clause to obtain the appropriate recourse for each party. The current standard forms each approach this differently:

- *EJCDC.* Owner has the ability to go forward with the project using the design-builder's documents, effectively paying only for the actual cost of the work actually performed by the design-builder through the date of termination, with no premiums to the design-builder.

- *AGC.* Owner has the ability to go forward with the project without using the design-builder's documents, and will pay design-builder an agreed-on premium to do so.

- *AIA.* Owner does not have the ability to go forward with the project with anybody's documents, and is potentially exposed to an unliquidated premium to do so.

The *DBIA Contracting Guide* concludes that a certain payment should be made to the design-builder by an owner for use of the design-builder's work product in a termination for convenience scenario and that the premium will be lower in the event of a project abandonment and substantial if the project is not abandoned. It also suggests that the owner's rights to use the design be conditioned on the indemnification principles set forth in Sec. 14.6.1 above.[22]

14.6.3   Termination for default of design-builder

The default of the design-builder creates special considerations relative to the ownership and use of the design. The current standard forms do not adequately and clearly address this issue, particularly given the manner in which they

---

[22]*DBIA Contracting Guide*, p. 8.

address ownership of documents. In fact, only AIA's form states that in the event of a design-builder default the owner will be given a license to use the work product, "conditioned upon the Owner's execution of an agreement to cure the Design/Builder's default in payment to the Architect for the services previously performed and to indemnify the Architect with regard to claims arising from such reuse without the Architect's professional involvement."[23]

If the default occurs before the drawings are signed and sealed, the owner may prefer to have the right—but not the obligation—to use the current design professional of record to complete the work, since obtaining another design professional will be time-consuming and expensive. The contract between the owner and the design-builder should require the design-builder to provide a contingent assignment of its contract with such design professional in the event of a design-builder default. This will create a practical problem if the design-builder is integrated, with in-house design capabilities, and the owner will likely have to reprocure design services elsewhere.

If a default occurs after the design is completed, the issue is less complex, since the owner should be able to complete the work without having to obtain another design professional. A contingent assignment clause would still be appropriate under these circumstances to give the owner the option of having inspection and other duties of the design professional performed by one already familiar with the project.

14.6.4 Termination for default of owner

In the event of an owner default, the design-builder should attempt to maintain control over the contract documents. Although this will not be a practical problem if the owner abandons the project, it is more likely that the design-builder and owner will be in dispute as to who, as between the owner and design-builder, has actually breached the contract and created the default.

Given this, it may be appropriate for the parties to deem the default of the owner in the same manner as a termination for convenience for purposes of the ownership of documents and use of the design. This will provide the design-builder with monetary compensation and indemnifications for use of the work product and enable the owner to use the documents to go forward with the project. Note that some parties may increase the financial consequences to the owner if the termination is ended as a result of owner default.

14.6.5 Use of design by owner on other projects

The final issue to consider in this area is the right of the owner to use the design on other projects without compensation or involvement of the design-builder. Owners who take this position claim that they have already paid for the design and should be free to use it whenever and wherever they please. Design-builders

---

[23]Subparagraph 1.3.3, Part 2 of AIA Document A191.

contend that their product is not intended to be a prototype and that they should not only be able to receive some financial benefit from the use of their efforts on other projects but also be protected from liability for issues arising on projects in which they have not been involved. Given that each site is different—particularly from an environmental and siting perspective—the concerns of the design-builder are reasonable.

The current standard forms do not uniformly address the issue of ownership of documents for this purpose.

- *EJCDC*. All documents give the design-builder an ownership interest in the work product, but do not preclude owner's use on other projects. If the owner does use the documents on other projects without adaptation or approval by the design-builder, the use will be at the owner's sole risk and will indemnify the design-builder from the consequences.[24]

- *AGC*. All documents are owned by the design-builder and are not to be used by the owner without the design-builder's consent.[25]

- *AIA*. All documents are owned by the design-builder's architect and are not to be used on other projects without permission of the design-builder.[26]

On the basis of the policy reasons associated with this issue, there are compelling reasons for the "ownership of documents" clause to at least provide the owner with the right to use the documents for the operation and maintenance of the facility. If the owner plans to use the work as a prototype for other projects, it would be appropriate for the parties to know this at the time of contracting and to have appropriate compensation given to the design-builder and its team to reflect their investment and the value added to the process.

14.7 Owner's Right to Review the Design

Under the design-bid-build and construction management systems of project delivery, the owner has the ability to be substantially involved in development of the design. The design professional works directly for the owner, enabling the owner to interact directly with the design professional's staff in assessing various design options. The contractor's involvement in design is limited to providing value engineering advice that the owner and design professional will consider as the design evolve..

Design-build poses some challenges in the area of owner involvement in design, particularly since the development of the design for the project will be progressive, taking place mostly after contract execution and the establishment of a price. Although some owners will take a turnkey type of approach

---

[24]EJCDC 1910-40, Paragraph 3.04.

[25]AGC 410, Subparagraph 3.1.7.

[26]Subparagraphs 1.3.1 and 1.3.2, Part 2 of AIA Document A191.

to this issue and play virtually no role in the design review, many owners are interested in having access to and evaluating the design. This helps them understand what they are buying, as well as giving them some ability to influence design decisions that are important to their overall goals.

On the other hand, excessive involvement by the owner can influence the design-builder's quality, cost, and schedule goals. This is particularly true if the design-builder was selected on the basis of price after competitively developing a proposal on the basis of a detailed RFP and owner's program. Moreover, owners need to be concerned about getting too actively involved with the design, since a design-builder could then use this involvement as an excuse to overcome the single point of responsibility feature of the process. As a result of these and other issues, it is important for the parties to understand how the owner and the design-builder will work together and communicate with each other in the development of the design.

The standard forms address the design review process similarly, although there are some subtle differences that should be noted. Part 2 of AIA Document A191 states that the design-builder will submit "Construction Documents for review and approval by the Owner."[27] No time period is established for this review and approval process. Moreover, since the term "construction documents" is a defined term, it does not appear that the AIA format allows for interim submissions of designs for review.

EJCDC 1910-40 provides for a preliminary design submission that will be reviewed and approved by the owner, after which the design-builder will proceed to prepare and submit final drawings and specifications for owner review and approval.[28] The EJCDC clause does require that the design-builder submit the design documents and review them with the owner within the time specified in the design-builder's progress schedule. The inference should be that the owner will also be required to act within the time set forth in the schedule, although the language of the scheduling clause does not specifically require this.[29]

AGC's clauses are more comprehensive, requiring three phases of submissions: schematic design documents, design development documents, and construction documents.[30] Each set of documents is subject to the owner's "written approval," with the owner having the general obligation to "review and timely approve schedules, estimates, Schematic Design Documents, Design Development Documents and Construction Documents furnished during the Design Phase."[31]

The *DBIA Contracting Guide* notes that the design process is an area where owners may choose to play an active role to ensure that their goals are being achieved, particularly when price is being established during the design process

---

[27]Subparagraph 3.2.3, Part 2 of AIA Document A191.

[28]Subparagraphs 6.01.B and 6.01.C, EJCDC 1910-40.

[29]Subparagraphs 2.04.A and 2.06.A, EJCDC 1910-40.

[30]Subparagraphs 3.1.4–3.1.6, AGC 410.

[31]Subparagraph 4.2.1, AGC 410.

on a cost-plus or GMP basis.[32] The *DBIA Contracting Guide* also notes that because the design-builder can be adversely affected if the owner does not timely review and approve the design submissions, the parties should establish a specific understanding, on a project-by-project basis, of the needs and goals of each party relative to this issue. DBIA's suggested language calls for specific design review workshops to be used for each major submission. This will allow an interactive dialogue that promotes clearer and more comprehensive understandings as to what has taken place relative to the design. The language is as follows:

1. Design-Builder shall submit to Owner interim design submissions as the parties may agree upon to support the overall project schedule. On or about the time of the scheduled submissions, Design-Builder and Owner shall meet and confer about the submissions, with Design-Builder identifying during the meeting, among other things, the evolution of the design and any significant changes or deviations that have taken place from previous design submissions. Minutes of the meeting will be maintained by the Design-Builder and provided to all attendees for review. Following the design review meeting, Owner shall review and provide its response to the interim submissions in a time that is consistent with the turnaround times set forth in Design-Builder's schedule.

2. Design-Builder shall submit to Owner, for Owner's review and approval, Construction Documents setting forth in detail the requirements for construction of the Work and which shall be consistent with the latest set of interim design submissions, as modified in the design review meeting. The parties shall have a design review meeting to discuss such Construction Documents consistent with Subparagraph 1 above and Owner shall review and approve the Construction Documents in a time that is consistent with the turnaround times set forth in Design-Builder's schedule. Construction shall be in accordance with the approved Construction Documents.

As is apparent from the above-referenced language, the key to DBIA's philosophy is that the parties should determine in advance of contract execution precisely how many design iterations will take place and whether the owner needs to review and approve such submissions. It is always wise, however, for the owner to have a direct role in reviewing and acknowledging the acceptability of the final contract documents.

It should be noted that many owners are content to have certain documents provided to them on an "informational" basis for their files. As opposed to documents furnished on an "action" basis—where the review, comment, or approval may take time and negatively affect the design-builder's progress—informational documents can likely satisfy an owner's needs in a variety of areas of the project.

14.8   Owner's Right to Make Changes to the Work

The ability of an owner to make changes to the work is a long-standing principle of construction contracting. This ability arises through what is commonly

---

[32]*DBIA Contracting Guide,* pp. 16–17.

known as the "changes" clause. A typical changes clause (1) gives the owner the right to make changes within the general scope of the work and (2) compels the contractor to make the changes, but (3) requires that the owner compensate the contractor for the time and cost of implementing the change. The right of the owner to require a contractor to perform changes is critical, since evolving owner needs (financial, technical, or other) and constraints by outside sources (statutory, regulatory, or lending institutions) could render the design insufficient before the construction even starts.

Although few question that the owner should be given this right to require changes to be made (and that the contractor should have the reciprocal right to adjustments in the contract price and contract time), the structure of a design-build relationship may require the parties to think through the changes process more carefully than under other project delivery methods. For example, if a project design is based on a standard prototype of the design-builder (as might be the case with public-assisted housing or medical office buildings), the design-builder may preclude the owner from making wholesale changes to the design. Likewise, if the design-builder's prototype design is not only proprietary but also essential to the ability of the design-builder to meet performance guarantees, one might expect the design-build contract to preclude owner discretion in design changes. This might be found in

- The power generation industry, where pieces of equipment, such as a turbine or the software associated with the turbines' operation, are critical to system performance
- The telecommunications industry, where telemetry configurations associated with the vendor's intelligence network are proprietary and form the entire basis for the system design

Given this, it is important for a design-builder to clearly establish what aspects of the design are off limits for changes.

Current standard forms do not address this issue, and give the owner the broad-based right to make changes. By way of example, AIA enables a change in the work to be made by

- Change orders executed by both the design-builder and owner reflecting the change and the price and time adjustment due to the change[33]
- Construction change directives, which enable the owner to direct a change in the work before agreement on price and time adjustments[34]
- Minor changes by the design-builder in the construction documents that are consistent with the intent of such documents and do not affect price or time[35]

---

[33]Paragraph 8.2, Part 2 of AIA Document A191.

[34]Paragraph 8.3, Part 2 of AIA Document A191.

[35]Paragraph 8.4, Part 2 of AIA Document A191.

Although EJCDC is similar in approach to AIA, there are several subtle differences. First, EJCDC gives the owner—not the design-builder—the right to make minor changes in the work.[36] Moreover, while AIA's contract specifies directly that amounts not in dispute can be submitted in payment applications, pending a final determination of the cost associated with a construction change directive, EJCDC does not address this in its work change directive process.

Despite these differences, however, both AIA and EJCDC are consistent in requiring the design-builder to perform changes required by the owner even if there is a dispute as to whether the owner will ultimately assume responsibility for the financial consequences of the change.[37] This differs from the 1993 edition of AGC 410, which states that the design-builder does not have the obligation to perform changed work until either a change order has been issued or the owner has issued a written order to proceed with the change, with any additional price to be paid to the extent additional cost is incurred.[38]

The *DBIA Contracting Guide* endorses the AIA change order approach, noting that while the design-builder should be given the right to make minor changes in the work, it should be done with prompt notice to the owner, as specified in Subparagraph 8.4.1 of Part 2 of AIA Document A191.[39] DBIA did note the problem that many have found with the changes process: that the failure of an owner to agree on changes can significantly impact a design-builder's cash flow and its ability to complete the work. Neither the AIA nor EJCDC clauses address payment when the owner disputes the design-builder's entitlement to a change or for other scope of work disagreements. DBIA suggested language to address this is as follows:

> In the event Owner and Design-Builder disagree upon whether Design-Builder is entitled to be paid for work required by Owner, or in the event of any other disagreements over the scope of work included within the Contract Sum, Owner and Design-Builder agree to negotiate in good faith to resolve the issue amicably. As part of the negotiation process, Design-Builder shall furnish Owner with a good faith estimate of the costs to perform the disputed work in accordance with Owner's interpretations. If the parties are unable to agree, and Owner expects Design-Builder to perform the work in accordance with Owner's interpretations, Design-Builder shall proceed to perform the disputed work, conditioned upon Owner issuing a written order to Design-Builder directing Design-Builder to proceed and specifying Owner's interpretation of the work that is to be performed. In such event, Design-Builder shall be entitled to submit in its payment applications an amount equal to fifty percent (50%) of its estimated cost to perform the work,

---

[36]Subparagraph 1.01.A.21, EJCDC 1910-40.

[37]AIA does this by inference through the construction change directive process. EJCDC specifically states that the design-builder is obligated to proceed through Paragraph 6.17 of EJCDC 1910-40.

[38]Paragraph 9.3, AGC 410.

[39]*DBIA Contracting Guide*, pp. 21–22.

and Owner agrees to pay such amounts, with the express understanding that such payment by Owner does not prejudice Owner's right to argue that it has no responsibility to pay for such work and that receipt of such payment by Design-Builder does not prejudice Design-Builder's right to seek full payment of the disputed work under the applicable terms of the Agreement in the event the Owner's order is deemed to be a change to the Work.

This would effectively place both parties at some risk for getting the issue resolved and avoid requiring the design-builder to finance the work.

In addition to the above, it is important for both the owner and the design-builder to address in the contract the pricing of changes, including overhead and profit markup on such work. This will enable the parties to set out the ground rules for how changes are to be administered and will avoid misconceptions about what either party expects. For example

- Will the design work associated with developing proposals for the owner on proposed changes be reimbursable to the design-builder?

- How will the design-builder be allowed to recover the costs of design services performed by in-house employees? Will it be on a cost basis or on preestablished rates?

- Will there be limitations on what subcontractors can charge for overhead and profit? If so, these should be established in the contract between the design-builder and owner and passed on to the subcontracts between the design-builder and its subcontractors.

- If work is to be routinely done on a time-and-materials basis, what does the owner expect in the way of paperwork?

Because these are all project-specific issues, they need to be considered independently from the standard forms.

14.9   Differing Site Conditions

One of the most important issues confronting owners during project development is the nature of the geotechnical conditions for the project.[40] Subsurface soil conditions can not only affect project design but also be a major factor in overall construction cost and schedule. Moreover, to the extent that subsurface conditions differ from those planned and budgeted, the financial viability of a project can be significantly jeopardized.

Notwithstanding the importance of identifying soil conditions prior to design, geotechnical site investigation is often treated in a haphazard and superficial manner. This is particularly true on design-build projects, where the owner may feel that the design-builder will determine site conditions once

---

[40]See generally M. C. Loulakis, B. P. Waagner, and H. C. Splan, "Differing Site Conditions," in *Construction Claims Deskbook,* Wiley Law, New York, 1996, Chapter 11.

under contract and limit its own investigation to only a few borings. However, even an owner that undertakes an extensive geotechnical investigation program does not ensure full understanding of subsurface conditions, since conditions can vary dramatically along any continuum.

As noted in Sec. 14.3 above, the responsibility for and allocation of risk of differing site conditions can be a contentious contract issue. The answer to the question is dependent largely on how the owner views risk. Some owners have adopted a procurement philosophy which shifts all the risk of site conditions to the contractor, regardless of whether the actual conditions encountered were foreseeable. Faced with this risk, a contractor must decide during the bidding process how much contingency to place in its bid to account for such risk.

Other owners take a more enlightened view of risk allocation. They conclude that because of the uncertainty of subsurface risk, they will incur higher bids for a problem that might never arise. Rather than prepay for such a problem, these owners believe that it is best to handle this exposure by permitting a contractor to recover additional costs and obtain a time extension under a differing site conditions clause.[41] This approach has long been used at the federal contracting level, for reasons stated in a well-recognized United States Claims Court decision:

> The purpose of the changed conditions clause is thus to take at least some of the gamble on subsurface conditions out of bidding. Bidders need not weigh the cost and ease of making their own borings against the risk of encountering an adverse subsurface, and they need not consider how large a contingency should be added to the bid to cover the risk. They will have no windfalls and no disasters. The Government benefits from more accurate bidding, without inflation for risks which may not eventuate. It pays for difficult subsurface work only when it is encountered and was not indicated in the logs.[42]

Numerous other public and private owners have followed suit, based on sound economic principles as well as recognizing that the absence of such a clause can result in an acrimonious relationship on the project.

While a design-builder's responsibility for site conditions can be more extensive than that of a general contractor working on a design-bid-build basis (particularly if the design-builder is taking on development risks), the substantial cost and time impact to the design-builder if the risk actually materializes requires that an owner carefully consider assuming the risk through a differing site conditions clause. EJCDC's differing site conditions clause adopts the traditional federal government approach to this issue.[43] These provisions provide that the design-builder shall receive an adjustment to the contract price and contract time for

---

[41]Differing site conditions are sometimes also referred to as "changed conditions," "latent physical conditions," and "concealed conditions," depending on the contract provision governing the issue.

[42]*Foster Construction, C. A. & Williams Bros. Co. v. United States,* 193 Ct. Cl. at 613–14, 435 F.2d at 887.

[43]Subparagraph 4.02.A, EJCDC 1910-40.

- Subsurface or latent physical conditions which differ materially from those indicated in the contract documents, commonly known as a "type 1 differing site condition"

- Unknown physical conditions, of an unusual nature, which differ materially from those ordinarily encountered and generally recognized as inhering in work of the character called for in the contract documents, commonly known as a "type 2 differing site condition"

This is virtually identical to AIA's "concealed conditions" clause[44] and AGC 410's "unknown conditions" clause.[45] Note, however, that AGC's clause provides relief in terms of conditions that are different from those which the design-builder reasonably anticipated, which is potentially a broader baseline than the conditions reflected in the contract documents. The intent of all of these clauses is to grant the design-builder monetary and schedule relief for conditions falling under the clause, thereby enabling the design-builder to avoid inflating its bid to account for the risk of dealing with such conditions.

Although the philosophy behind using the differing site conditions clause in a design-build project is clear, the selection process of the design-builder may impact how the clause is used in practice, particularly regarding the appropriate baseline for determining anticipated site conditions. If the owner has conducted an investigation prior to selection, this should be the appropriate baseline, since the design-builder's assumptions will be based on this information. If the design-builder is selected before sufficient information has been obtained, the parties could agree that a postaward geotechnical investigation will be undertaken by the design-builder and that this will be the baseline for the differing site conditions clause. This is particularly logical if the parties are attempting to jointly develop a GMP or have used allowances or unit prices for foundation and/or sitework (e.g., piles versus spread footings).

If the parties elect to use a differing site conditions clause, they should ensure that the contract does not have any extraneous language that could defeat the intent of the clause. Disclaimers (i.e., unclassified site conditions) should not be used unless the parties specifically agree that the limited risk contained in the disclaimer will be assumed by the design-builder. In addition, broad site investigation and inspection clauses should be used only if applicable and not as a substitute for overriding a differing site conditions recovery.

14.10 Changes of Law

Under other forms of project delivery, there is little issue associated with changes of law that affect the design or construction. Owners routinely bear this risk and will process change orders to both the design professional and the

---

[44]Paragraph 8.5, Part 2 of AIA Document A191.

[45]Paragraph 9.5, AGC 410.

contractor to implement the change. This practice also holds true in the design-build process, with the owner continuing to bear responsibility for changes in law. There are, however, some questions that the parties need to consider:

- What is the benchmark date for establishing when a change of law occurs (proposal submission, contract execution, etc.)?

- Does the design-builder have the obligation to design and construct the facility to meet laws and regulations in existence as of the benchmark date or the date the project is substantially complete?

- Should the design-builder be entitled to recover for any changes of law that affects its work, including those that change commercial terms (such as taxes)?

The answers to these questions can vary from owner to owner and project to project. For example, some parties shift this risk entirely to the owner effective on the proposal date, on the theory that the design-builder's price is based on assumptions made as of that date. Others require that the design-builder assume financial responsibility for changes in law that affect construction, such as new OSHA administrative interpretations, on the theory that this is a normal risk that a construction contractor assumes and is in the best position to manage. Still other owners do not want to be "nickel and dimed" by changes of law and require the design-builder to assume the financial risk for changes in law that do not have "a major affect on construction or design."

Each of the standard forms treats these issues somewhat differently. AIA Document A191 establishes the date of the design-builder's proposal as the benchmark date, since this is the date when the price will be provided to the owner for the work.[46] AIA's documents also state that the design-builder will comply with all laws, ordinances, rules, and regulations relating to the project.[47] Thus, under this document, it appears that the design-builder would be required to accommodate any changes in law, regardless of when they occur. As to the third question, compensation is limited to "changes in the construction." Therefore, there is a possible gap in who bears responsibility for design, administrative, or economic changes (permit filings, public hearings, taxes) that might be required by changes of law.

EJCDC treats the issue more broadly than does AIA, since it does not limit compensable changes of law to those that simply affect construction. Instead, EJCDC allows an adjustment for any law change that has an effect on cost or time of performance: "Changes in Laws and Regulations not known or foreseeable on the date of receipt of Proposals having an effect on the cost or time of performance may be the subject of a claim under Article 9."[48] However, the

---

[46]Subparagraph 8.6.1, Part 2 of AIA Document A191.

[47]Subparagraph 3.2.11, Part 2 of AIA Document A191.

[48]Subparagraph 6.08.C, EJCDC 1910-40.

"foreseeability" language can create a constraint on design-builder recovery, since this is a fairly loose standard. Although it was likely intended to deal with issues such as the Americans with Disabilities Act, which was in draft form for months before it was enacted, this could be problematic in interpretation after contract award.

The *DBIA Contracting Guide* sets forth a philosophy that is similar in form to both AIA and EJCDC. However, DBIA's suggested language expressly addresses situations where the design-builder is under contract but will be providing a price over the course of an evolving design—as is often the situation with GMP contracts. In these situations, DBIA's position is that the key date for establishing a benchmark for the changes of law risk is the date when firm pricing has been provided to the owner. Its clause reads as follows:

> The Contract Sum and/or Contract Times shall be adjusted to compensate the Design-Builder for the effects of any changes in Laws and Regulations affecting the performance of the Work not known or foreseeable on the later of (1) the date of Design-Builder's Proposal for the Work or (2) the date negotiations on this Agreement have concluded and the parties have agreed upon a lump sum or GMP, as applicable.[49]

Although each of the above-referenced contracts appears to give the design-builder recourse for virtually all changes of law, many owners do not actually feel that it is appropriate to bear the risk of changes in law that affect the general business of the design-builder in performing the work. Such potential areas might include increases in payroll withholding taxes, changes to existing (or the implementation of new) gross receipts taxes, and income taxes. Likewise, if the change of law affects reporting requirements, there is a possible, but more remote, connection to the design builder's costs. The parties need to think through these issues during contracting and carve out any exceptions to the general changes of law philosophy.

Parties should be aware that international construction projects are handled very differently from domestic construction projects relative to changes of law. Given the risks that are associated with working in a foreign country, many design-builders have successfully convinced the owner to accept most of the risk of changes in law, even with respect to economic issues. This is particularly true in developing countries, where major changes in legislation or political control can occur with little warning.

14.11 Permit responsibility

Although design-builders are generally willing to assume more responsibility for permits and governmental approvals than is a typical general contractor, the issue remains one that is important for owners and design-builders to address

---

[49]*DBIA Contracting Guide,* p. 18.

during contracting. Numerous (environmental, site-related, occupancy-related, etc.) permits may be more efficiently obtained by the owner rather than the design-builder. Consequently, it may be appropriate for the contract to have a specific permit and approvals list that identifies which party has primary responsibility for obtaining those permits and approvals associated with the project. This approach is endorsed by DBIA, which suggests model language as follows:

> Unless otherwise provided in Owner's Permit List, attached hereto as Exhibit ___, Design-Builder shall, with the assistance of the Owner, obtain and pay for all necessary permits, licenses, government charges and inspection fees necessary for the prosecution of the Work.

This language is similar to the approach used by EJCDC, which specifically makes the design-builder responsible for obtaining permits except as stated elsewhere in the contract documents.[50]

AIA's position is more rigid than this, in that it imposes on the design-builder the obligation to file for and obtain "documents required to obtain necessary approvals of governmental authorities having jurisdiction over the Project,"[51] and that the owner must provide cooperation with respect to the design-builder's obligations to secure "building and other permits."[52] On the other hand, AGC imposes the responsibility on the owner to obtain the building permit and requires the design-builder to merely assist in this endeavor,[53] without regard to the fact that in many design-build facility contracts, the owner anticipates that this service will be accomplished by the design-builder.

Regardless of who has primary responsibility for obtaining permits and approvals, it is in the best interests of both parties to provide reasonable cooperation and assistance to each other in facilitating this effort. This is particularly true for approvals that are sensitive or problematic, as might be the case with an unreasonable fire marshal or local building inspector. In fact, it may be beneficial for the owner to determine at the outset of the project which permitting agencies create the greatest potential for difficulty and determine how to make them more comfortable with supporting the process.

14.12   Time-Related Issues

Given that expedited project delivery is one of the primary factors driving many owners to select the design-build project delivery method, the parties must carefully consider time-related issues in the contract. These issues include (1) scheduling requirements, (2) time-extension relief, and (3) liquidated damages and early completion bonuses, each of which is discussed below.

---

[50]Subparagraph 6.07.A, EJCDC 1910-40.

[51]Subparagraph 3.2.4, Part 2 of AIA Document A191.

[52]Paragraph 2.3, Part 2 of AIA Document A191.

[53]Subparagraph 3.3.5, AGC 410.

14.12.1 Scheduling requirements

Because project scheduling can take many forms, it is important for the owner to identify in the contract what it reasonably needs from the design-builder. Issues to consider include

- The type of scheduling software that will be used by the design-builder and whether it needs to be compatible with the owner's scheduling software (if the owner will in fact be independently monitoring the schedule)

- The form of and level of detail within the schedule—including decisions on whether the schedule will be a bar chart or CPM (critical-path method), the minimum number of activities within the schedule, and the maximum duration of any activity

- The frequency of schedule updates

It is important for owners to limit contractual scheduling requirements to those that are reasonably needed by the owner to manage their role on the project. Imposing substantial and unnecessary requirements will have a cost impact on the design-builder. Moreover, since design-builders are motivated to properly manage the schedule—to meet completion dates and ensure profitability—they will typically have internal scheduling controls that will be used on the project.

In addition to the format of scheduling, the parties should also have a clear understanding of what each must do to support the project schedule. If the design-builder is selected directly, or based on proposals submitted under Part 1 of AIA Document A191 or AGC 400, it is likely that a preliminary schedule (bar chart or simplified CPM) will be developed before the contract is executed. If the design-builder is selected competitively, it is more likely that the preliminary schedule will be a deliverable after contract award. Regardless, the schedule should be treated as a dynamic tool that will be used to assist both parties in administering the project, and not merely a paperwork exercise for the design-builder.

It is also critical for the owner to advise the design-builder early in the process as to any constraints that the owner will have in meeting its obligations, including such items as delivery of owner furnished equipment, interconnections, and permits. Likewise, if the design-builder expects owner responses within specific periods of time for design and other submissions, these should be made known early and established in the baseline schedule. This is supported by the philosophies expressed in the *DBIA Contracting Guide*. The *DBIA Contracting Guide* also notes the importance of having both design and construction activities scheduled.[54] This is different from the

---

[54]*DBIA Contracting Guide,* pp. 18–19.

[55]Subparagraph 3.3.4, AGC 410.

philosophies expressed in AGC 410[55] and Part 2 of AIA Document A191,[56] which focus on the schedule for only the construction portion of the work.

Finally, the parties should carefully evaluate the guaranteed dates for project completion. While most contracts have a single date for substantial completion of the entire project, on some projects it may be more prudent to specify milestones for completion of portions of the work that would enable the owner early access to part of the project. If used, milestone dates should be clearly identified in the contract, with an appropriate definition as to what constitutes completion of the milestone.

14.12.2   Time extensions

One of the central risks that needs to be addressed by the parties in the contract is how to treat delays to the guaranteed completion dates in the contract. The conventional approach to this risk has been to give time extensions to contractors for events beyond their reasonable control. This approach has been adopted by the standard form design-build contracts as well. For example, Part 2 of AIA Document A191 states:

> If the Design/Builder is delayed at any time in the progress of the Work by an act or neglect of the Owner, Owner's employees, or separate contractors employed by the Owner, or by changes ordered in the Work, or by labor disputes, fire, unusual delay in deliveries, adverse weather conditions not reasonably anticipatable, unavoidable casualties or other causes beyond the Design/Builder's control, or by delay authorized by the Owner pending arbitration, or by other causes which the Owner and Design/Builder agree may justify delay, then the Contract Time shall be reasonably extended by Change Order.[57]

While most owners are willing to grant this relief to design-builders, several key issues need to be addressed in the contract relative to the right to obtain the time extension. These include

- The time within which the design-builder is obligated to provide notice of the delay
- The supporting documentation needed to seek recovery, including any requirements for fragnet schedules or CPM analyses
- The right of the design-builder to be compensated for time extensions

The latter issue is particularly significant, since many owners prefer to grant time extensions as the sole relief for delays. In these cases, the contract will have the design-builder waive, through a "no damages for delay" clause, its relief to extended general conditions and other time-related costs. Although this may be immaterial to some design-builders, it can have a major financial

---

[56]Paragraph 4.4, Part 2 of AIA Document A191.

[57]Paragraph 4.5, Part 2 of AIA Document A191.

impact on those design-builders who perform work with major pieces of equipment themselves.

14.12.3 Liquidated damages and bonuses

Liquidated damages have been used for many years on construction contracts. Owners find it an appropriate technique to create an incentive for contractors to meet contractual schedule obligations—establishing a "penalty" to the contractor for late completion. They also find it as a convenient way to recover the anticipated losses to the owner for late project delivery. Many contractors have shifted their policy to require the use of liquidated damages, since the exposure to actual damages for delay could be substantial and jeopardize the company.[58]

Notwithstanding the widespread use of liquidated damages, few owners and contractors feel that they are appropriate. Their primary argument is that contractors and owners may actually spend more time positioning themselves through letter writing to avoid or recover liquidated damages than in working together to overcome project delays. Contractors also argue that owners use liquidated damages as a tool to extort them into giving up their legitimate claims at the end of the project.

Neither AIA Document A191 nor AGC 410 addresses the subject of liquidated damages. EJCDC 1910-40-A and 1910-40-B do address the subject and contain relatively standard language indicating how such damages are to be assessed.[59] The *DBIA Contracting Guide* directly addresses the policy issues behind liquidated damages and provides a sample clause for parties who decide that it is appropriate to use them.[60]

If liquidated damages are used, it is important for the parties to consider several points:

- The owner should limit the amount of the liquidated damages to only those costs reasonably necessary to compensate the owner for delays.

- The owner should pay attention to market conditions and not specify liquidated damages that are commercially unreasonable given the contract value. For example, it will be difficult to obtain any design-builder who will accept a $1,000,000/day liquidated damages clause—regardless of the support for the legitimacy of the calculation behind the amount.

- The owner should consider a grace period (7 to 30 days) on the assessment of liquidated damages to avoid having the parties fight over who bears responsibility for only a few days of delay.

---

[58]See generally M. C. Loulakis, J. G. Gilmore, and S. B. Hurlbut, *Contracting for the Construction of Power Generation Facilities,* Federal Publications 89-5, April 1989.

[59]Article 3, EJCDC 1910-40-A.

[60]*DBIA Contracting Guide,* pp. 20–21.

The owner may also evaluate whether a liquidated damages clause should be limited to only those damages that are difficult to prove with certainty (such as lost revenue or profits), with those that can be defined (such as the extended costs of owner consultants) being directly claimable by the owner outside the liquidated damages clause.

Early-completion bonuses should also be viewed by the owner as a tool for incentivizing (motivating) the design-builder to expeditiously complete the project. The prospect of receiving such a bonus will give the design-builder an incentive to manage the project differently—such as spending its own money to work overtime or add resources—to give it the best chance of achieving the bonus. This works in the best interests of the project, since it will ensure that all project participants are doing all that is reasonably possible to achieve an expedited project delivery. To protect the owner's interests, it may be appropriate for the owner to establish (1) a grace period (similar to that proposed above with liquidated damages) and (2) a ceiling on the amount of bonus that can be obtained.

14.13   Warranty-Related Issues

A variety of warranty obligations are imposed under typical design-build contracts. Many of these warranties are no different from those contained in traditional design-bid-build construction contracts, such as those requiring that all materials and equipment furnished under the contract be new unless otherwise specified. Such warranties are almost viewed as boilerplate by the parties and do not create either contracting problems or significant risk to the design-builder. There are, however, a number of warranty-related issues that are unique to design-build contracts that should be considered by the parties, all of which are discussed below.

14.13.1   Warranty for defective construction

It is typical for design-build contracts to specifically address a warranty for the correction of any defective construction work. The AIA, EJCDC, and AGC standard design-build forms each contains a specific warranty clause requiring the construction to be free from defects. For example, AIA's version states that "The Design/Builder warrants to the Owner that...the construction will be free from faults and defects."[61]

AGC's form discusses this warranty in terms of the "materials and equipment furnished under the Construction Phase [being] free from defective workmanship and materials."[62] EJCDC states this in terms of "all Construction [being] in accordance with the Contract Documents and will not be *defective.*"[63]

---

[61]Subparagraph 3.2.9, Part 2 of AIA Document A191.

[62]Subparagraph 3.7.1, AGC 410.

[63]Subparagraph 6.18.A, EJCDC 1910-40.

All three documents limit the design-builder's "callback" obligations to defects that occur during the first year after substantial completion of the work.

Some design-build contracts provide extended warranties for work that is corrected during the original warranty period. Sometimes called "evergreen" provisions, these clauses will establish a new 12-month warranty on any replaced or corrected work, giving the design-builder a longer exposure for callbacks. If evergreens are used, it is prudent for the design-builder to establish some outside date that extinguishes its obligations to return to the site.

14.13.2 Warranty for defective design

The responsibility for defects in design after substantial completion has been made more difficult to understand than might be expected, given the language of current design-build contracts and the differences with how this issue is handled under other forms of project delivery.

The design professional working directly for an owner generally does not contractually warrant that its design will be free of defects or fit for the purposes intended. Instead, its obligation is to provide a design that meets prevailing industry standards. If the design professional fails to meet this standard, it will be held accountable to the owner under a negligence theory of liability. Likewise, the contractor working directly for an owner on a design-bid-build approach does not face liability if the project does not function as the owner anticipated, since it is obligated to construct the work only in accordance with the design furnished by the owner. Thus, unless the A/E has been negligent, the owner on a design-bid-build approach ultimately faces the risk of design defects.

The single point of responsibility feature of design-build might be expected to offer more than this. However, the design-build documents sponsored by EJCDC and AGC adopt philosophies on this issue that parallel those found in a design-bid-build setting. EJCDC specifically states that the standard of care for design professional services will be the "care and skill ordinarily used by members of the engineering profession practicing under similar conditions and at the same time and locality."[64] AGC does not specifically discuss design standard of care, but requires that the design-builder "procure the design...and construction...and shall exercise reasonable skill and judgment in the performance of its services."[65] This appears to be identical in substance to the EJCDC approach relative to design. AIA does not have a "standard of care" clause in its design-build family of documents. The problem is exacerbated by the fact that the AIA, EJCDC, and AGC standard design-build forms limit warranties and the one-year "callback" to construction defects, with no mention of defects in the design.

Because many owners will be choosing design-build to obtain a single point of responsibility relative to the integration of design and construction, the

---

[64]Subparagraph 6.01.A, EJCDC 1910-40.

[65]Article 3, "Contractor's Responsibilities," AGC 415.

DBIA Contracting Guide stated that the issue of design defects should be treated more precisely than the language offered by AIA, AGC, and EJCDC. The *DBIA Contracting Guide* offered the following suggested language to address responsibility for the design defect issue[66]:

> **Standard of Care.** The standard of care for all design services performed under this Agreement shall be the care and skill ordinarily used by members of the architectural or engineering professions practicing under similar conditions at the same time and locality. Notwithstanding the above, in the event that the Contract Documents specify that portions of the Work be performed in accordance with a specific performance standard, the design services shall be performed so as to achieve such standards.
>
> **Warranty.** Design-Builder warrants to the Owner that (a) the construction shall be new unless otherwise specified, of good quality, in conformance with the Contract Documents and free of defects in materials and workmanship, and (b) the design shall be performed in accordance with the Contract Documents. Construction or design not conforming to these requirements shall be corrected in accordance with Article _____ hereof.

This language is based on DBIA's philosophy that performance specifications should be used in design-build contracting, since this enables the owner to define its needs and lets the design-builder do what it does best: develop solutions to those needs. When the contract establishes specific performance requirements that can be objectively measured, the design-builder should be obligated to meet such performance requirements, regardless of what the ordinary standard of care in the industry is relative to such issue. DBIA's language is consistent with AIA, EJCDC, and AGC in establishing an industry standard of care when the contract does not have objective performance specifications.

14.13.3 Exclusivity of remedies

Some design-builders attempt to place requirements in the contract that limit the owner's recourse for postcompletion problems to the specific remedies set forth in the warranty section of the contract. These "exclusivity of remedies" provisions generally have the practical effect of

- Creating a time limitation (often one year) on the liability of the design-builder for any failure of the facility to perform
- Limiting the design-builder's obligations for postcompletion defects to the correction of such defects by the design-builder
- Precluding the owner from exercising its right to correct defects on its own

Although these provisions are not used frequently in facility contracts, they are often part of the commercial terms in industrial projects (e.g., power

---

[66]*DBIA Contracting Guide,* pp. 23–24.

generation projects) when the successful performance of the entire project will be based on a performance test.

14.13.4 Waiver of UCC warranties

An area that is related to the "exclusivity of remedies" issue is whether the contract will waive either party's recourse under the Uniform Commercial Code (UCC). As discussed more fully later in this book (Sec. 15.4.2, subsection on contractor's contingent E&O liability coverage), Article 2 of the UCC addresses the sales of goods and has a number of rights and remedies that substantively affect the parties to the sale, including implied warranties and remedies for breaches of such warranties. The nature of design-build makes the application of the UCC inapplicable and can seriously jeopardize the rights and recourse that the parties specifically negotiated into the contract. Consequently, it is prudent for the contract to specifically state that the UCC is not to be applied to design-build contracts and for the parties to disclaim any recourse under the UCC.

14.14 Indemnification

Indemnification provisions are standard tools for distributing the risk of loss on a contract. Sometimes referred to as "hold harmless" provisions, these clauses define a party's obligations to reimburse another party for the losses it incurs or the damages for which it may be held liable. They have a long history of use in the construction industry, and are designed to transfer risks to a party who may be factually more culpable or who has the ability to insure for such a risk. Subject to considerations of public policy or statutory restrictions, courts routinely uphold and enforce indemnity clauses in a variety of contexts.[67] Three major areas of indemnity issues typically arise in a design-build context: general indemnity, patent indemnity, and hazardous waste indemnity.

14.14.1 General indemnity

The design-builder typically provides a general indemnity to the owner. There are three basic types of general indemnity arrangements:

- *Broad form.* These clauses result in the indemnitor assuming any and all liability associated with the event, regardless of fault, and even if the damage is due to the indemnitee's sole negligence. In essence, the indemnitor becomes the insurance carrier for risk of loss to the owner on the project. Because of the severe nature of this risk transference, numerous jurisdictions have adopted statutory provisions which prohibit the indemnification of a party for its sole negligence or intentional misconduct.[68]

---

[67]See Loulakis and Shean, *Risk Transference,* endnote 5, pp. 9–10.

[68]See, e.g., Alaska Stat. § 45.45.900 (1995); *Ga. Code Ann.* § 13-8-2 (1995); *Md. Cts. & Jud. Proc. Code Ann.* § 5-305 (1995); *Va. Code Ann.* § 11-4.1 (Michie 1993).

- *Intermediate form.* Some state statutes forbid indemnification except to the extent that the indemnification relates to damage or loss caused by the indemnitor.[69] The indemnitor assumes any and all liability if it is partially negligent—even if such partial negligence is as little as 1 percent.

- *Limited form.* This is a form that limits one's indemnity obligations to the extent of one's own negligence.

A reasonable indemnity clause should relate to third-party damages or claims for personal injury or property damages, both of which are readily insurable and capable of being reasonably flowed down to subcontractors and suppliers.

The general indemnity provisions contained in all three standard forms are virtually identical in substance and follow the limited form of indemnity approach. For example, AIA's indemnity, which is similar to the time-tested AIA Document A201 form of indemnity,[70] states as follows:

> To the fullest extent permitted by law, the Design/Builder shall indemnify and hold harmless the Owner, Owner's Consultants, and agents and employees of any of them from and against all claims, damages, losses and expenses, including attorneys' fees, arising out of or resulting from the performance of the Work, provided that any such claim, damage, loss or expense is attributable to bodily injury, sickness, disease or death or to injury to or destruction of tangible property (other than the Work itself), but only to the extent caused in whole or in part by negligent act or omission of the Design/Builder, anyone directly or indirectly employed by Design/Builder or anyone for whose acts the Design/Builder may be liable, regardless of whether or not such claim, damage, loss or expense is caused in part by a party indemnified hereunder.[71]

AGC and EJCDC have similar provisions.[72] The *DBIA Contracting Guide* endorses the language on general indemnity contained in all three standard forms, with the addition of an obligation that the design-builder "defend" as well as "indemnify and hold harmless."[73]

In addition to the general indemnity provided by the design-builder to the owner, there is often discussion about whether the owner should provide reciprocal indemnification coverage to the design-builder to cover the negligence of the owner or its agents or separate contractors. Although this can be a contentious issue during contract negotiations, it is not typical for the owner to provide such indemnity, since any negligence of the owner is likely to be "passive" when compared with the negligence of the contractors

---

[69]See, e.g., *Gen. Oblig. Law* § 29C (1994); *Minn. Stat.* § 337.02 (1993).

[70]Case law seems to support the position that Subparagraph 3.18.1 of AIA Document A201 (1987) is a limited form of indemnity. See, e.g., *Dilliard v. Shaughnessy, Fickel, and Scott, Inc.,* 884 S.W. 2d 722 (Ct. App. Mo. 1994) (A201 indemnity provision requires contractor to indemnify owner or architect for that portion of damage caused by general contractor or its subcontractors).

[71]Subparagraph 11.5.1, Part 2 of AIA Document A191.

[72]Subparagraph 6.19.A, EJCDC 1910-40; Subparagraph 11.1.1, AGC 410.

[73]*DBIA Contracting Guide,* pp. 24–26.

working on the site. Owners who perform significant work at the site with other contractors should be willing to provide appropriate coverage for the wrongful acts of those separate contractors. Additionally, some owners are willing to provide indemnity once they take over the facility—as might be the case when testing will be done by the owner's operators under the direction of the design-builder.

14.14.2   Patent indemnity

Patent indemnity is an issue that frequently arises on design-build projects where the engineer is the lead. In a direct design-build situation, this indemnity is typically provided by the design-builder to the owner. If, however, the owner has specified a particular process or uses prescriptive design specifications in its RFP, there are logical reasons to require the owner to provide a cross-indemnity to the design-builder.

AIA, AGC, and EJCDC forms all contain provisions requiring the design-builder to indemnify the owner for patent right violations. The primary differences among the forms is in how they treat the owner's role in possibly creating the patent violation. For example

- *EJCDC*. EJCDC calls for indemnity for "any infringement of patent right or copyrights incident to the use in the performance of the Work or resulting from the incorporation in the Work of any invention, design, process, product or device not specified in the Conceptual Documents."[74]

- *AIA*. AIA is more limited indemnity than EJCDC, specifically stating that the indemnity does not apply "when a particular design, process or product of a particular manufacturer is required by the Owner."[75] This recognizes that the owner's role may extend beyond that specified in its RFP or program.

- *AGC*. AGC is similar to AIA, focusing indemnity on work "selected by the [Design-Builder] and incorporated into the Work."[76] However, it adds a requirement that the owner indemnify the design-builder if a patent violation occurs on materials, methods or systems specified by the owner.

On engineered projects with a heavy process orientation, the preceding clauses, although workable, do not comprehensively or practically address what happens in the event of a patent problem. The *DBIA Contracting Guide* recommends the following patent and copyright indemnity for those projects where more specificity is appropriate.[77]

---

[74]Subparagraph 6.06.A, EJCDC 1910-40.

[75]Subparagraph 3.2.12, Part 2 of AIA Document A191.

[76]Paragraph 3.5, AGC 410.

[77]*DBIA Contracting Guide*, pp. 26–27.

Patent and copyright indemnification

1. Design-Builder shall defend any action or proceeding brought against the Owner based on any claim that the Work, or any part thereof, or the operation or use of the Work or any part thereof, constitutes infringement of any United States patent or copyright, now or hereafter issued. Owner agrees to give prompt notice in writing to Design-Builder of any such action or proceeding and to provide authority, information and assistance for the defense of same. Design-Builder shall pay all damages and costs, including attorneys' fees, awarded against the Owner or Design-Builder in any such action or proceeding. Design-Builder further agrees to keep Owner informed of all actions and the defense of such actions.

2. In the event the Owner is enjoined from the operation or use of the Work, or any part thereof, in connection with any patent suit, claim or proceeding, Design-Builder shall at his sole expense take reasonable steps to procure the right to operate or use the Work. If Design-Builder cannot so procure the aforesaid right within a reasonable time, Design-Builder shall then promptly, at Design-Builder's option and at Design-Builder's expense: (i) modify the Work so as to avoid infringement of any patent or copyright, or (ii) replace said Work with Work which does not infringe or violate any said patent or copyright.

3. The provisions of Subparagraphs 1 and 2 above shall not be applicable to any action or proceeding based on infringement or violation of a patent or copyright (i) relating solely to a particular process or the product of a particular manufacturer specified by Owner and such processes or products which are something other than that which has been offered or recommended by Design-Builder to Owner, or (ii) arising from modifications to the Work by Owner or its agents after acceptance of the Work. If the action or proceeding is based upon events set forth by the preceding sentence, Owner shall indemnify Design-Builder to the same extent as Design-Builder is obligated to defend Owner in Subparagraph 1 above.

4. The provisions set forth in this "patent and copyright indemnification" Paragraph shall constitute the sole agreement between the parties relating to liability for infringement or violation of any patent or copyright.

14.14.3   Indemnity for hazardous materials

The issue of hazardous waste on construction projects has long provided for indemnification flowing from the owner to the contractor for hazardous materials, as well as prescribing a mechanism for addressing how the contractor is to respond when confronted with such materials.[78] Both AGC and EJCDC treat the hazardous waste obligations of the owner and design-builder in a manner similar to that taken under design-bid-build contracts.[79] The most significant aspects of these clauses include[80]:

---

[78]Subparagraphs 10.1.2–10.1.4, AIA Document A201 (1987); Paragraph 4.06, EJCDC 1910-8 (1996).

[79]AIA Document A191 is silent on this issue. However, AIA Document A491 (*Design/Builder—Contractor Agreement*) incorporates AIA Document A201, which has such a clause.

[80]Paragraph 3.4, AGC 4.10; Paragraph 4.04, EJCDC 1910-40.

- The owner being contractually responsible for hazardous materials discovered at the site

- The design-builder being obligated to stop the work on discovery of any area affected by hazardous materials

- The owner being required to remediate the condition, with the design-builder resuming the work after the area has been mutually agreed on as being remediated

- The design-builder being paid for costs associated with the hazardous materials

- The owner indemnifying the design-builder for any claims or losses arising out of or relating to the hazardous materials

Note that EJCDC specifically states that the owner will not be responsible for any hazardous materials brought to the site by the design-builder,[81] whereas AGC is silent on this issue.

14.15   Suspension and Termination

It is important for the parties to discuss within the contract the rights either party may have to suspend or terminate the work. Except for one significant area, the rights and remedies associated with suspension and termination involve the same principles as those in design-bid-build and other contracting systems. The one exception is the ownership of interest in and use of the design by the owner, which issue is discussed extensively in Sec. 14.6. The following sections will note a few key concepts about the principles of suspension and termination.

14.15.1   Suspension by owner

The owner may need the right to suspend the work if there is a change in circumstances that could affect the design or development of the project. Although there are other contractual vehicles for delaying the project (including the changes clause), it is more typical for the parties to specifically place a suspension of work clause in the contract that will address these issues directly. Some of the questions the parties should consider are

- What costs will the design-builder be entitled to recover in the event of a suspension? Note that most clauses allow the design-builder to receive the costs of demobilization and remobilization.

- Whether the design-builder can receive profit on its costs? In the federal construction contracting arena, there is no profit allowed on a suspension of work claim.

---

[81]Subparagraph 4.04.A, EJCDC 1910-40.

■ How long the suspension can last, in the aggregate and individually, before the design-builder can declare the contract to be terminated?

Each of these questions requires the owner to have a clear understanding of the project development risks it is confronting and to discuss these openly with the design-builder.

14.15.2 Suspension by design-builder

The right of the design-builder to suspend the work is generally tied into the failure of the owner to make payments when due. In fact, the right to suspend is a much more powerful and practical right of the design-builder than the right to terminate for default, since an owner will often be willing to let the design-builder suspend much more quickly than to allow the design-builder to terminate the contract. The most important aspect of this clause, therefore, is the time required to pass before the design-builder can trigger its rights to stop work.

14.15.3 Termination by owner for default of design-builder

The right of owner to default a design-builder is critical to protecting the owner's ability to timely and properly complete the work. The most critical issues associated with this clause are

■ The events of default by design-builder

■ The time given the design-builder to cure the default

■ Any certification requirements or notices to surety required to be given by the owner before effecting the termination

■ The financial exposure to the design-builder for liquidated damages and performance deficiencies when the work is completed by others

Design-builders should pay careful attention to the first issue and attempt to avoid cases where they can be terminated because of a single breach of contract. Because default is such an extreme remedy, an event which triggers a default should be material and/or persistent failures of the design-builder to perform.

14.15.4 Termination by design-builder for owner default

Most design-builders have limited rights to terminate because of owner default, given the extreme hardship to the owner if it has to go elsewhere to complete the work. Events of default which many owners will accept as a basis for termination include (1) failure of the owner to pay within a specific period of time and (2) breach of any representations that the owner has the ability to pay. As with the suspension of work remedy, the keys to this clause are how quickly the design-builder can terminate and the financial recovery to the design-builder in the event it is required to terminate.

14.15.5   Termination by owner for convenience

As discussed in Secs. 14.4 and 14.6, the ability of the owner to terminate the design-builder's services for convenience is an important issue to the owner, since it enables an owner who is confronted with an unviable project to mitigate its costs without being in breach of contract. Although this right is important, the process must consider the interests of the design-builder, since the design-builder's valuable design concepts and management approaches are often the key to the project's success should it be resurrected. It is this reason that makes design-build termination for convenience language different from that found in design-bid-build contracts.

Although AIA, AGC, and EJCDC recognize the owner's right to terminate for convenience, each approaches the process differently. For example, the AGC forms calculate the cost of a convenience termination in two ways, depending on whether the cost-plus AGC contract (AGC 410) or the lump-sum AGC contract (AGC 415) is used. AGC 410 specifies that on such termination the design-builder will receive, in addition to its proven costs (inclusive of demobilization), all of its design phase compensation plus part or all of its construction-phase fee. If the termination occurs prior to commencement of construction, the design-builder receives 25 percent of its construction-phase fee. If it occurs after construction is under way, then the design-builder receives 100 percent of its construction-phase fee.[82] Because there is no separate fee in a lump-sum contracting method, AGC 415 provides that the design-builder will receive the balance of its unpaid design costs, as reflected in the schedule of values, plus a percentage of the contract price—which is a blank in the contract to be filled in by the parties when they execute the agreement.[83]

EJCDC and AGC are similar in that they each recognize the right of the owner to terminate for convenience for any reason. Each requires the owner to pay the design-builder for its costs incurred to date, plus demobilization.[84] However, EJCDC does not specify any premium to be associated with the termination, requiring only that the owner pay reasonable overhead and profit on completed work.[85] Unlike either AGC or EJCDC, AIA expressly limits the ability of the owner to terminate the design-builder's services for convenience to situations where the project is abandoned:

> This Part 2 Agreement may be terminated by the Owner upon 14 days' written notice to the Design/Builder in the event the Project is abandoned. If such termination occurs, the Owner shall pay the Design/Builder for Work completed and for proven loss sustained on materials, equipment, tools, and construction equipment and machinery, including reasonable profit and applicable damages.[86]

---

[82]Paragraph 12.3, AGC 410.

[83]Paragraph 11.3, AGC 415.

[84]Subparagraph 14.03(A), EJCDC 1910-40.

[85]Subparagraph 14.03(A), EJCDC 1910-40.

[86]Subparagraph 12.1.1, Part 2 of AIA Document A191.

As indicated by this referenced language, AIA's reimbursement structure is similar to those of AGC and EJCDC in that it is cost-based. However, because the term "applicable damages" is not defined, the owner could potentially be exposed to the design-builder's expected overhead and profit on the unperformed work, thus creating a controversy in the event of a termination. AGC, by liquidating the premium to an agreed percentage, avoids the controversy. EJCDC does not recognize the right of the design-builder to obtain overhead or profit on unperformed work.[87]

14.16   Limitations of Liability

Clauses which limit the liability of the design-builder to the owner are commonly used in certain industry sectors of the design-build market, particularly when the risks are so great (1) as to place the design-builder's capital structure in significant jeopardy and (2) that the best and most responsible contractors would be unwilling to submit proposals for the project. These clauses are generally enforceable if they are agreed on between parties of relatively equal bargaining positions, are reasonable in amount, and clearly define what liability is being limited. It should be noted, however, that some states have found these clauses to be void as against public policy when they attempt to disclaim liability for a party's negligence.[88] See Fig. 14.4.

There are several types of limitations of liability clauses. Three of the most common in design-build contracting are discussed below.

14.16.1   Overall caps on liability

Normal contracting practices in certain industries will have a maximum limit on the contractor's total, aggregate liability, however caused. The theory to support this limitation is that the risks arising from these projects are so great that contractors are unwilling to expose their companies to these risks in order to gain a limited opportunity.

Most maximum or aggregate liability limitations are expressed as a percentage of the contract price—with ranges all over the lot, even exceeding the contract price. If agreement is reached at all, the amount is often determined according to what the owner and project lender will permit when compared to how much uncovered exposure a contractor will have from the performance of a major vendor or subcontractor.

The parties may develop variations from a blanket maximum or aggregate liability provision by carving out certain types of risks or conduct from the operation of the limitation, or by establishing sublimits. For instance, fraud, gross negligence, and willful misconduct of the contractor might be expressly excluded from the blanket limitation. Also, liabilities of the contractor which

---

[87]Subparagraph 14.01.B, EJCDC 1910-40.

[88]See, e.g., *City of Dillingham v. CH2M Hill Northwest, Inc.*, 873 P.2d 1271 (Alaska 1994).

COMMON FORMS OF LIABILITY LIMITATIONS			
Form of Limitation	**Expression in Contract**	**Benefit to Owner**	**Benefit to Design-Builder**
1. Liquidated damages for schedule delays	Expressed in terms of daily rate for each day of delay beyond a contractually fixed date	Owner does not have to prove exact amount of actual damages and can understand what it will receive if the project is delayed.	Design-builder understands what it will pay if the project is delayed and has limited its liability accordingly.
2. Liquidated damages for performance failures	Usually expressed in terms of a rate for shortfall in performance (e.g., $/kilowatt)	Owner does not have to prove exact amount of actual damages and can understand what it will receive if performance guarantees are not met.	Design-builder understands what it will pay if it fails to achieve performance guarantee and has limited its liability accordingly.
3. Overall liability cap	Usually expressed in terms of a dollar amount	Should reflect more competitive pricing and willingness of certain design-builders to consider the project.	Design-builder can assess overall risk of performing work on the project and recognizes its risk is not unlimited.
4. Waiver of consequential damages	Contract provision	Should reflect more competitive pricing and willingness of certain design-builders to consider the project. Owner should be able to deal with risk through its own insurance.	Design-builder recognizes its risk is not unlimited for damages that are remote.
5. No damages for delay	Contract provision limiting rights of design-builder for delays to a time extension only	Protects owner from claims for indirect costs, such as home office overhead.	There is no benefit to design-builder from this clause.

Figure 14.4

are covered by insurance—such as indemnification or patent liability—may be carved out of the blanket limitation. On the basis of the particular dynamics of the project, other adjustments may be made in order to negotiate an acceptable provision.

14.16.2   Waiver of consequential damages

Consequential damages are the indirect result of an alleged failure and typically include such items as loss of use of a facility, loss of market position, or economic losses faced by customers of the owner. These damages are so significant that many parties to commercial ventures (regardless of whether they are in construction) will require that they be waived by contract. Most owners recognize that contractors are not in a position to be the insurers on a project, and that one catastrophic economic loss could potentially bankrupt a construction company. It is also widely viewed that the owner has a broader opportunity to earn profits over the course of the expected useful life of the facility and is in a much better position to absorb, through third party insurance or self-insurance, the catastrophic economic loss.

It should be understood that consequential damage disclaimers are not intended to totally absolve contractors from liability for consequential damages. The principal, if not exclusive, elements of actual damages which are covered by liquidated damages for late completion and failure to achieve performance guarantees are consequential in nature. Thus through liquidated damages clauses, the owner is able to keep the design-builders responsible for the two most significant and probable sources of consequential loss. These risks, however, have defined parameters and are quantifiable. To address this situation, the *DBIA Contracting Guide* recommends the following language[89]:

> Neither Owner nor Design-Builder shall be liable to the other for any consequential loss or damages, whether arising in contract, warranty, tort (including negligence), strict liability or otherwise, including but not limited to loss of use and loss of profits. Notwithstanding the above, this limitation as to consequential damages is not intended to and shall not affect the enforceability of the liquidated damages set forth in Article _____, which both parties recognize have been developed, in part, to reimburse Owner for some damages that might otherwise be deemed consequential.

The last sentence of this clause is obviously intended to be used when the parties have agreed on a liquidated damages formula for project delays.

14.16.3   No damages for delay

A commonly used limitation of liability clause is the "no damages for delay" clause. This clause attempts to limit the owner's liability to a contractor for

---

[89]*DBIA Contracting Guide,* p. 31.

delays by precluding the contractor from recovering additional compensation for the delay and only giving the contractor an adjustment to the contract performance time. Although there are several recognized exceptions to the enforceability of these clauses, many courts have enforced the clauses to the detriment of the contractor.

Parties to the design-build contract should attempt to balance the legitimate interests of both parties relative to this issue. As opposed to treating this an "all or nothing" issue, the parties could

- Make certain delays, such as those beyond the control of either party, subject to the clause and noncompensable.

- Agree to limit the clause to a certain number of days of delay.

- Agree to a specific cost per day for delay to avoid an unwarranted inflation of the claim by the design-builder and its subcontractors.

The parties should further recognize that some states have statutes that expressly find "no damages for delay" clauses to be unenforceable as against public policy.[90]

14.17 Dispute Resolution

The construction industry has long used alternative dispute resolution (ADR) procedures, largely on the basis that construction demands real-time, cost-effective ways to resolve conflict. Design-build is particularly conducive to ADR given the high level of teamwork that is associated with the owner, design, and construction teams. As a result, many design-build projects will use innovative ADR techniques, including dispute review boards, project arbitration panels, and project mediators.

Although it is relatively easy to draft a dispute resolution clause, the parties are advised to think through several issues:

- How will disputes be formally resolved—through arbitration or litigation? There is a growing trend to having disputes over certain dollar amounts resolved in court through a bench trial rather than through arbitration.

- Will mediation be mandatory before initiation of formal dispute processes? Although mediation is an excellent means of letting the parties amicably resolve disputes, it is not right for every dispute. Sometimes the parties are better served by saving mediation until after there has been a certain amount of discovery and commensurate understanding of the respective positions.

- Will the loser pay attorneys' fees?

- What will be the venue of the dispute process?

---

[90]See B. B. Bramble and M. T. Callahan, *Construction Delay Claims,* 2d ed., Wiley Law Publications, New York, 1991, p. 78.

Finally, the parties should carefully evaluate whether there should be a mandatory cooling-off period before formal dispute processes are initiated. Using *stepped negotiations*—project-level and then senior-level executives—as a condition precedent to filing any formal actions has been proven to be quite effective in resolving disputes early and cost-effectively. Most importantly, any process that preserves the relationship between the parties should be considered for inclusion in the contract.

14.18   Conclusion

The subjects identified in this chapter have been discussed in detail given that they are some of the most critical design-build contracting issues. Of course, many other clauses are important in a design-build contract, including

- Choice of law provisions
- Payment mechanisms
- Confidentiality
- Lender requirements
- Key personnel
- Subcontractor listing and selection

To ensure that nothing is omitted, both owner and design-builder should consult with competent counsel and insurance experts before embarking on the negotiation of any major design-build endeavor.

Insurance and Bonding Issues in Design-Build

15.1 Introduction

As the advantages of the design-build process become more widely recognized, an ever-increasing number of owners, design professionals, and contractors are becoming interested in using and/or participating in the process. The dynamics of today's construction world suggest that the design-build market share will continue to grow at the expense of other delivery systems. However, there are some challenges to the growth and successful use of the design-build process. One of these challenges has been, and continues to be, the implementation of integrated insurance and surety programs that address some of the unique risk issues associated with design-build.[1]

Until the late 1990s, the insurance and surety industries were reluctant to embrace design-build. They feared that the risks resulting from the merger of design and construction created far greater exposure to claims and potential losses than those resulting from traditional project delivery systems where the architect or engineer (A/E) is retained directly by the owner.[2] Under traditional delivery systems, liability for design defects traditionally falls on the A/E, while liability for construction defects traditionally falls on the contractor, who has no privity of contract relationship with the A/E. Construction insurance and bonding products have long addressed liability exposures under traditional project delivery systems and provide fairly predictable coverage.

---

[1]See generally M. C. Loulakis and O. J. Shean, *Risk Transference in Design-Build Contracting,* Construction Briefings, 2d series, 96-5, Federal Publications, April 1996; M. C. Loulakis, "Single Point Responsibility in Design-Build Contracting," in *Design-Build Contracting Handbook,* R. F. Cushman and K. S. Taub, eds., Wiley Law Publications, New York, 1992, p. 29; B. L. Hum, "Insurance Aspects of Design-Build Construction," in Cushman and Taub, *Design-Build Contracting Handbook,* pp. 173–191.

[2]See generally D. G. Gavin and M. C. Loulakis, "The Effect of the New Design Build Contracts upon the Contract Surety," address presented at the 19th Annual Surety Claims Institute, June 23, 1994.

The primary distinction between design-build and all other project delivery systems is the merger of design and construction into a single point of responsibility, making the design-builder accountable for the functional performance of the construction project. This has many advantages to the owner because, among other things, the owner is no longer required to prove whether the A/E or contractor is responsible for a defect in the work.[3] However, if a problem does arise under a design-build project, there is the potential for gaps in coverage, since conventional insurance and bonding are built around the concept of separate responsibility for design and construction.

Notwithstanding initial concern about design-build exposure and risk of loss, the environment for design-build insurance and surety products has rapidly evolved to keep pace with industry needs and owner demands. As those underwriting coverage for professional liability insurance and bonds have become more familiar with how the design-build process works, they have also become more amenable to providing products that can be used by their clients in the design-build sector. In fact, since 1998, a flurry of new professional liability insurance [often referred to as *errors and omissions* (E&O) insurance] policies have been issued to specifically provide coverage for design-build risks. While this shift in philosophy is largely a function of market demand and the reality that design-build is here to stay, it is also based on relatively low rates of problems and claims on design-build projects, as addressed in Chap. 16.

Given the importance of insurance and bonding to the construction industry and the growth of design-build, it is critical for owners and those providing design-build services to have a strong working knowledge of how design-build insurance and bonds operate. This chapter reviews these insurance and bonding fundamentals and discusses, among other things, what risks are covered by existing products. Remember that these markets are quite dynamic and that products are constantly being developed and changed to meet perceived and actual industry demands and underwriting concerns. Consequently, prudent owners and design-builders should be vigilantly informed about the design-build insurance and surety environment and determine what is being done to respond to their individual, as well as industry, needs and concerns.

15.2   Risk Management Overview

Myriad risks can arise on a specific construction project. These can include the risk of

- Economic loss caused by poor performance, such as a contractor's ineffective scheduling and coordination or the failure of an owner to timely perform its duties
- Personal injury and property damage to third parties or employees

---

[3]See Loulakis, "Single Point Responsibility in Design-Build Contracting," (fn. 1, above).

- Force majeure events, such as unusual weather, governmental actions, or labor strikes
- Unforeseen site conditions
- Design deficiencies that affect the economic viability of the project

Some risks (e.g., project financing, currency exchange rates, labor strikes, weather, site conditions) are purely business risks that are inherent to the development or execution of a construction project.[4] Business risks are usually managed by assigning the risk to one of the parties to the contract, with the willingness of the party to accept the risk dependent on its risk tolerance and profit potential. Careful and thoughtful contract administration and planning can help prevent many of these risks from actually arising on the project.

Most business entities are unwilling to take direct responsibility for certain risks, since a potential loss can jeopardize their financial viability. These risks are managed by transferring them to a third party, such as an insurance carrier or bonding company. Some of the most typical insurance and bonding products that are found in the construction industry include the following:

- *Professional liability (E&O) insurance.* This protects the insured against liability arising from negligence, errors, and omissions in rendering professional services.
- *Commercial general liability (CGL) insurance.* This protects the insured from claims by third parties for bodily injury or property damage.
- *Builder's risk insurance.* This protects the insured's property interests in the work during construction.
- *Performance bond.* This provides a guarantee to the obligee (such as the project owner) that a contractor will perform its contractual obligations.
- *Payment bond.* This provides a guarantee to the obligee (such as a project owner) that the contractor will fulfill its payment obligations to subcontractors and suppliers and is written to make subcontractors and suppliers beneficiaries to the bond.

Additional insurance and bonding products are widely available and used in the construction industry, such as pollution insurance, automobile liability coverage, efficacy and business interruption insurance, and bid bonds. These products are discussed briefly in Sec. 15.9.

The sections that follow address several specific insurance and bonding strategies. It is critical to remember at the outset, however, that insurance and bonding are simply components of an overall risk allocation strategy for parties to the construction process. Although insurance is available for almost every conceivable risk, most prudent parties will perform a cost/benefit analysis to

---

[4]See R. J. Smith, "Allocation of Risk—The Case for Manageability," paper presented at the World Conference on Construction Risk III, Paris, France, April 26, 1996.

determine whether the premiums associated with the insurance coverage make business sense. The determination of what, if any, insurance or bonding may be needed should be based on the specific needs and risk tolerance of the party seeking protection, as well as the complexity and difficulty of the underlying project.[5]

For example, some owners may not need the design-builder to provide either E&O insurance or performance bonds, as the reputation and creditworthiness of the design-builder may be sufficient to address any losses covered by such instruments. On the other hand, some owners have the philosophy that the protection afforded by readily available insurance and bonding products makes good business sense. They believe that the security it provides justifies the additional cost to the project, even if the reputation and balance sheet of the design-builder make it appear unnecessary.

It is also important to remember that the contract requirements between the owner and design-builder should not be the sole factor in determining what, if any, insurance or bonding the design-builder needs from those with whom it contracts. The design-builder faces its own set of risks arising from the performance of its contract, and needs to engage in a risk analysis to determine the insurance and performance security that is appropriate for its design subconsultants and trade subcontractors. For example, many contractor-led design-builders will require their A/E to have specific levels of E&O coverage to protect the design-builder's own interests relative to the A/E's responsibility for an acceptable design. Similarly, sureties providing bonds to contractor-led design-builders may require contractors to obtain appropriate E&O coverage from the contractor's A/E to minimize the surety's potential for direct loss.

15.3   Insurance Coverage—Generally

Insurance policies have some common features that can be important to understanding coverage on a design-build project.[6] Some of the more significant features are discussed in this section.

15.3.1   Exclusions and endorsements

The *insuring agreement* of an insurance policy identifies the risks covered by the policy. To understand the full extent of the protection provided by the policy, it is, however, often more critical to review the risks that are excluded from the policy. Exclusions, which are set out specifically and clearly in the policy,

---

[5]Design-Build Institute of America, *Design-Build Contracting Guide,* Washington, DC, 1997, pp. 28–29.

[6]For an excellent treatment of insurance products and coverage issues in construction contracting, see O. J. Shean and D. L. Patin, *Construction Insurance: Coverages and Disputes,* The Michie Company, Charlottesville, VA, 1994; W. J. Palmer et al., *Construction Insurance, Bonding, and Risk Management,* McGraw-Hill, New York, 1996, p. 23.

are an insurance underwriter's way to ensure that the insuring agreement is not construed as covering certain risks. Exclusions are particularly important to consider when evaluating how standard E&O policies and commercial general liability (CGL) policies relate to design-build relationships. As discussed in the sections below, each of these policies has several exclusions that restrict coverage for design-build risks.

Exclusions can be removed through an *endorsement*. Typically underwritten and priced separately by the insurance carrier, endorsements are the most common way of broadening the scope of coverage in an underlying policy without obtaining a totally new policy. An A/E or contractor should not hesitate to seek work on design-build projects simply because it believes that its insurance carrier will rigidly adhere to the standard list of exclusions in underlying policies. Because of the sharp increase in design-build and the interests of insurers in being competitive and responsive, many of the classic exclusions related to design-build are being added back into coverage through endorsements with little, if any, increase in premiums.

15.3.2 Coverage triggers

Coverage under insurance will be triggered either on a claims-made basis or an occurrence basis (see Fig. 15.1). *Claims-made insurance* covers claims made during the policy term (generally one year) for acts that occurred within the policy term. This is obviously quite restrictive, since a claim may not be made until several years after the act giving rise to the damage. To address this issue, most claims-made policies can be expanded to cover acts and claims beyond the term of the policy. Acts or claims occurring prior to the policy date can be covered by making the coverage retroactive to such earlier time. Likewise, "tail coverages" extend the term during which claims may be discovered or reported. Retroactive dates and tail coverages are often subject to additional underwriting criteria and may result in an additional premium depending on the length of the extension beyond the initial policy.

COVERAGE TRIGGERS	
Claims-made	**Occurrence Basis**
Covers claims made during policy term	Covers acts *occurring* during policy period
Can be expanded and can apply retroactively	When claim occurs is irrelevant
Associated with professional liability policies	Associated with commercial general liability, builder's risk and property and casualty insurance

Figure 15.1

Occurrence-based policies do not trigger coverage on the claim, but instead focus on the act (referred to as an *occurrence*) giving rise to the loss. As a result, as long as the act occurs during the policy period, occurrence-based policies will cover the loss regardless of when the claim is made. This offers a substantial benefit over the claims-made policy, since it provides the insured with protection that will never expire.

Parties should recognize that these coverage triggers play an important role in the overall management of risks through insurance vehicles. Because of the differences between claims-made and occurrence-based policies, it is possible that gaps in coverage will be created in administering policies that operate on different coverage triggers.[7] Some circumstances that may create gaps include the following:

- The insured cancels or refuses to renew a claims-made policy and cannot buy a replacement policy.

- The insured cancels or does not renew and chooses not to buy a replacement policy.

- The insured switches from a claims-made to an occurrence-based policy.

- A claims-made policy is renewed subject to an advanced retroactive date.

- The insurer attaches an exclusion for a specific product, accident, or project to the renewal of a claims-made policy.

Consequently, if any member of the design-build team anticipates the protection of risks through a claims-made policy, as will be the case with E&O coverage, it is important to take some careful contract administration steps to ensure that these, and similar, events will not impede coverage.

15.3.3 Policy limits

All insurance has specific policy limits for the insurable event. The limits of the policy, which depend on the nature of the business and the type of project, are expressed in terms of a limit per claim and an aggregate limit over the life of the policy. Keep in mind that these limits are applicable not only to the actual damages recoverable from the insured but also to the costs of defending the claim. In design-build liability cases, the costs of defense can be substantial and can reduce the monies available to pay the actual claim.

15.3.4 Waiver of subrogation

Insurance policies often contain clauses that enable the insurer to recover the losses it pays by recovering from the responsible party. This right is commonly referred to as *subrogation,* where the insurer will stand in the shoes of the

---

[7]Shean and Patin, *Construction Insurance* (fn. 6, above), Section 6-4.

insured with respect to its rights against the responsible party. The trend in construction contracting is to require parties to a contract to waive their rights, and their insurer's rights, to subrogation on the theory that if a loss is caused by an insurable event, the insurance coverage should benefit all parties to the transaction.[8] This trend to require waivers of subrogation is also present in the standard design-build contracts published by the American Institute of Architect (AIA), the Associated General Contractors of America (AGC), the Engineers Joint Contract Documents Committee (EJCDC), and the Design-Build Institute of America (DBIA).

15.4 Professional Liability (E&O) Insurance

As discussed briefly in Sec. 15.3, professional liability (E&O) insurance is designed to protect those providing design services from liability arising out of negligence (i.e., malpractice), that results in personal injury, property damage, or economic losses.[9] In addition to covering the economic damage resulting from the negligence, the E&O policy also covers attorneys' fees incurred in defending the action. Although most E&O coverage is written for A/E firms, some owners, developers, and contractors who have in-house designers also obtain such coverage.

Because E&O policies are limited in coverage, there are several key points to consider about E&O coverage as it relates to design-build risks.[10] See Fig. 15.2.

CHARACTERISTICS OF A TYPICAL E&O POLICY	
1.	Claims-made trigger which covers claims made during the policy period
2.	Policies renewed on an annual basis
3.	Negligence standard used to identify coverage; this measures the design professional's conduct against that of a similarly situated professional
4.	If insured contracts for a higher standard of care, damages resulting from a breach of the higher standard will not be covered by the policy
5.	Economic losses are covered, including failure of the design-builder to properly perform the design or late completion caused by design-builder's professional negligence
6.	Generally there is no additional insured status

Figure 15.2

[8]See Deutsch et al., *Construction Industry Insurance Handbook,* Wiley Law Publications, New York, 1991, Section 9.6.

[9]Hum, "Insurance Aspects of Design-Build Construction" (see fn. 1, above), pp. 173–191.

[10]See generally M. V. Niemeyer, "Risk Management and Surety for Design-Build Projects," *CFMA Building Profits,* p. 6 (Sept./Oct. 1996).

Section 15.4.1 first addresses some of the typical features of E&O insurance, including both exclusions and coverage issues. Specific strategies for reducing the gaps in E&O insurance to enable parties to obtain coverage for design risks in a design-build setting are discussed in Sec. 15.4.2.

15.4.1 Typical features of E&O insurance

One of the most important features of an E&O policy is that it is written on a claims-made basis. Occurrence-based policies have historically not been used in E&O insurance, since the liability tail (the period in which claims could be made for occurrences arising during the policy period) is too long for an E&O carrier to adequately underwrite and price. Thus, the claims-made policy requires that the claim, "a demand for money or services, or the filing of a lawsuit or institution of arbitration proceedings naming the insured and alleging an error, omission, or negligent act,"[11] be made within the policy period, which may be extended by the retroactive date or tail coverage. As a practical matter, the claims-made nature of E&O insurance requires that these policies be renewed on an annual basis in order to effectively maintain coverage for prior acts.

Coverage under an E&O policy is triggered on a negligence-based standard of care. This standard measures the design professional's conduct against the degree of care and skill exercised by similarly situated professionals performing similar tasks on similar projects in the same geographic region as the project.[12] If the insured contracts to provide a more severe standard of care, such as "the highest standard of care in the industry," damages flowing from a breach of this standard will not be covered by conventional E&O policies.

As with all insurance policies, exclusions are dependent on the underwriting philosophies of the insurer and custom in the industry. Typical E&O exclusions include the following[13]:

- Warranties and guarantees beyond the negligence standard
- Faulty work
- Services or activities not normally provided by a design professional
- Failure to advise about insurance
- Failure to complete drawings, specifications, or other instruments of service, or failure to process shop drawings, on time or within a defined period of time
- Providing or revising estimates or statements of probable construction costs or cost estimates

---

[11]Hum, "Insurance Aspects of Design-Build Construction," (see fn. 1, above) pp. 175–176.

[12]See Deutsch et al. (fn. 8, above), Section 10.2 [citing *Carter v. Deitz*, 556 So.2d 842, 867 (La. Ct. App.), *writ denied*, 566 So.2d 992 (La. 1990)].

[13]Hum (fn. 1, above) p. 176; Deutsch et al. (fn. 8, above), Appendix A.

- Pollution
- Projects where the insured performs professional services and where construction is performed, in whole or in part, by the insured or an entity under common ownership with, or management and control of, the insured

The last exclusion (commonly known as the *design-build exclusion*) is just one of several exclusions that relate directly to issues that can arise under design-build projects. It can be particularly problematic if the A/E is in a lead position on the design-build team.

One of the reasons that E&O insurance is so important is that it not only covers personal injury and property damage arising from negligent acts but— unlike the coverage afforded under CGL policies—also covers economic losses. This can be important in a design-build setting, where the damages that might arise from professional negligence include the failure to (1) develop complete design documents, which results in greater cost to the design-builder, and (2) specify proper equipment to enable performance to be achieved. Each of these breaches may give rise to direct damages, such as the additional labor, materials, and equipment costs associated with the design deficiency, as well as to delay damages that may be payable as a result of the deficiency.

Although few judicial cases deal with coverage under an E&O policy for design-build issues, a 1995 case, *Bell Lavalin, Inc. v. Simcoe & Erie Gen. Ins. Co.*,[14] provides some guidance. *Bell Lavalin* involved a dispute arising over a subcontractor walking off the project at a point in time when it had completed 87 percent of its work. The subcontractor claimed that it was commercially impractical to complete its work because the design-builder refused to grant its time extension request after more than 6 months of delays on the project. The design-builder terminated the subcontractor for default and finished the work itself. The subcontractor sued the design-builder on a variety of theories and was awarded $7 million in damages by the jury.

The design-builder sought to recover the $7 million award from its E&O carrier, claiming that the failure to grant a time extension was determined by the jury to be a negligent act and the cause of its damages. The 9th Circuit Court of Appeals disagreed, finding that the "cause of the damages...was the simple fact that the [subcontractor] performed 87% of its contract work for [the design-builder] and [the design-builder] refused to pay." In essence, the court concluded that the dispute between the design-builder and its subcontractor involved a breach of contract and other theories and had nothing to do with professional liability.

---

[14]61 F.3d 742 [9th Cir. (Circuit Court) 1995].

15.4.2 Strategies for reducing the design-build gaps in E&O insurance

As is evident from the standard E&O coverage discussion in Sec. 15.4.1, several issues need to be considered in determining how to implement a viable E&O program on a design-build project. Some of these issues relate specifically to the terms and structure of the E&O coverage. Other ways to reduce the risk of gaps for E&O liability coverage are tied into CGL coverage and bonding issues, which are addressed in other sections of this chapter.

Modifying the exclusions to fit the role of the A/E. As discussed in Sec. 15.3.1, the exclusions to an insurance policy are the key to understanding coverage. Common E&O exclusions create some practical problems for A/Es to engage in the practice of design-build. Therefore, the A/E who is involved in design-build, as either a lead or a subconsultant, should have a strong sense of what exclusions need to be deleted or modified and of the willingness of its carrier to engage in this process. The role of the A/E within the design-build team will have a major impact on how such exclusions are viewed.[15]

Consider, for example, the typical E&O design-build exclusion, which generally states that

[A]ny claim based upon or arising out of a project for which the assembly, construction, erection, fabrication, installation or supplying of materials was provided in whole or in part by:

1. The Insured, or
2. A subcontractor of the Insured, or
3. Any enterprise and/or any subsidiary of any enterprise that any Insured controls, manages, operates or holds ownership in or by any enterprise that controls, manages, operates or holds ownership in the Named Insured.[16]

This provision effectively nullifies coverage if the A/E is involved (1) as the lead design-builder, (2) as an employee in an integrated design-build firm, or (3) in a joint venture type of relationship. Although most underwriters claim that the primary intent of the exclusion is to avoid covering faulty construction work, the broad wording of the exclusion covers far more than defective work. Deleting this exclusion completely, or restricting its coverage to faulty construction work, is a strategy for addressing this issue.

Additional problems arise if the A/E is involved in a joint venture relationship. Most E&O policies preclude the A/E from joint venturing with anyone other than another design professional. Therefore, the A/E's policy will not provide coverage to the A/E if the design services performed by the A/E are within a joint venture relationship with a contractor. The problem is compounded by the fact that

---

[15]For a discussion of how the role of the design professional will affect its overall risk, see *Design/Build Issues and Insurance Coverage*, Schinnerer Expert Risk Management Opinion/Note, Victor O. Schinnerer & Company, Inc., Chevy Chase, MD, 1996.

[16]Deutsch et al. (fn. 8, above), Section 10.6.

the joint venture itself, since it contains a contractor, will not be able to obtain its own insurance. The exclusion precluding the A/E from joint venturing must be bought back through an endorsement or be otherwise modified.[17]

Similar to the joint venture exclusion is an equity interest exclusion that does not allow claims to be made by any business enterprise that is wholly or partly owned, operated, or managed by the A/E. If the business deal results in the A/E having an equity position in the project, this limitation should be either deleted or modified. The underwriter may be willing to limit this equity exclusion to those situations where the A/E has more than a certain ownership position (such as 15 to 25 percent).

Because of the problems mentioned above, several of the major E&O carriers have issued new design-build endorsements that eliminate most design-build coverage gaps. These policies are particularly helpful to those organizational structures where the A/E is the lead on the design-build team or in a joint venture with a general contractor.[18] In these cases, the newer policies are permitting the design-build entity—whether it is a joint venture, limited liability company (LLC), limited liability partnership (LLP) or corporation, to be the named insured on the policy.

Note, however, that even the most broadly defined design-build E&O policy does not cover damages caused by faulty work, guarantees, or warranties. These remain issues that are considered beyond the scope of professional responsibility and are either pure business risks (in the case of guarantees or warranties) or the responsibility of others (in the case of faulty work).

Contractor's contingent E&O liability coverage.   When the A/E is a subconsultant to the general contractor on a contractor-led design-build team, the contractor relies on the A/E's E&O policy for coverage of design defects. This creates some potential problems since (1) there is no guarantee that the A/E's E&O policy will be renewed and available when a claim is made, (2) the A/E's E&O policy's aggregate limits may be exhausted from claims on other projects, and (3) the A/E will not typically have sufficient assets to pay for a loss in the absence of insurance. As a result, if a claim is made against the contractor for design defects and there is no coverage, the assets of the contractor are at risk.

Several insurance carriers have addressed this risk by developing products to cover the contractor's vicarious liability for negligent design.[19] Although each product is different, all essentially provide the contractor with coverage in the event that the A/E's E&O policy is either not available or insufficient to cover the actual losses arising out of the negligent act or omissions (commonly called *excess* or *surplus* coverage). These policies also provide coverage

---

[17]Loulakis and Shean, *Risk Transference* (see fn. 1, above), p. 14.

[18]See, e.g., A. L. Abramowitz, "Professional Liability Insurance in the Design/Build Setting," *The Construction Lawyer* **16**(3):3–7 (Aug. 1995).

[19]The insurance companies providing a program for contractor's professional liability coverage are constantly changing. See Niemeyer, "Risk Management and Surety" (fn. 10, above), p. 10.

on a primary basis to the contractor for the negligent acts or omissions of those design professionals who are employees of the contractor, or for services that the contractor performs that are construction management in nature.[20]

Project E&O insurance.   In addition to E&O policies that are underwritten directly for a company (called *practice policies*), the insurance market has developed project-specific E&O policies.[21] These policies offer several advantages, including

- *Coordinated insurance coverage.* These policies should cover all design professionals performing services on the project, regardless of whether they work for the A/E, owner, or contractor. This ensures that there is adequate professional coverage and limits for all design services. It also avoids having insurance companies fight among themselves as to accountability for the damages giving rise to the claim.

- *No coverage erosion from other projects.* Since the policy is dedicated to a specific project, there is no concern that coverage will be eroded by claims on other projects.

- *Longer and predictable terms of coverage.* These policies extend the term of E&O coverage for a contractually specified period, avoiding the issue of retroactivity periods and whether the individual design professionals will maintain their coverage sufficiently after the project is completed.

- *Broader coverage.* Project-specific policies may include coverage for a wider variety of claims and with greater limits of coverage than a practice policy.

Of course, the cost of a project-specific policy may not justify the benefits. The policies tend to be expensive, although prices do fluctuate downward in a soft insurance market. Moreover, because an A/E's overhead will contain a factor for the cost of its practice E&O policy, there is a double payment for design liability protection. Finally, it should be noted that even though project policies limit the named insureds to design professionals, it is rare that either the owner or the contractor-led design-builder is afforded such status. While this may not have significant practical ramifications (since the policy is project-specific), it is often perceived as a loss of control over the carrier and can be problematic to owners and contractors who expect to be protected under the policy.

15.5 Commercial General Liability (CGL) Insurance

Commercial general liability (CGL) insurance protects an individual or business entity from losses due to (1) bodily injury and property damage to a third party, (2) personal and advertising injury, and (3) medical payments. All prudent businesses will have CGL insurance as a central component of their overall risk management program. Thus, unlike E&O coverage, which is obtained only by

---

[20]It should be remembered that there are many ways other than insurance to address the liability of the contractor and its subconsultant A/E for design defects. These are addressed in Chap. 16.
[21]Hum (see fn. 1, above), pp. 178–179.

those performing design services, CGL coverage will be in place for all parties involved on a construction project, including the owner, design-builder, A/Es, contractors, and subcontractors. CGL coverage works in the same manner as discussed in Sec. 15.3.1, and *coverage* is defined by the insuring agreement and modified and further defined by exclusions and endorsements. See Fig. 15.3.

15.5.1 Exclusions

As with E&O policies, the list of exclusions to a CGL policy can be extensive and should be carefully examined. Although a detailed review of such exclusions as they pertain to the construction industry is not appropriate for this chapter,[22] it is worthwhile to note some of the most common exclusions in these policies:

- *Workers' compensation.* There is no coverage for damages that are specifically addressed by worker's compensation laws.

- *Employer's liability.* This excludes employee-related injuries, since they are covered by workers' compensation. Note, however, that this does not preclude coverage in "action over" cases, where the injured employee of a design-builder sues the owner, who in turn sues the design-builder on an indemnification theory.

- *Pollution.* There are broad pollution exclusions in CGL policies.

- *Professional liability.* This excludes coverage for professional acts or omissions by the insured or those working under the insured.

- *Work product.* This excludes coverage for work performed by or on behalf of the insured once such work has been completed.

CHARACTERISTICS OF A TYPICAL CGL POLICY	
1.	Occurrence-based trigger
2.	Covers bodily injury and property damages to third parties, medical payments, and personal and advertising injury
3.	Does not cover pure economic losses
4.	Additional insured status can be conferred on any number of parties, enabling them to have direct right of access to carrier
5.	Waivers of subrogation are a significant contractual issue

Figure 15.3

---

[22]See Shean and Patin, *Construction Insurance* (see fn. 6, above), Section 6-1 for a more comprehensive review of these policies.

- *Care, custody, and control.* This excludes coverage for work that was under the care, custody, or control of the insured.

- *Joint ventures.* Unless modified, the insured's activities in a joint venture are not covered.

Keep in mind that the exclusion language in a typical CGL policy is quite comprehensive, with limitations and exceptions often embodied within the exclusion itself. It should never be assumed automatically that something is excluded or covered without careful review of the policy and consultation with an insurance expert.

15.5.2   Key differences between CGL and E&O coverage

Although CGL and E&O insurance cover very different types of claims, the two policies have a number of structural differences that should be considered:

1. *Occurrence-based trigger.* Unlike E&O insurance, CGL insurance policies are generally written on an occurrence basis, triggered by an event that occurs during the policy period. This means that claims made after the policy period are covered.[23]

2. *Additional insureds.* A design-builder's CGL policy will have a number of additional insureds, including the owner, owner's consultants, and owner's lender. This gives these insureds a direct right of recourse against the carrier. E&O policies rarely permit anyone other than the A/E to be afforded status as an insured. It is vital to remember that while it is common to have multiple insureds in a CGL policy, these additional insureds must be specifically identified in the certificate of insurance issued for the project. Except for this requirement, the CGL policy will limit who can make claims under the policy.

3. *Scope of coverage.* CGL policies do not cover pure economic losses, such as delay damages, loss of productivity, or increased expense associated with leasing alternative facilities.[24] E&O policies cover these damages if they arise out of the negligent acts or omissions of an insured.

15.5.3   Strategies for reducing the design-build gaps in CGL insurance

As with E&O coverage, some gaps in CGL coverage may arise because of the nature of design-build services and the fact that most CGL coverage is not written with design-build in mind. Some of the most significant gaps are discussed below.

---

[23]F. J. Baltz and G. E. Bundschuh, *Insurance & Bonding for a Design-Build Project,* Construction Briefings, 2d series, 97-11, Federal Publications, Oct. 1997, p. 4.

[24]T. R. Tennant, *Design-Build Professional Liability Insurance: Are You Covered?,* The American Institute of Constructors, St. Petersburg, FL, Dec. 1997.

Professional liability exclusion.   As noted earlier, most CGL policies specifically exclude from coverage the design portion of the design-build contractor's services.[25] This exclusion is significant to those design-build teams that are led by a general contractor, as well as to those situations where the contractor and A/E team through a joint venture, LLC, LLP, or similar corporate or partnership structure.

The typical professional services exclusion in the CGL insurance policy provides, in pertinent part, that[26]

> This insurance does not apply to "bodily injury," "property damage," "personal injury," or "advertising injury" arising out of the rendering or failure to render any professional services by or for you, including:
>
> 1. The preparing, approving, or failing to prepare or approve maps, drawings, opinions, reports, surveys, change orders, design, or specifications; and
>
> 2. Supervisory, inspection or engineering services.

As is evident by the language of the policy, this exclusion eliminates coverage for design work performed by a contractor's personnel, A/Es performing work on behalf of a contractor, and for other professional services, including construction supervision, inspection, and engineering services. The rationale for this exclusion is that E&O insurance is better suited to provide coverage for these risks.

The broad language of the above-mentioned exclusion gives insurers the ability to deny coverage for bodily injury and property damage losses arising out of activities that are commonly performed by the contractor rather than design professionals. This argument was successfully made in the case of *Harbor Ins. v. Omni Construction.*[27] In this case, property adjacent to the construction site was damaged as the result of excavation at the construction site. Omni (the general contractor) sought recovery from its CGL carrier for the costs of repairing the adjacent property. The CGL carrier denied coverage because the damages were caused by errors in the design of the sheeting and shoring system prepared by Omni's subcontractor—thus triggering the professional services exclusion in the policy.

Omni argued that these services were not professional services within the meaning of the exclusion, but were rather "means and methods" of construction. The Washington, D.C. Circuit Court of Appeals rejected Omni's position, concluding that preparation of sheeting and shoring designs were professional services. The court was persuaded not only by the clear language of the exclusion but also by the fact that the sheeting and shoring plans were signed and sealed by a professional engineer for the trade subcontractor.[28]

---

[25]Shean and Patin, *Construction Insurance* (see fn. 6, above) Section 8-3.

[26]Idem, p. 258.

[27]912 F.2d 1520 (DC. Cir. 1990).

[28]Several cases address the professional services exclusion in the context of a CGL policy. For example, in *National Union Fire Insurance Co. v. Structural Systems Technology, Inc.,* 756 F. Supp. 1232 (E.D. Mo. 1991), the insured, the general contractor, had contracted with a steel fabrication subcontractor for work on a radio tower. During the repair of some steel rods furnished by the subcontractor, the structure collapsed. The owner's claim was based on the negligence in

The *Omni* case created some major concerns for both contractors and design-builders. As a result, two new endorsements were issued in 1996 to address the professional services exclusion. One, known as *Exclusion—Contractor's Professional Liability Endorsement* (ISO Form CG 22 79), allows coverage for incidental design services performed as part of the contractor's means and methods of construction. This would have addressed the problem that Omni faced in dealing with its policy. It does, however, specifically exclude damages resulting from professional design services, whether performed by the insured or one of its subcontractors, on a global basis. In short, its intent is to provide limited coverage for situations where trade contractors may take on peripheral design work as an adjunct to the trade work—as opposed to allowing coverage for wholesale and substantial design services.

The most important endorsement affecting design-builders is the *Limited Exclusion—Contractor Professional Liability Endorsement* (ISO CG 22 80). This allows coverage for bodily injury or property damage from professional design services in connection with a project that the contractor is also constructing and was developed specifically to address design-build relationships.

Work product exclusion.   As noted above, typical CGL policies restrict coverage of damages to the property caused by the removal and/or replacement of the faulty work performed by the named insured. The purpose of this exclusion is to preclude coverage of losses incurred as a result of defective construction that requires the removal of conforming work in order to perform the necessary repairs. This is done to encourage high-quality work on the part of the contractor and ensure that the risk of poor work is borne by the party doing the work—not simply shifted to the insurance company.[29] As a result, unless the design-builder is able to "buy back" this exclusion, this problem will generally be considered a business risk and the design-builder will be required to address it through effective contract administration and performance.

Integrating CGL with other coverages.   While it is possible to adjust the exclusions and endorsements on a CGL policy to broaden the CGL coverage, it is important to remember that CGL does not cover economic losses. Thus, even

the design, manufacture, and assembly of the radio tower. As a result of these allegations, the insurer denied coverage under the applicable CGL policy because of the professional services exclusion. See also *Womack v. Travelers Insurance Co.*, 251 So. 2d 463 (La. App. 1971), where a CGL claim was denied on the basis that failing to transmit plans to the appropriate utility constituted the negligent performance of a professional service. The *Womack* court was not persuaded that the failure to transmit was simply an administrative function as opposed to a professional function. See Shean and Patin, *Construction Insurance*, Section 8-3.

[29]On a similar issue, there is a fair body of substantive law interpreting the CGL policy and determining that it does not indemnify the insured for the costs of correcting or replacing the work performed by the insured, or an entity under its control, under the theory that poor work quality should be a risk borne by the contractor, and not the insurance company. Other courts have interpreted the CGL policy to be ambiguous in connection with coverage for these breach of warranty claims. Accordingly, this ambiguity would be construed against the insurance company, as the drafter of the insurance agreement, and a breach of an insured's warranty of high-quality work would likely be covered. See Hum (fn. 1, above), p. 188.

with modifications to the standard policy, neither contractors nor A/Es who lead the design-build team should rely on CGL as a catchall for any claims of professional liability. This policy, if applicable at all, will only cover the risk of personal injury and property damage.

Since a majority of the claims based on professional liability are for economic losses, the risk protection afforded by CGL should be augmented by practice policies for E&O (either with the A/E or contractor), project E&O policies, and some of the other strategies identified in Sec. 15.4.2. Moreover, it is important to remember that indemnification clauses and other contractual risk allocation vehicles will play an important role in managing these risks.

15.6   Builder's Risk Insurance

Builder's risk insurance protects the named insureds against physical damage to the structure and all materials to be incorporated into the structure.[30] Although builder's risk insurance is not unique to the design-build delivery system, it is used on a high percentage of design-build projects. Typically, a builder's risk property insurance policy is purchased by either the owner or the design-builder (at the expense of the owner) and covers the replacement or repair value of the structure and materials to be used in the structure. Sometimes builder's risk insurance may be available as an endorsement under the owner's existing standard property policy.

Builder's risk insurance policies are typically written on either (1) an "all risk" basis or (2) a "named perils" basis. All-risk policies are intended to protect against loss incurred as a result of all the potential risks to the property except those expressly excluded, while named-perils policies, for the most part, cover only the risks identified within the policy. However, named-perils policies can be supported by interlocking insurance policies commonly known as "difference-in-conditions" policies. The difference-in-conditions policy covers all perils except those covered by the named-perils policy, which is specifically excluded by this supplemental policy. The integration of these two particular insurance policies essentially creates an all-risk policy.

15.6.1   Exclusions

Regardless of whether the form is all-risk or named perils with the addition of a difference-in-conditions policy, numerous exclusions are applicable. Although exclusions can differ greatly among different insurers, some common exclusions are[31]:

- Loss of use or occupancy
- Penalties for delay in completing the project

---

[30]Baltz and Bundschuh, Insurance and Bonding (see fn. 23, above), pp. 11–12; Deutsch et al., *Construction Industry Insurance Handbook* (see fn. 8, above), Section 9.1.

[31]See Deutsch et al. (fn. 8, above), Sections 9.5–9.7.

- Loss caused by war, civil rebellion, insurrection
- Mechanical breakdown
- Loss caused by freezing or frost
- Loss due to faulty work or design

Obviously, the latter exclusion is important to a design-build risk management program and needs to be carefully interwoven into other insurance vehicles. Careful attention must be paid to what property may be excluded from coverage as well. Often items such as automobiles, landscaping, and tools and equipment not incorporated into the completed structure are not included.

15.6.2   Rights of subrogation

A key issue in connection with builder's risk insurance policies used on a design-build project is the waiver of subrogation. As noted in Sec. 15.3.4, after the insurance company pays a claim, it has the option to pursue the party responsible for the loss. This is the insurance company's method of recouping the amount paid on the claim and maintaining the culpability of the party at fault for the loss. If a party partly responsible for the loss obtains a waiver of subrogation from the insurance company or the insured, the waiver will protect the party from incurring any loss from the carrier. The waiver of the right of subrogation has the effect of shifting the loss to the insurance company.

15.6.3   Strategies for using builder's risk on design-build

Parties engaged on a design-build project should consider several questions with regard to builder's risk insurance:

- *Who should obtain the policy?* The entity that obtains the policy controls the risks that are being covered by the policy and the extent of the limits and deductibles associated with a loss. Therefore, it is often beneficial for the design-builder to take the lead in this area to ensure that appropriate recourse is available for specific project risks that concern the design-builder.
- *Who should be identified as an insured?* Because of the importance of control and having direct discussions with the carrier when a loss occurs, it may be prudent for the owner, design-builder, and major subcontractors to be identified as insureds on the policy. Unlike E&O coverage, which generally does not permit coverage for additional insureds, most builder's risk carriers do not have a problem with a broad inclusion of parties as additional insureds under the policy.

- *What happens if the loss is caused by design errors or defective work?* Because exclusions for design errors or defective work are present in most builder's risk policies, the design-builder will not be paid under the policy for the costs of correcting the negligent design or faulty work. However, coverage will likely be afforded to protect the insureds from the consequential damages resulting from the incident, notwithstanding that design errors or defective work may have contributed to the problem.

- *Does the design-build contract play a role in builder's risk?* Many of the rights and responsibilities of the parties relative to builder's risk coverage can be defined by the design-build contract. This can include coverage, deductibles, waivers of subrogation, and how monies obtained from the carrier are to be handled.

- *Can the scope of coverage be extended?* There is a specific endorsement available to the builder's risk policy for "soft cost" coverages which would cover items such as (1) A/E fees, (2) insurance premiums, (3) professional fees (i.e., attorney's fees), (4) interest expenses in connection with construction finance costs, and (5) real estate property taxes. However, most soft-cost endorsements exclude factors such as necessary improvements to correct original deficiencies, weather delays, liquidated damages, licensure problems, excess reprocurement delays, and labor strikes.

15.7 Other Insurance Products

A variety of other insurance products are used on construction projects. Some of this insurance, such as worker's compensation, is well known by those in the construction industry and not unique to design-build risks. Other insurance, such as efficacy insurance, is not typically used and can be beneficial for capitalizing certain risks on a design-build project, particularly if the design-builder takes an ownership position in the project.

The following sections briefly identify other insurance products that are available for those working on the design-build team. Please note that the short discussion of these insurance products is not intended to be a commentary on their use or importance. Rather, it is a reflection of the fact that little is unique or problematic about this insurance relative to the design-build process.

15.7.1 Workers' compensation

Workers' compensation insurance is a form of no-fault insurance that provides replacement wage benefits and medical care for injured employees. This insurance is intended to act as a ceiling to an employer on liability to an employee for injuries suffered while in the course of employment. However, because of indemnification clauses, it is common for an employer to be brought into an action filed by an injured worker against others (such as the owner, A/E, or subcontractors) and to bear the consequences of the liability for such parties.

Although the construction industry is familiar with workers' compensation policies, it should be noted that for those design-builders who are structured as either a joint venture or LLC, LLP, or other special-purpose corporation, there may be an issue as to who is technically the "employer" of an injured worker. Care should be taken to address this in developing insurance coverage for the project, as well as in formulating contracting strategies (such as indemnification responsibilities) among the parties.

15.7.2   Automobile coverage

Most CGL policies exclude coverage arising out of claims from automobiles. As a result, coverage is afforded through a specific business auto policy.

15.7.3   Pollution liability

CGL and E&O policies frequently exclude many pollution-based claims from coverage. To cover this gap, pollution liability policies are used. Their primary purpose is to insure the costs associated with cleaning up unknown, preexisting environmental damage and to address environmental damage that may occur during the duration of the insurance policy. Although pollution liability insurance was originally offered on a claims-made basis only, insurance companies have started to offer pollution coverage on an occurrence basis.

It should be noted that there has been a trend in the E&O market to include pollution coverage within the standard E&O policy. This is well suited to the design-build environmental remediation industry, since a single product should eliminate any gaps or disputes between insurers over which coverage applies in the event of a problem.

15.7.4   Excess liability

Excess liability insurance, as its term suggests, provides protection for major damages that an insured could not reasonably expect. This policy essentially provides additional limits for all major policies, including automobile, CGL, and sometimes E&O insurance.

15.7.5   Efficacy coverage

Efficacy insurance is a type of coverage that protects against losses from contract provisions that impose economic liability on the insured. It is intended to cover risk of loss for liquidated damages due to schedule delay, performance shortfalls, and other warranties or guarantees. Because this insurance is generally considered quite costly, its use is typically subject to a severe cost-benefit analysis.

15.7.6   Wrapups

A "wrapup" insurance program is intended to centralize insurance and loss control for the entire project. These policies are typically written for a single project and controlled by the owner, although contractor-controlled wrapups are becoming more common.

The benefits to a wrapup program are similar to those described in 15.4.2, subsection on project E&O insurance, relative to project-specific E&O policies, except that the breadth of the coverage is far greater. The primary advantage is that a wrapup enables the owner to eliminate the inefficiency and cost associated with separate insureds and separate insurers in all areas where insurance is obtained, including general and professional liability. Such programs are not used mainly because they can be quite costly and administratively burdensome. As a result, it is generally believed that the most appropriate project for a wrapup insurance program is one of significant cost and complexity.

15.8   Performance Security—Generally

Performance security has long been used in the construction industry to ensure that there is appropriate recourse in the event a party fails to perform its contract obligations. The most typical form of performance security in the United States is the performance bond, although the use of letters of credit and corporate guarantees is increasing.

Unlike insurance, performance security is tied directly into the business arrangements between the parties. Consequently, the best management of performance risk (which is the primary purpose of these instruments) is to ensure that contract terms are negotiated in an appropriate manner. Parties should not believe that their performance needs on a project will be achieved simply because a party has offered a performance security. One should also note that performance security on design-build projects is often contingent on having the risk of losses for design defects covered by E&O insurance. The sections that follow review the typical forms of performance securities and how they apply to design-build relationships.

15.9   Surety Bonds

A *surety bond* is a credit transaction between a party providing construction services (called the *principal*) and the surety for the benefit of another party (called the *obligee*). The surety lends its financial backing to the project and guarantees to the obligee that the obligations of its principal will be met. The surety also guarantees that, in the event that the principal defaults or fails to perform in accordance with the contract, it will step in and perform on the principal's behalf.[32] Design-builders, contractors, and subcontractors are

---

[32]See Palmer et al., *Construction Insurance, Bonding, and Risk Management,* (see fn. 6, above), p. 23.

typically principals, while owners and those who contract with subcontractors are obligees.

Generally, there are three forms of surety bonds issued on construction projects, with each guaranteeing a different element of performance:

- *Bid bond.* This bond is used to qualify those design-builders submitting bids or proposals on a project. It guarantees that, if the bid or proposal is accepted, the design-builder submitting the bid or proposal will enter into a contract with the owner to perform the work specified on the project. The bid bond also ensures that the design-builder will provide whatever additional guarantees (i.e., bonds) are necessary to comply with the requirements specified in the contract documents.

- *Payment bond.* This bond guarantees the design-builder's contractual obligations to pay specified subcontractors on the project. Payment bonds are especially prevalent in "no lien" jobs, such as public works projects, because they provide subcontractors with a means of recovery when they do not possess lien rights.

- *Performance bond.* This bond guarantees to the project owner the performance of the design-builder's obligations under the contract in accordance with the terms and specifications incorporated into the same. Importantly, the surety's guarantee of performance is no greater or less than that of its principal as defined by the contract documents.

Because the surety bond is a credit transaction, as opposed to a transfer of risk, the surety possesses both a contractual right to indemnification and an equitable right of subrogation against its principal. The surety's indemnification rights—typically set forth in an instrument called a *general agreement of indemnity*—contemplate that the surety will be reimbursed and held harmless by its principal for all monies expended by the surety in connection with its performance on the bond. This means that the assets of the principal, as well as the assets of any individual executing the General Agreement of Indemnity to induce the surety to provide bonding, will be at risk in the event of a loss by the surety. The surety's right of subrogation entitles the surety to funds payable to its principal in the event that the surety is called on to perform under the bond.

15.9.1 Design-build considerations in the surety market

The design-build concept has been particularly challenging for the construction surety industry.[33] This is largely because the three types of bonds provided by

---

[33]See generally K. Ryan, "Bonding Design-Build Contracts," Chap. 5 in Cushman and Taub, *Design-Build Contracting Handbook*; J. Carter, "Construction Exposures From Professional Liability," *CFMA Building Profits,* p. 45 (Nov./Dec. 1995).

sureties were originally developed in contemplation that the plans and specifications would be completed before the construction contractor provided a price and that the contractor had no responsibility for design risks. This enabled surety underwriters to focus on their clients' abilities to perform and manage the construction process—not on an integrated design and construction program.

The single point of responsibility feature of design-build, coupled with the role of the A/E as part of the principal's team, changes the paradigm by which sureties have evaluated risk and provided credit. Because the surety's liability to the owner is coextensive with that of the principal, the surety is concerned about the contractor's assumption of design responsibilities, warranties of design, and the absence of the design professional as a neutral third party who can capably monitor the project performance of the contractor.

As a result of this paradigm shift, many sureties originally took the position that they would not provide performance bonds for design-build projects. Today, because an increasing number of contractors are engaged in design-build and the delivery system has been quite successful, sureties have relaxed their position and demonstrated a greater willingness to provide performance bonds that guarantee all the work, including design.

Part of this change in philosophy arises out of the fact that owners who require performance security for their design-build projects will often insist that any surety bond provide full coverage for design and construction services. For example, the DBIA's *Design-Build RFQ/RFP Guide for Public Sector Contracts*[34] makes the following recommendation regarding bonds:

> The form of the contract bond form must specify that the bond shall secure the design-builder's faithful performance of the entire contract, including any and all professional design services necessary to complete the contract. Separate bonds, or construction only bonds, must not be allowed to secure a design-build contract.[35]

Because DBIA espouses best practices in design-build contracting and procurement, this approach is being viewed favorably by public and private owners, requiring sureties to respond with products that offer this service.

This does not mean, however, that the surety industry has embraced either design-build or the notion that bonds should cover design services. A more in-depth discussion of surety concerns about design-build is set forth in the following sections.

Design risk.   As noted above, the primary concern for the surety in connection with the bonding of design-build firms is the increased exposure created by the joining of the design portion and construction portion of the project under the bond. This increased risk arises from a variety of factors, including (1) contractual performance guarantees, (2) functional failure of the design, (3) shop drawing review and analysis, (4) coordination of the plans and specifications

---

[34]Design-Build Institute of America, *Design-Build Manual of Practice,* Document 210, 1996.

[35]Idem, p. 40.

for subcontractors, and (5) permitting requirements. The surety's problem with this risk is compounded by the fact that the bond covers 100 percent of the contract price, which theoretically means that liability for such amount could be due entirely to defective design. E&O coverage, which is specifically designed to deal with design risk, typically represents only a small fraction of the overall contract price on large projects, since it is generally written for amounts between $1 and $15 million. Thus, the surety's fear is that it is being subjected to far more risk than is commercially practical.

The surety's increased responsibility is evidenced by the case *Nicholson & Loup, Inc. v. Carl E. Woodward, Inc.*,[36] where the design-builder was an integrated firm with an in-house A/E. When the project's foundation failed, the design-builder was found liable for professional negligence. The surety, a party to the action, attempted to extricate itself from liability by arguing that the bond covered only the construction phase of the project and that the surety was not responsible for errors in design services.

The court disagreed with the surety's position, looking to the broad language of the surety bond to find the surety financially responsible for the defects in the project performance arising out of the faulty design. Specifically, the court said that "the design responsibilities were...a part of that contract. As [the surety] guaranteed all aspects of the project, the surety agreement covered [the A/E's] responsibilities."[37]

On the basis of the *Nicholson* case, and the broad language of performance bonds, some sureties have attempted to reduce their exposure for design liability by doing the following:

- *Requiring separate contracts.* Have the design-builder and owner execute separate contracts for design and construction, with only the construction contract bonded.

- *Excluding design from the bond.* In the event that the design-build is performed under a single contract, have only construction services bonded and specifically state in the contract that neither the penal sum of the bond nor the bond premium is based on design costs.

- *Limiting liability to construction.* Have the design-build contract limit the design-builder's liability to only construction defects, with the owner looking to the E&O policy for any damages caused by defects in design.

- *Assigning rights against the A/E.* If the design-builder is contractor-led or a joint venture, limit the contractor's liability and have the contractor assign its rights for design defects against the A/E to the owner.

All of these suggestions presuppose that there will be a viable recourse to the A/E's E&O policy on the project.

---

[36]596 So. 2d 374 (La. Ct. App. 1992).

[37]596 So. 2d 374 at 398 (La. Ct. App. 1992).

Two other points should be noted. First, liability for design risk is purely a function of what contract obligations are assumed by the design-builder. If the design-builder is able to reduce its liability for design, this will directly translate to lower risk to the surety and, perhaps, create a greater willingness of the surety to provide a bond. Second, it should be remembered that the issue of exposure to design risk is principally a concern for sureties that bond contractor-led design-builders. Integrated design-build firms should be capable of easily obtaining bonds that cover design, since their businesses are focused on offering design-build services and the surety can underwrite the risk more readily.

Loss of A/E protection.   Under the traditional design-bid-build delivery system, the A/E is typically responsible for overseeing the work performed by the general contractor. This oversight role includes payment certifications and inspection of the work for conformance with the plans and specifications. While these services are intended to benefit the owner, they also benefit the surety, who could feel comfortable in knowing that its principal was being paid only for work actually and properly in place, thereby reducing the surety's risks in the event of a default.[38]

In a design-build project, the A/E is part of the construction team and, in theory, becomes an unreliable source of protection for overpayment and quality. This problem can be compounded when an integrated design-builder is involved and the A/E is an employee of the same firm performing construction. Moreover, regardless of the design-build organizational structure, the A/E will typically not be contractually responsible for providing oversight services to the design-builder for payment applications. These are services performed directly by the design-builder.

Although these concerns are appropriate theoretical issues, they can be rectified by considering how the design-build project is actually managed. First, although the design firm will not be providing oversight to the owner, it is likely that the owner itself will have oversight done by either in-house personnel or an outside consultant (often either an A/E or a construction manager). This oversight will evaluate the progress payment process and ensure that there is neither front-end loading nor payment for work not properly in place. Second, there is a strong incentive for the full design-build team to maintain quality over the completed work, since they will have to correct any problem under the single point of responsibility concept.

Incomplete construction documents.   The fast-track nature of most design-build projects can have an impact on bonding. If the project is written on a cost-plus basis, or will have a GMP or lump-sum provided at some point after contract execution and after the design is sufficiently developed, the surety may be

---

[38]The primary risk is that if the principal defaults after being overpaid, there will be insufficient funds remaining in the contract balance to complete the work, thereby exposing the surety to the differential.

asked to provide bonding on the basis of estimates on incomplete construction documents. Scope changes that affect either contract price or project complexity can impact a surety's exposure. Although this can be inherent in any design-build contract where price is to be determined after contract execution, it may be possible to structure the relationship so that the bond is provided after the GMP or lump sum is developed. The surety should provide to its principal appropriate assurances at the time of contract execution that it will provide such bonds (unless, of course, the financial capacity of the principal materially deteriorates during the interim).

Bonding the A/E.   Most design-build teams are currently led by either integrated design-build firms or general contractors who subcontract for A/E services. This is primarily because these firms have the capital base and bonding capacity to effectively stand behind their contractual obligations to the owner. As a result, the primary bonding issues currently affecting the design-build industry revolve around the exposures of contractors and their sureties for design-related risks.

As more A/Es express interest in being the lead design-builder, one of the first questions that they must answer is how to best satisfy the owner's need for performance security. Although the process for meeting the underwriting needs of a surety is beyond the scope of this chapter,[39] it should be remembered that three primary criteria—often referred to as the "3 Cs"—drive a surety's decision to establish a relationship with a potential principal:

- *Character*—the qualitative aspects of a principal's business, focusing on personal backgrounds, relationships, reputation, and integrity
- *Capacity*—the ability of the principal to do the work (includes factors such as prior experience, management structure, location of the work and size of the project)
- *Capital*—the financial integrity of the principal, including net worth, liquidity, and any other factor that will indicate the principal's ability to pay in the event of a loss

Simply put, the A/E who is interested in leading the design-build team must demonstrate that it has the financial ability and talent to manage the process fully and absorb the risk of loss. Since many A/Es are not capitalized as fully as general contractors, this becomes problematic unless there is a concerted effort on the part of the A/E to establish and capitalize a company that will have designer-led design-build at the core of its business plan.

---

[39]See generally Palmer et al., *Construction Insurance* (fn. 6, above).

Design-builder's interest in reducing bond exposure. As noted above, bonds are not insurance, but rather credit vehicles. Therefore, the interests of the principal—whether as design-builder, contractor, A/E, or subcontractor—in reducing exposure is virtually identical to those of the surety. These interests include having reasonable contract terms with the obligee and ensuring that the business risks assumed in the contract reflect industry custom and the profit/loss potential for the contract. As discussed in Sec. 15.9, any loss incurred by the surety will be passed on to the principal through the general agreement of indemnity or otherwise.

By the same token, it is important for those providing design-build services to remember that the design-build process does anticipate greater responsibility by the design-builder than a typical general contractor or A/E would have in its individual capacity under other delivery systems. Because this single point of responsibility is often the primary reason why owners select design-build, attempts to reduce bond exposure by shifting design or construction risk away from the design-builder—and thereby converting the design-build system to one that looks more like a traditional delivery system—will likely be rejected.

15.10   Alternative Forms of Security

While surety bonds are the most common forms of guarantee within the United States, other forms of performance security are available. In fact, in the international construction market, these alternative forms—most notably third-party guarantees and irrevocable letters of credit—are the preferred and traditional way of securing performance.

15.10.1   Third-party guarantees

A *third-party guarantee* is a guarantee by someone other than a party to a contract that ensures that the contract will be performed and, if not, that the guarantor will be held responsible for the damages. It is often used with LLC or LLP structures, since the entity acting as the lead design-builder lacks the corporate assets necessary to provide the owner with assurances that it can live up to its financial commitments in the event of a nonperformance. These guarantees can be provided by anyone, but are most often provided by either a parent corporation or, in the case of a special-purpose corporation, members of the LLC or LLP who have strong balance sheets.

15.10.2   Irrevocable letters of credit

An *irrevocable letter of credit* is a written instrument whereby an issuer (generally a bank) guarantees that it will pay to the beneficiary (generally a project owner) a certain sum of money on presentation to the issuer of appropriate documentation. In the context of construction projects, the letter of credit provides security to the owner that in the event that the design-builder fails to fulfill any specific obligation under the contract, the owner can draw on the letter of credit. The irrevocable letter of credit represents funds that are available "on demand" by the project owner, provided the documentation required by the "drawing clause" has been furnished to the issuer of the letter of credit.

Letters of credit have significant advantages to an owner. They provide a ready source of funds to compensate the owner for damages suffered, generally without the need to pursue litigation. Unlike a surety bond or guarantee, the rights to draw against a letter of credit are more relaxed, although this is entirely dependent upon what is identified in the drawing clause as a condition to presentation to the issuing bank. For example, some drawing clauses may require the owner to present an executed arbitration award as a condition to receiving its funds. Alternatively, a more liberal drawing clause may enable the owner to draw on it only by presenting a certification from the owner that it has suffered damages attributable to the design-builder and that it has the right under the contract to recover monies from the design-builder.

Most letters of credit are issued in amounts far less than the contract amount—often in the range of 10 to 20 percent of the contract amount. Therefore, in the event of a major problem that exceeds the amount of the letter of credit, the owner would have to rely on the creditworthiness of the design-builder or other forms of performance security, such as third-party guarantees.

15.11 Conclusion

As noted at the outset of this chapter, the insurance and bonding markets are a "work in process" relative to design-build. Parties should not assume that any rule is rigorously applied without exception or that products being offered by one company are identical to products being offered by others. As a result, this is clearly a segment of the construction market where knowledgeable insurance and surety brokers can do a great service to their clients, with respect to available products as well as risk allocation in general.

Design-Build Liability

16.1 Introduction

The rights and duties of those contracting under traditional project delivery methods—shaped by years of judicial precedent and practical experience—are relatively well defined and predictable. The same cannot be said for the rights and responsibilities of those participating in the design-build process. Although the design-build concept dates back thousands of years, it is a relative newcomer to the modern construction world. Consequently, the construction industry and the court system are still in the process of determining what is expected from parties to design-build contracts.

Additionally, given the realities of the design-build market, it is likely that design-build liability will remain a "work in process" for the foreseeable future. This is due largely to the fact that

- Standard form design-build contracts have been in use for only a short time, resulting in few cases interpreting this type of contract language.
- Most design-build contracts resolve disputes through processes other than litigation, which results in few reported decisions describing liability.
- There is a comparatively small incidence of conflicts and disputes between parties to a design-build contract.

Consequently, cases that do address design-build issues are frequently ones of "first impression," in which decisions are typically based on general construction law principles and commentary from design-build experts.

Despite the absence of extensive judicial precedent specifically addressing design-build liability, enough information is available to provide strong indications of the most fertile areas for potential problems. This chapter reviews the most critical problem areas, such as the design-builder's single point of responsibility and the liability of the design professional for defective design. Before doing so, however, the following sections provide a brief historical overview of liability in the construction industry.

16.2 Historical Review of Construction Liability

As described in Chap. 2, the history of design-build dates back to the ancient civilizations of Egypt, Greece, and Rome, where the master-builder concept was used in the design and construction of most projects. A review of this history not only provides examples of some of the greatest construction works ever executed but also chronicles the development of construction industry liability.[1] See Fig. 16.1.

16.2.1 Strict liability of the master builder

The concept that the master builder was the true single point of responsibility was first expressed in the ancient Babylonian Code of Hammurabi. Hammurabi's Code provided "eye-for-an-eye" accountability to the master builder for both personal injury and economic losses that were caused by design or construction defects. For example, the Code imposed personal injury liability as follows:

> If a builder has built a house for a man, and his work is not strong, and if the house he has built falls and kills the householder, that builder shall be slain. If the child of the householder be killed, the child of that builder shall be slain.[2]

This strict liability concept applied to anyone who was affected by the master builder's defective work, regardless of whether there was a contractual relationship—known as *privity of contract*—between the parties.

The Romans were the first to introduce the concept that privity of contract should exist between the master builder and an injured party in order for the master builder to be subject to liability. The requirement for privity was rigidly applied under Roman law, regardless of whether the loss was for personal injury or economic injury. However, although the application of the privity of contract doctrine under Roman law did have the effect of reducing the potential number of parties to whom the master builder could be held liable, the strict liability feature of Hammurabi's Code still applied to the master builder when it was in privity of contract with an injured party.[3] In fact, it was not until the Renaissance, when the architectural profession began to develop a design expertise independent from building, that the strict liability concept for design or construction defects finally began relaxing somewhat.[4]

---

[1]For an excellent perspective on the history of architect's liability, see J. A. Felli, "The Elements of Ohio's Liability Provisions for Contemporary Design-Build Architects: An Unwillingness to Expand the Plan," *U. Dayton L. Rev.* **17**, 109, 111–115 (1997).

[2]H. G. Block, "As the Walls Came Tumbling Down: Architects' Expanded Liability under Design-Build/Construction Contracting," *J. Marshall L. Rev.* **17**:1 (note 2) (1984).

[3]See Felli, *Ohio's Liability Provisions* (see fn. 1, above), at 112.

[4]See B. J. Miller, "Comment, The Architect in the Design-Build Model: Designing and Building the Case for Strict Liability in Tort," *Case W. Res. L. Rev.* **33**, 116, 117 (1982).

Present — Negligence Standard of Care for Design Defects

Early 1970s — Privity Defense Totally Eroded for Personal Injury
- Series of personal injury cases against A/Es demonstrate that the privity of contract defense was no longer necessary for action against A/E for personal injury or property damage arising from A/E's negligence

1957 — Beginning of End of Privity Defense to A/E Negligence
- Inman case determined that A/E owed a duty of care to those who could foreseeably be injured as a result of A/E negligence
- Other courts quickly followed for personal injury cases

Early 1900s — Privity Defense Erodes in Manufacturing Sector
- Product liability cases expand liability against manufacturers
- Privity requirement erodes more slowly for design professionals

Mid 1800s — Privity of Contract and Negligence Standards Become Norm
- English and U.S. courts begin to adopt Roman requirement of privity
- Privity becomes coupled with requirement of negligence for recovery against architects & engineers

1350-1500 Renaissance — Relaxation of Strict Liability Doctrine
- Architectural profession begins to develop design expertise apart from building
- Professional standards begin to develop

0-100 A.D. Ancient Rome — Privity of Contract Concepts First Introduced
- Privity needed to claim either economic loss or personal injury
- Strict liability concepts remain applicable if privity existed

2000 B.C. Ancient Babylon — Strict Liability of Master Builder
- Responsible for personal injury & economic loss caused by design or construction defects
- No privity of contract required

Figure 16.1 Liability for design defects: a historical perspective.

16.2.2 Designer's liability: privity and negligence concepts

The Industrial Revolution of the mid-1800s resulted in several developments which gave rise to the modern system of design and construction liability. First, courts in England and the United States began to adopt the Roman concept of privity of contract as a basis for liability.[5] Then, these courts began to require proof of negligence in order to recover against architects and engineers (A/Es).[6] As a result of these developments, the liability of A/Es gradually became limited to claims made by those who had contracts with the A/E. Even then, recovery was predicated on a showing that the applicable standard of care had been breached.

The privity defense in American society began eroding in the early 1900s through product liability cases that expanded liability against manufacturers of products.[7] However, the privity defense in lawsuits against design professionals did not fade as quickly as it did against manufacturers. A typical example of the view of A/E liability under early 1900s law is provided in *Ford v. Sturgis*.[8] This case involved the death of a patron at the Knickerbocker Theater in the District of Columbia from the collapse of the theater's roof and balcony due to unusually heavy snow. The victim's heirs sued the architect for negligent design. The court denied the claim based on the lack of privity, stating: "It is elsewhere given as a better ground that the negligence of the Owner in maintaining the defective building...is the true proximate cause of the third person's injury."[9] The opinion of the *Sturgis* court—that an injured plaintiff's recourse for negligent design was against the owner of the property, not the architect—prevailed for another 40 years after the 1916 *McPherson* case.

The advancement of tort concepts in the American legal system finally brought the end of the strict privity defense to A/E negligence. The first ruling of this kind was handed down in *Inman v. Binghamton Housing Authority*,[10] a 1957 New York case involving a suit against an architect by a child injured in a housing development designed by the architect. The court concluded that the A/E did owe a duty of care to those who could be foreseeably injured as a result of the A/E's negligence, regardless of whether or not a contract existed. After *Inman,* other courts around the country quickly followed suit in situations where a plaintiff was personally injured as a result of A/E negligence.

By the early 1970s, courts around the country concluded that privity of contract was not necessary in bringing an action against an A/E for personal injury or property damage arising from an A/E's negligence. At present, how-

---

[5]See Felli (fn. 1, above), at nn. 31–32; *Winterbottom v. Wright,* 152 Eng. Rep. 402 (Ex. Ch. 1842) (finding that without privity there would be excessive, unlimited liability, and that it would be decidedly more difficult for plaintiff to prove cause and effect).

[6]See Felli (fn. 1, above), at 114.

[7]The first product liability case that eliminated privity as a defense was *McPherson v. Buick Motor Co.,* 111 N.E. 1050 (N.Y. 1916) (involving a personal injury caused by a manufacturing defect).

[8]14 F.2d 253 (D.C. Cir. 1926).

[9]Idem, at 255.

[10]143 N.E. 2d 895 (N.Y. 1957).

ever, A/E liability for economic losses in the absence of privity is still very much in flux.[11] Decisions in many states find that A/Es owe a duty of care, and can be liable, to contractors, suppliers, and other third parties who suffer economic losses associated with delays, disruption, and inadequate plans and specifications. Others take an opposing view, strictly forbidding such claims on the basis of lack of privity and limiting exposure for economic losses to the party with whom the plaintiff has contracted.

Although privity is no longer an absolute condition to A/E liability in the United States, the concept of negligence as a basis for A/E liability has not eroded. Courts that have considered the question apply a standard-of-care analysis and do not expect the A/E to guarantee its results. One frequently cited expression of this view is found in *Surf Realty Corp. v. Standing,*[12] where the court defined the A/E's standard of care as follows:

> He [the architect] must possess and exercise the care of those ordinarily skilled in the business, and in the absence of a special agreement, he is not liable for fault in construction resulting from defects in the plans because he does not imply or guaranty a perfect plan or a satisfactory result.[13]

Notwithstanding this general rule, the evolution of the A/E's role under more modern project delivery systems—and the lack of familiarity by courts as to how A/Es actually practice—has created some confusion as to whether the A/E should be held to a higher standard of care than negligence.[14] This is discussed more fully in Sec. 16.5 herein.

16.2.3 Contractor's liability

Although the history of A/E liability has evolved substantially over time, the liability of the construction contractor who does not design the project has

---

[11]See discussion in Sec. 14.4 (this book) on the issue of the A/E's responsibility for economic losses to parties with whom it is not in privity of contract.

[12]195 Va. 431, 78 S.E.2d 901 (1953).

[13]Idem, at 908.

[14]Consider, for example, this passage from *Waddington v. Wick,* 552 S.W.2d 147, 150 (Mo. App. 1983):

> An architect is one who makes plans and specifications for designing buildings and other structures, and supervises their construction. In lay person's terms, a property Owner tells an architect basically what kind of structure is desired, and the architect takes it from there, designing the specifications which are later used by engineers and builders.

This rendition omits the critical issue in A/E relations with owners—that the contract defines what services the A/E will be providing. But see *Moundsview Independent School Dist. No. 621 v. Beutow & Assocs., Inc.,* 253 N.W.2d 836, 839 (Minn. 1977), where the court understood the importance of the contract:

> It is the general rule that the employment of an architect is a matter of contract, and consequently, he is responsible for all the duties enumerated within the contract of employment....An architect, as a professional is required to perform his services with reasonable care and confidence and will be liable in damages for any failure to do so.

remained consistent. Similar to the Egyptian and Roman concepts of law, the contractor is strictly accountable to the owner in the performance of its work in accordance with the contract documents, regardless of applicable industry standards. Moreover, unless the contract documents specify otherwise, the contractor is generally expected to provide guarantees that it will execute the design of the A/E for a specific price and within a specific time period.[15]

Importantly, since it has no responsibility for the design, the general contractor does not warrant that the A/E's design, when executed, is sufficient to meet the owner's needs. Instead, the owner, as the party who contracts with the A/E and contractor, is held to impliedly warrant the sufficiency of the plans and specifications to the contractor. This implied warranty, often referred to as the *Spearin* doctrine, serves two purposes: (1) it supports the premise that if there is a design defect which prevents the project from functioning in accordance with the owner's expectations, the contractor will not be liable; and (2) it enables the contractor to be paid for any changes to or defects within the design documents.[16] Absent negligence on the part of the A/E, these costs are to be borne by the owner and are not recoverable from the A/E.

16.2.4   Snapshot of liability under traditionally delivered project

The movement away from the master-builder concept to separately procured design and construction services resulted in a major transfer of accountability for design and construction. This shift has placed the owner at the center of the universe for liability purposes, with A/Es and contractors having more limited responsibilities. Consequently, the liability issues fall out as follows:

- *Owner liability.* The owner will be responsible to the contractor under the implied warranty of specification standard, requiring the owner to deal with overruns caused by defects in the design. Additionally, the owner will be liable for additional costs that the contractor incurs from A/E acts or omissions.

- *A/E liability.* The A/E is responsible to the owner for providing a design that meets the owner's requirements. However, unless there is a contract provision to the contrary, the A/E does not warrant that the plans and specifications are error-free and will only be accountable to the owner for defects if

---

[15]Liability of the contractor to third parties injured as a result of the contractor's negligence has historically been handled on the same basis as A/E liability, with privity of contract remaining the critical issue. Note, however, that because the contractor typically assumes more responsibility for liability to third parties—primarily through the indemnification clause—its potential liability to third parties is far broader than that of A/Es to third parties.

[16]*United States v. Spearin*, 248 U.S. 132, 39 S.Ct. 59, 63 L.Ed. 166 (1918).

there is a breach of industry standards. The A/E will also be liable to third parties for negligence in accordance with the privity laws of the applicable jurisdiction.

- *Contractor liability.* The contractor on a traditionally delivered project is strictly responsible to the owner for meeting the plans and specifications provided by the owner. The contractor does not, however, have responsibility for design errors, either in the completed project or in executing the work. As with the A/E, the contractor is liable to third parties for negligence in accordance with applicable privity laws.

It must be noted that these general theories of liability may be modified if a contract shifts responsibility for design among the parties, as often happens when the traditional system is used with construction management techniques.

16.3   Design-Build Liability—Generally

Liability under a design-build contract is different from that in a traditional contract setting. As was the case with the master-builder concept, a single entity, the design-builder is responsible for both the design and construction functions and thus warrants the sufficiency of the design. This generally renders the design-builder responsible for design problems that affect the scope of work (which typically involve duties owed to the trade contractors relying on the plans and specifications). They are also responsible for design problems that affect the ability to achieve performance requirements (which typically involve duties owed to the owner of the facility). However, many questions remain unanswered as to whether the extent of the design-builder's liability will move to the strict liability concepts associated with the master builder of yesteryear or whether the industry will adopt standard of care and negligence principles.

The remainder of this chapter will review the legal theories on which design-build liability is founded. As will be noted, the *potential liability* exposure to members of a design-build team is indeed substantial. However, before reviewing this potential liability, it is important to review the liability exposure that is actually experienced.

16.3.1   Potential versus actual liability under design-build

Industry experience suggests that while the *potential* for major liability exposure exists on design-build projects, *actual* liability to date on design-build projects has been quite low. This view is supported by a report published in the 1990s by Victor O. Schinnerer, the underwriter for CNA's professional liability program. The report indicates that the claims frequency of insureds working on design-build projects was almost 50 percent of those providing services on traditional projects. Figure 16.2, based on data for 1985 to 1992, demonstrates this experience. Although the claim incident rate is lower for design-build, experience shows that the magnitude of the claims in a design-build setting is higher than in a traditional setting.[17]

Although there is no other specific published data of which the authors are aware relative to the actual difference in liability between design-build projects and projects delivered under other methods, there appears to be a significant reduction in conflicts among the parties involved in design-build projects.

Perhaps the most important factor in reducing overall liability is that the design-build team is organized far differently from a traditionally delivered project. In design-build, the construction and design professional project team members work closely together with several common goals:

- Developing a high-quality design that meets the owner's needs
- Taking steps to ensure that the design is balanced with the cost and schedule objectives for the project
- Ensuring that the legitimate goals of both the design and construction team members are understood and agreeing on mutually acceptable ways to achieve such goals

To that end, designers and contractors are motivated to communicate with each other frequently, effectively, and cordially. This can be very different from communications on a traditional project, where design and construction personnel have conflicting interests and frequently blame each other for problems.

On a design-build project, the designer relies heavily on the construction staff, not only to evaluate the design and suggest value engineering alterna-

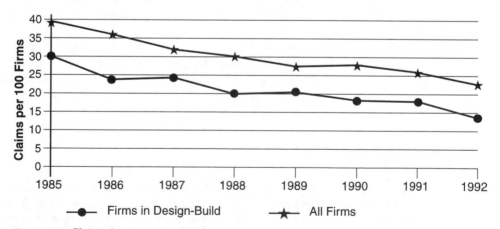

Figure 16.2 Claims frequency per 100 firms.

---

[17]See A. J. Abramowitz, "Professional Liability Insurance in the Design-Build Setting," *The Construction Lawyer* **16**(3): 3–7 (Aug. 1995).

tives that can be effectively implemented, but also to point out constructability problems. This generally results in a more cost-effective design than in a situation where the designer is working in isolation from the contractor. Additionally, depending on the relationship between the design and construction teams, it may be possible to remove the A/E from the process of reviewing certain submittals, concentrating its focus on those submittals that are critical to producing a high-quality project.

In addition to being recognized for positive interaction with each other, the organizations currently working in the design-build sector are characterized by strong reputations for their quality and integrity. This can be attributed to the manner in which owners select design-build teams. Many owners use a best-value approach with shortlisted firms that have strong credentials. Moreover, many of the teams have long-standing experiences and common bonds with one another.

Finally, design-build offers the owner a single point of responsibility, a deterrence to conflict within the design-build team and between the owner and the design-builder. The design-builder is fully aware of what commitments it has made to the owner and the consequences of failing to satisfy those commitments. A properly drafted design-build contract should put the design-builder in control of the process for achieving these commitments. While this control creates a potential liability, experience has shown that vesting responsibility for achieving a goal in a single entity creates a greater likelihood that the goal will be properly reached.

These factors, among others, create an environment which is not conducive to conflict and that provides the team with an incentive to amicably resolve disputes, giving each the best opportunity to achieve their respective goals. Furthermore, to the extent that conflicts do arise, there is a strong incentive to resort to find creative, informal processes to resolve the conflicts as opposed to resorting to arbitration or litigation.

16.3.2 Major areas of design-build liability

Despite the relatively few reported cases of conflict on design-build projects, there are several areas where claims have arisen and where we have some guidance as to the respective duties of the parties to the design-build process. The most significant deal with the liability of the design-builder under the single point of responsibility theory (discussed in Sec. 16.4) and the expanded liability of the A/E for design defects (discussed in Sec. 16.5). These areas are important not only because of the large liability exposure but also because they reflect changes from how liability would be decided under other project delivery systems.

Several other areas of the design-build process present major areas of liability exposure. These arise through contract and procurement issues, including contract formation, copyright, and the resolution of differing site conditions (discussed in Sec. 16.6). Job-site injury exposure has also been clarified by judicial precedent (discussed in Sec. 16.7).

16.4 Single Point of Responsibility

One of the most important factors leading an owner to choose design-build is the merger of design and construction functions into a single point of responsibility. This feature not only eliminates the conflicts that can result from separation of the design and construction functions but also enables the owner to obtain warranties that the facility will meet certain specific performance requirements. In fact, in some construction sectors, such as power generation, this feature virtually mandates the use of design-build in order to obtain financing of the project.[18]

Despite the importance of the single point of responsibility in the design-build contract, there are only a few cases which address this topic. These cases suggest that, to date, the courts are not reluctant to find the design-builder, and its team members, liable for failing to meet the higher duties that the single point responsibility entails, including the satisfaction of performance requirements. The legal theories that are used to justify these findings are broadly based, with courts having evaluated the single point of responsibility under theories related to breach of contract, the *Uniform Commercial Code* (UCC), and strict liability. There has also been little room for design-builders to use defenses such as impossibility of performance or owner interference. A discussion of these issues follows.

16.4.1 Breach of express contract duties

Breach of express contract duties is the most common legal theory used by owners who have claims against design-builders. This is a relatively easy theory to use when the owner employs objective performance specifications and has testing and guarantees built around these specifications, as would be the case, for example, in a power-generation project where the design-builder will be responsible for achieving specific-capacity and heat-rate requirements.[19] It can also be used where commercial facilities do not work as intended.

Consider, for example, *Rivnor Properties v. Herbert O'Donnell, Inc.,*[20] where the owner of the project contracted for the design and construction of an office building in the greater New Orleans area. The design-builder subcontracted the design to an architect who had no contractual supervisory obligations incident to construction. Shortly before completion of the project, the owner complained about water leaking into the building. Over a period of about 3 years,

---

[18]See M. C. Loulakis et al., *Contracting for the Construction of Power Generation Facilities,* Construction Briefings 89-5, Federal Publications, April 1989.

[19]A performance specification differs from a design specification in that performance specifications set forth certain quantified, objective standards to be achieved without dictating the design—in essence, identifying the problem but not the solution. In a design specification, a design professional has already engineered and specified a "cookbook" approach to the design which the contractor is required to follow.

[20]633 So.2d 735 (La. Ct. App. 1994).

the design-builder attempted to remedy the leaks. Its subcontractor performed patchwork on the cap flashing, installed caulking around the building, and drilled holes in the exterior curtainwall system. Despite these efforts, the leaks continued, resulting in, among other things, glass breakage and discoloration.

Finally, the owner hired its own experts to determine the causes of the various problems. The owner performed the work recommended by its experts and then filed suit against the design-builder for breach of contract and express and implied warranties. Various third-party claims and counterclaims were filed by the design-builder and its subcontractors, including the architect. The trial court found the design-builder fully liable to the owner, and the Louisiana Court of Appeals affirmed this decision. The court held that the design-builder "was charged by contract with the sole responsibility for all construction means, methods, techniques, sequences and procedures and for coordinating all portions of the work under the contract," as well as quality control inspections. The court also found that several members of the design-build team, namely, the architect and several trade subcontractors, contributed to the leakage problem and were liable to the design-builder for a share of the damages.

Design-build cases addressing breach of contract frequently do so within the context of liquidated damages. For example, in *CIT Group/Equipment Financing v. ACEC Maine*,[21] the design-builder of a power facility was required to meet two sets of performance tests as a condition to acceptance. One set of performance tests established substantial completion and the start of commercial operation. The second set of performance tests, which would determine the plant's efficiency and reliability, were to be undertaken on the one-year anniversary of substantial completion (with a 30-day grace period). Shortfalls in electrical capacity or heat rate were to be handled through liquidated damages.

The plant achieved substantial completion and was in commercial operation when, 9 months after substantial completion, one of the turbine generators failed. The turbine was finally repaired one year and 56 days after the original substantial completion date and, as a result, the second set of performance tests could not be undertaken on the date scheduled in the contract. The owner alleged that this was a breach of contract and triggered $32,276,440 in liquidated damages.[22] The design-builder attempted to defend on the basis that the liquidated damages were not contemplated for this type of defect and were unenforceable as a penalty.

The court found for the owner, describing the allocation of risk in the following manner:

> The parties agreed that if the Facility performed at the Guaranteed Performance Level the Contractor would be relieved of its liability under the Construction Contract for the plant's failure to perform at the specified levels over the plant's

---

[21]782 F. Supp. 159 (D. Me. 1992).

[22]This figure represented the actual contract price, which was the limit of the design-builder's liability. Note that the actual calculation of liquidated damages amounted to over $167 million.

lifetime, including incidental and consequential damages. On the other hand, if the Facility was unable to meet the performance standards, as it was unable to do in this case, the Owner would be compensated by a one-time payment according to a formula.[23]

The court rejected the notion that the liquidated damages were a penalty, concluding that the clause was enforceable because the damages (1) were difficult to estimate accurately and (2) served as a reasonable forecast of the actual loss at the time the contract was executed.[24] Importantly, the court found it difficult to put a price on the lack of reliability of the plant given that, in its first year of operation, a turbine failed.

Likewise, the court was not persuaded by the design-builder's argument that the liquidated damages were not intended to be used for a total failure as a result of a temporary problem that could be corrected through warranty repairs. The court noted that the clause was clear on its face and, if the parties had intended for the liquidated damages to apply only to permanent shortfalls in performance, the contract could have been clearly drafted to reflect this intention.

A case which reached a similar result is *Royal Ins. Co. of America v. CNF Constructors, Inc.,*[25] where CNF, the design-builder of a combined-cycle cogeneration facility, was sued by the plant's owner and its insurance company for consequential damages arising out of the temporary shutdown of a steam turbine. The contract contained typical requirements for performance testing and allowed the owner to go on line and sell energy commercially before final plant completion.

Three days after the plant achieved provisional acceptance and went on line commercially, the steam turbine experienced a catastrophic failure (which CNF acknowledged was due to defects in design, material, work quality, or installation). As a result, the plant completely shut down for 3 months while CNF repaired the turbine. CNF completed the repairs and ultimately was able to achieve final completion of the plant within the time required by the contract, notwithstanding the shutdown and lack of power generation during the period between provisional acceptance and final completion.

CNF argued that the contract precluded it from being liable for the owner's lost revenue during the shutdown. Although CNF conceded that the turbine failure caused a breach of the warranty provision, CNF contended that it complied with its contractual obligation to "repair or replace the defective part" and that this obligation did not expressly include consequential damages as a potential liability during the warranty repair. The court rejected this argument, citing the fact that the contract neither (1) imposed any limitations on the available remedies to the owner nor (2) expressly waived the owner's rights to recover consequential damages.

---

[23]782 F. Supp., at 163.

[24]Idem, at 162.

[25]1995 U.S. Dist LEXIS 112 (S.D.N.Y. Jan. 3, 1995).

CNF was also unsuccessful in its argument that the liquidated damages for late-delivery provisions of the contract precluded recovery for consequential damages. These provisions addressed liability for failure of CNF to achieve provisional and final acceptance by certain dates; they did not address what would happen if the temporary failure of a component part caused an interim shutdown.

Design-builders should also exercise caution in how they use marketing statements in terms of the single point of responsibility issue. It is not unusual for design-builders (or, for that matter, anyone trying to sell a service) to make representations about their superior expertise, skill, or suitability to perform a particular type of project. While the design-builder may believe that it is only marketing its organization and that such statements should be considered mere "puffing," courts and arbitrators may not always agree. The design-builder may face liability for its failure to fully meet the standards established in the contract documents and the design-builder's representations. For example, if the design-builder indicates that it possesses superior expertise or skill in a particular area, the design-builder may be held to a standard commensurate with its representations as opposed to the normal professional standard of care.

16.4.2 Implied warranties

Parties to a design-build contract are typically careful to ensure that any duties owed by the design-builder to the owner are expressly stated in the contract documents. However, in the event the contract does not clearly limit the duties of the design-builder to what is stated in the contract, it is possible for the owner to argue that the design-builder is liable for a breach of implied warranties.[26] An owner may need to make this type of claim to overcome specific contract limitations—such as a limitation of liability or warranty limitation— or statute of limitations constraints that preclude its use of breach of contract theories. Implied warranties are created by both judicial precedent and application of the UCC, each of which is discussed below.

Judicially imposed implied warranties. Two types of judicially created implied warranties can affect a design-builder: the implied warranty that the work will be performed in a professional manner and the implied warranty that the work will be fit for the purposes for which it is intended.

There is little case law on the subject of implied warranty of work quality. Because work quality is typically a specific obligation of the design-builder

---

[26]Various implied obligations can govern the obligations of the owner to the design-builder. These include the implied duties of the owner to (1) cooperate with the contractor and (2) not hinder the contractor's performance. See J. Cibinic, Jr. and R. C. Nash, Jr., *Administration of Government Contracts,* 2d ed., George Washington University Press, Washington, DC, 1985, pp. 320–356.

under its contract with the owner, the existence of this implied warranty may not, in and of itself, create any expansion of the single point of responsibility concept. However, this implied warranty may affect the relationship of the design-builder to parties other than the owner.[27] It may also affect an A/E's liability to an owner under a designer-led design-build process if the contract does not expressly deal with construction work—although there are no cases on point as of the date of this publication.

There is more case law that addresses the implied warranty that the work performed by the design-builder will be fit for the purposes for which it is intended. This warranty has primarily been used in cases involving the sale of products and has major potential liability ramifications to the design-build team, since it might imply that the design-builder is providing a product and not a professional service.

Most of the cases finding design-builders liable for breach of the fitness of purpose warranty arise in the context of residential homebuilders.[28] There are, however, some cases that place liability for breach of this warranty on builders of commercial and public projects. For example, in *Kellogg Bridge Co. v. Hamilton*,[29] the U.S. Supreme Court held that insufficient pilings designed and built by a bridge company were impliedly warranted to be fit for a particular purpose, stating

> The law therefore implies a warranty that [the defective piling] was reasonably suitable for such use as was contemplated by both parties. [The bridge] was constructed for a particular purpose, and was sold to accomplish that purpose; and it is intrinsically just that the company, which held itself out as possessing the requisite skill to do work of that kind...should be held liable.[30]

Other courts have found implied warranties of fitness for a particular purpose in design-build projects involving a warehouse,[31] a refrigeration system for an ice skating rink,[32] and a grain storage facility.[33]

Although Sec. 16.8 describes some cases finding A/Es on design-build teams liable for economic damages based on implied warranties, the majority of courts refuse to recognize implied warranty claims against A/Es based on the

---

[27]See *McMillan v. Brune-Harpenau-Torbck Builders, Inc.*, 455 N.E. 2d 1276 (Ohio 1983) (concluding that the original design-builder could be found responsible for breaching its implied warranty that the work would be done in a professional manner).

[28]See Felli (fn. 1, above).

[29]110 U.S. 108 (1884).

[30]Felli (fn. 1, above), at 136 [quoting *Kellogg Bridge Co. v. Hamilton*, 110 U.S. 108 (1884)].

[31]*Kennedy v. Bowling*, 4 S.W.2d 438 (Mo. 1928).

[32]*Prier v. Refrigeration Eng. Co.*, 442 P.2d 621 (Wash. 1968).

[33]*Robertson Lumber Co. v. Stephen Farmers Cooperative Elevator Co.*, 274 Minn. 17, 143 N.W. 2d 622 (1966) [applying an implied warranty of fitness because (1) the design-builder held itself out to be competent, (2) the owner had no expertise in design or construction, (3) the owner provided no plans and specifications, and (4) the owner intended to rely on the design-builder's expertise after explaining the specific purpose of the facility].

professional nature of the A/E's services.[34] As courts are confronted with lawsuits alleging breaches of implied warranties, it is likely that they will carefully evaluate the logic of how A/Es are treated—and may ultimately conclude that the professional nature of the design-build relationship does not justify looking beyond the contract and industry standards in finding design-builder liability.

UCC implied warranties.   Article 2 of the *Uniform Commercial Code* (UCC) is applicable to the sale of goods and establishes some commercial terms governing the rights and responsibilities of the buyer and seller of goods. Included in the scope of the UCC are warranties by the seller that its goods will be fit for the purposes intended and of merchantable quality.[35]

Most courts are reluctant to apply UCC warranties to construction projects, citing that (1) construction is a service-intensive process and is not the sale of "goods" as defined by the UCC and (2) any "goods" that might be furnished on the construction project are not capable of being severable from the project without being harmed—specifically, they are non-mobile fixtures not fitting the definition of "goods" as defined by the UCC.[36] However, courts do not automatically conclude that the UCC is inapplicable. Rather, the court will attempt to balance the amount of goods and services provided by the contract to determine whether the project is arguably a product.[37]

The case of *Smith v. Arcadian Corp.*[38] provides a perspective of how courts look at the "goods/services" issue in the context of the UCC.[39] This case involved claims against a design-builder arising out of the failure of a high-pressure reactor used in a fertilizer plant. The owner's lawsuit was for indemnification from the design-builder for personal injury and property damage resulting from the failure of the high-pressure reactor constructed nearly 30 years before the lawsuit was filed. A state statute barred liability in regard to construction contracts more than 10 years after acceptance of

---

[34]See *Johnson-Voiland-Archuleta v. Roark Assocs.*, 572 P.2d 1220 (Colo. App. 1977) (no implied warranty of fitness for particular purpose to engineer's services); *Borman's Inc. v. Lake State Dev. Co.*, 230 N.W.2d 363 (Mich. App. 1975) (no implied warranty for design of drain system); *Sears, Roebuck & Co. v. Enco Assocs., Inc.*, 372 N.E.2d 555 (N.Y. 1977) (no implied warranty for design of snow melting pipes in a ramp that endangered the ramp's structural integrity); *Smith v. Goff*, 325 P.2d 1061 (Okla. 1958) (supervisory architect not liable for implied warranty when contractor's work was insufficient); *Ryan v. Morgan Spear Assocs., Inc.*, 546 S.W.2d 678 (Tex. Civ. App. 1977) (no implied warranties for the design of a foundation system).

[35]Every state except Louisiana has adopted the Uniform Commercial Code in some form. See UCC §§ 2-314, 2-315 (1977).

[36]See R. J. Bednar et al., *Construction Contracting* 828, George Washington Univ., National Law Center, Government Contracts Program, Washington, DC, 1991.

[37]*Bonebrake v. Cox*, 499 F.2d 951 (8th Cir. 1974) (applying the UCC, services were determined incidental); *Aluminum Co. of Am. v. Electro Flo Corp.*, 451 F.2d 115 (10th Cir. 1971).

[38]657 So.2d 464 (La. 1995).

[39]*Arcadian*, 657 So.2d, at 465, discussed in M. C. Loulakis, *Lessons Learned: How 1995's Design-Build Cases May Affect You*, 1995.

the work by the owner. Therefore, the owner needed to establish that the contract was not a construction services contract, but rather a contract for the sale of goods under the UCC.

In analyzing the issue, the court found that the following characteristics were present in a construction services contract:

- The purchaser has some control over the specifications.
- Contract negotiations take place before the object is built.
- The builder is to furnish its skill and labor to build the desired object, rendering the contract a service contract.

Because each test was resolved in favor of the design-builder,[40] the court dismissed the lawsuit. Interestingly, the court noted that even though the reactor could, standing alone, be viewed as a product, it was simply one component of an overall project and should not be looked at in isolation.

Despite the general rule that design-build contracts will be considered services contracts and outside the scope of the UCC, there is at least one case where a design-builder was found liable for breaching UCC warranties. In *Omaha Pollution Control Corp. v. Carver Greenfield,*[41] the court concluded that the contract, which called for the design and construction of a sewage processing plant to deal with the City of Omaha's slaughterhouse waste problems, was for the sale of a product. As a result, when the plant did not work, the court found liability due to breaches by the design-builder of the UCC's implied warranty of merchantability and the implied warranty of fitness for purpose and awarded the owner in excess of $3.3 million.[42]

The court's opinion in the *Omaha Pollution* case is particularly interesting in that the focus was on the finished plant rather than the services and products that went into designing and constructing the plant. The court appeared to disregard in its opinion arguments related to the design-builder's exercise of due care or breach of work quality under the contract. This result may have been reached because the design-builder was integrally involved in financing the project and had a proprietary method for treating the waste.

Given the result in *Omaha Pollution,* it is important for the parties to determine at the time of contracting whether UCC warranties will be applicable. It is generally in the interest of both owners and design-builders to fashion contractual remedies for performance problems, rather than allow remedies to be established by a statute developed for the sale of goods. As a result, many design-build contracts contain a clause which conspicuously excludes the UCC warranties and provides for the contract to dictate rights and responsibilities

---

[40]Idem, at 469.

[41]413 F.Supp. 1069 (D. Neb. 1976).

[42]Idem, at 1086.

for these issues.[43] Courts, however, may refuse to recognize the disclaimer for public policy reasons.[44] This issue of disclaiming warranties through the use of merger clauses has not been effective.[45]

16.4.3   Defenses to the single point of responsibility

There are questions as to how far the design-builder's single point of responsibility will actually extend when the design-builder is confronted with factors beyond its reasonable control. Although there are few cases on this subject, two lines of defense have surfaced: impossibility of performance and owner interference. Each of these is discussed in the following sections.

Impossibility of performance.   Because design-build contracts are often based on performance guarantees, there is an issue as to what liability the design-builder has if it finds that it is impossible to meet the performance guarantee. By accepting a performance-based specification, it can be argued that the design-builder has represented to the owner that the specifications are attainable. However, courts have considered this matter more precisely and have evaluated (1) the precise contract terms agreed on by the design-builder and (2) the relative knowledge of the owner and design-builder regarding the "impossible specification."[46] In doing so, courts appear quite willing to find that the design-builder has assumed responsibility for all damages as well as increased costs that it may experience as a result of impossibility.

In *Colorado-Ute Electric Association v. Envirotech Corp.,*[47] the design-builder (Envirotech) agreed to meet certain performance requirements in its contract to provide the owner (Colorado-Ute) with a hot-side electrostatic precipitator at a coal-fired electric power plant. Specifically, Envirotech agreed to comply with state air quality standards requiring that emissions opacity not exceed 20 percent and warranted that it would bear the cost of all corrective measures and field tests until continuous compliance could be achieved. Envirotech failed to achieve continuous compliance with the performance

---

[43]UCC § 2-316(2) states "to exclude or modify the implied warranty of merchantability or any part of it the language must mention merchantability and in case of a writing must be conspicuous, and to exclude or modify any implied warranty of fitness the exclusion must be by a writing and conspicuous" (1994).

[44]See *Melody Home Manufacturing Company v. Barnes,* 741 S.W.2d 349, 355 (Tex. 1987).

[45]See *O'Dell v. Custom Builders Corp.,* 560 S.W.2d 862, 869 (Mo. 1978); *Omaha Pollution Control Corp. v. Carver-Greenfield,* 413 F.Supp. 1069, 1086 (D. Neb. 1976). Merger clauses such as "this agreement supersedes any and all prior agreements or understandings between the parties" do not exclude implied warranties, since implied warranties are not expressly included in the terms of the contract.

[46]See generally T. Richelo, "Legal Implications of Design-Build Method in High Technology Projects," in *Design/Build: Issues for the 90's and Beyond,* A.B.A. Forum on Construction Industry, American Bar Association, Chicago, IL, 1990.

[47]524 F. Supp. 1152 (D. Colo. 1981).

requirements and claimed that such compliance was "impossible" to accomplish. The court held that Envirotech had expressly warranted that it could provide Colorado-Ute with a satisfactory precipitator and thus assumed the risk of impossibility. The court stated, "[Envirotech's] impossibility defense is inconsistent with its express warranties and cannot be employed to avoid liability."[48]

Similarly, in *Aleutian Constructors v. United States,*[49] the Court of Claims held that by altering the owner's initial design specifications for the design features at issue, the contractor had impliedly assumed the risk that performance under its proposed specifications may be impossible.[50] In this case, the contractor, Aleutian, agreed to construct an airplane hangar and dormitory building for the Air Force's Optical Aircraft Measurement Program at Shemya Air Force Base, Alaska. The area is known for its extreme weather conditions and high winds.

During construction, Aleutian obtained the government's approval to change the design of the roofing system provided that it warrant the materials and work for a 5-year period and verify that the proposed design would withstand a wind uplift pressure of 80 lb/ft^2. Soon after installation, the roofing system failed and Aleutian was forced to make substantial repairs and modifications to the roofing system. Aleutian filed a claim to recover the repair costs, alleging defective specifications and impossibility. The court rejected the claim and reasoned that when the contractor persuades the owner to change its design to one proposed by the contractor, the contractor assumes the risk that performance under its proposed design may be impossible.[51] Accordingly, by assuming responsibility for the design, the contractor assumed liability for all damages and losses arising from the inability of the design to meet the owner's performance goals.

Yet another instructive case in this area is *J.C. Penney Company v. Davis & Davis, Inc.*[52] where the issue involved the quality of work of certain sheetmetal and coping work. The project specifications provided that the work must "be true to line, without buckling, creasing, warp or wind in finished surfaces."[53] The owner refused to accept the work because it did not comply with the specifications. The design-builder did not dispute the assertion that the work did not comply with the specifications, but instead claimed that it was impossible to comply with the specifications. The court found that impossibility is not a basis to allow the design-builder to recover its additional costs from the own-

---

[48]Idem, at 1159 (citing *Gulf Oil Corp. v. Federal Power Commission*, 563 F.2d 588, 599 (3d Cir. 1977), certification denied, 434 U.S. 1062, 98 S.Ct. 1235, 55 L.Ed.2d 1762 (1978)); *City of Littleton v. Employers Fire Insurance Co.*, 169 Colo. 104, 453 P.2d 810, 814 (1969).

[49]24 Cl. Ct. 372 (1991).

[50]Idem, at 378.

[51]Idem.

[52]158 Ga. App. 169, 279 S.E.2d 461 (1981).

[53]279 S.E.2d, at 463.

er for attempting to comply with the specifications. The court reasoned that the specifications, although impossible to meet, were negotiated by the parties at arm's length.[54] Therefore, the owner was totally within its rights in refusing a product that did not meet all of its bargained-for specifications.[55]

Owner involvement and interference.   Design-builders are often provided with information from the owner that is to be used in the design and construction of the project. Depending on the owner's needs and level of sophistication, the preliminary information provided by the owner may vary from a verbal description to several hundred pages of technical specifications accompanied by drawings that are virtually complete. This latter concept—often referred to as "bridging"—has been promoted by some members of the A/E profession as the most appropriate means of undertaking a design-build procurement.[56]

Although there is little case law on this point to date, it appears that the information provided by the owner will be subject to the same *Spearin*-type warranties that are found in design-bid-build contracting, thereby subjecting the owner to cost overruns that the design-builder may experience in its reliance on information furnished by the owner. This is one of the primary reasons that many owners have refrained from using the bridging concept, since it can potentially destroy the single point of responsibility advantage to design-build.[57]

The owner's potential liability for furnishing defective preliminary design information is clearly illustrated in the case of *M.A. Mortenson Co.*[58] This case involved a design-build contract awarded by the Corps of Engineers to Mortenson for a medical clinic replacement facility at Kirkland Air Force Base, New Mexico. The solicitation contained design documents that were approximately 35 percent complete, with the solicitation informing proposers that such documents expressed the minimum requirements for the project. The Corps' design criteria informed all proposers that "[these] requirements may be used to prepare the proposals." The design documents furnished by the

---

[54]Idem, at 464.

[55]Idem.

[56]See Design-Build Institute of America, *Manual of Policy Statement, Policy No. 23, Design/Design-Build* (October 1998).

[57]When a contract is properly written in terms of a clear performance specification, courts will not be hesitant to find the design-builder liable for failing to meet such specifications. See *Utility Contractor, Inc. v. United States,* 8 Cl. Ct. 42 (1985). There, a contractor was to design and build a flood control system to collect rainwater along a creek in Oklahoma. Rainstorms caused the creek to overflow temporary cofferdams installed to keep the construction area dry. The contractor alleged that the government had failed to identify detailed procedures in the contract for protection of the permanent work during the construction phase. The court rejected the claim following its reading of the contract, taken as a whole, as requiring the contractor to possess sufficient hydrological expertise and construction skills to protect its unfinished work.

[58]ASBCA 39978, 93-3 BCA, Paragraph 26,189 (1993). This case is discussed in Loulakis, *Lessons Learned,* 1995 (see fn. 39, above).

Corps contained a number of options for structural systems, including calculations for these systems.

Mortenson's estimators, in originally pricing the work, did a takeoff of the structural concrete and rebar quantities indicated in the solicitation design documents. The final design was similar to that shown in the solicitation documents and was approved by the owner. Mortenson ultimately submitted a request for equitable adjustment based on the increased quantities of concrete and rebar associated with building to the final design. The Corps rejected the claim, believing that Mortenson assumed the risk of any cost growth due to these quantities because of the fixed-price nature of the design-build contract.

However, the Board agreed with Mortenson, finding that, although the solicitation did not require that the proposers use the information in the drawings, it also did not indicate "that the information was to be used at the proposer's risk."[59] The Board held that (1) the design-builder acted reasonably in relying on the technical information provided by the government and (2) the changes provision applied to the changes in the structural concrete and rebar. It specifically found that the doctrine of contra proferentum (construing the ambiguity against the drafter) applied and the government had warranted the adequacy of information on the solicitation design documents.[60]

In addition to furnishing defective information, the owner can potentially jeopardize the single point of responsibility by interfering with the design or construction process. Consider, for example, *Armour & Company v. Scott*,[61] which arose out of a design-build contract for the construction of a meat packing plant. The court found that the owner became so actively involved in the design process by modifying the electrical and mechanical systems and ultimately increasing the facility size that the owner assumed the role of a de facto partner of the design-builder. These substantial interferences constituted a breach of contract by the owner.

Sometimes, despite the best efforts of the owner to develop a performance specification and enable the design-builder to meet it, circumstances related to owner involvement can impact the single point of responsibility. Consider, for example, *Allen Steel Co. v. Crossroads Plaza Associates*,[62] which involved a commercial facility in Salt Lake City, Utah. In response to an owner's solicitation of design-build proposals for structural steel work, a contractor submitted in its proposal three structural design alternatives. However, the proposal specifically stated the following:

> This proposal is offered for the design, fabrication, and erection of the Structural Elements only for the tower and mall....Owner's engineer is to check this design

---

[59]Idem, at 130, 367.

[60]Idem, at 130, 368.

[61]360 F.Supp. 319 (W.D. Pa. 1972), affirmed, 480 F.2d 611 (3d Cir. 1973).

[62]1989 Utah LEXIS 124, at *1 (Utah Oct. 6, 1989), opinion withdrawn, 1991 Utah LEXIS 30 (Utah Apr. 10, 1991).

and make changes if necessary to enable him to accept overall responsibility for the design. Changes that effect [sic] quantity, weight, or complexity of structural members will require an adjustment in price.[63]

The proposal was accepted, and the contractor was directed to prepare detailed plans for steel fabrication based on its plans.

During the course of performance, however, inspectors from Salt Lake City stopped construction because of what they perceived as structural defects. The owner retained its own engineer to correct the defect. Steel had to be torn down to remedy the problem, resulting in delays to the project and substantial cost overruns. The owner backcharged the contractor for such costs, prompting litigation between the parties.

The sole issue in the case was whether the contract had effectively disclaimed responsibility for design defects by placing responsibility for the design within the control of the owner through its proposal. The court found that although the owner had only provided general design parameters for the structural steel, the contractor had effectively disclaimed its responsibility, since it had provided a design for purposes of the bid and transferred the risk of verifying adequacy of the design to the owner.

16.5   Architect-Engineer Liability for Design Errors

As is evident from the discussion in Sec. 16.4, the single point of responsibility feature of design-build will impose liability on the design-builder for design problems. The more challenging question that arises, however, is the extent of liability of the A/E as a member of the design-build team. The few cases that have dealt with this subject to date suggest that the potential liability may be substantial. Importantly, the cases have based liability findings on breaches by the A/E of express and implied warranties of performance.

One of the major cases establishing A/E liability for design errors based on express contractual warranties is *Arkansas Rice Growers v. Alchemy Industries, Inc.*[64] This case involved the construction of a pollution-free rice hull combustion plant capable of generating steam and marketable ash from the rice hull fuel. Rice hulls were required to be the sole fuel for the plant furnace. The contract between the developer of this process and the contractor (who was also the ultimate owner of the plant) required that the engineer provide

the necessary engineering plant layout and equipment design and the onsite engineering supervision and start-up engineering services necessary for the construction of a hull by-product facility capable of reducing a minimum of $7\frac{1}{2}$ tons of rice hulls per hour to an ash and producing a minimum of 48 million BTU's per hour of steam at 200 pounds pressure.[65]

---

[63]*Allen Steel,* 1989 Utah LEXIS 124, at *5.

[64]797 F.2d 565 (8th Cir. 1986).

[65]Idem, at 566.

The plant never performed properly, repeatedly shutting down because of the buildup of hulls in the furnace. As a result, the plant was unable to comply with state air pollution control standards and did not pass performance tests. The plant was closed three years after completion of construction. The contractor/owner successfully sued both the engineer and the developer for breach of contract and negligence, arguing that these parties had failed to design a plant capable of meeting the performance requirements.

The 8th Circuit Court of Appeals affirmed, stating that the design professional had warranted the performance of the design and, since the performance standards were not met, the warranty was breached.[66] The primary reason for the plant's failure to operate as warranted was the engineer's faulty design of the furnace system, since it could not support combustion at a temperature low enough to produce quality ash without the aid of fuel oil. Combined with the performance warranty, this justified a finding of liability against the engineer. Significantly, the court never looked at the issue of negligence from a standard-of-care perspective, only from the failure of the facility to function as warranted.

C.I. Maddox, Inc. v. Benham Group, Inc.[67] involved a design-build project for a coal processing system for an electric power plant. Maddox, a general contractor and the proposed design-builder, entered into an oral agreement with Benham, an engineering firm. Under the agreement, Benham was to prepare drawings, specifications, and equipment information to enable Maddox to prepare a lump-sum construction cost for the project. Having relied heavily on the quantity estimates prepared by Benham, Maddox submitted its formal proposal to the owner and was ultimately awarded a design-build contract.

After the contract was awarded, Maddox and Benham entered into a written agreement. As part of Benham's basic services, Benham was to keep Maddox "informed of the progress and quality of the Work" and endeavor "to guard [Maddox] against defects and deficiencies in the Work of [Maddox]."[68] These services did not include the compilation or preparation of bidding information, but did require that Maddox "furnish all cost estimating services required for the Project." The contract contained an integration clause stating that all prior agreements were superseded.

From the beginning, Maddox claimed problems with Benham's performance, including the fact that the drawings were often late and insufficient; Benham had underestimated the amount of work needed to complete the final design; and because prints for the electrical components of the project were not available, Maddox ended up having to install part of the wiring without plans. Maddox successfully convinced the jury to award it over $5 million in dam-

---

[66]Idem, at 570.

[67]88 F.3d 592 (8th Cir. 1996) discussed in Loulakis, *Lessons Learned: How 1996's Design-Build Cases May Affect You*, 1996.

[68]*C.I. Maddox, Inc. v. Benham Group, Inc.*, 88 F.3d at 595.

ages, including over $2.7 million for bidding errors, engineering errors, and delays caused by Benham.

Benham appealed to the 8th Circuit Court of Appeals. Among other things, it argued that the oral contract was inadmissible, since the written contract's integration clause superseded "prior negotiations, representations or agreements." The court disagreed, finding that the oral contract for preliminary bidding was a separate and independent contract which was bargained and paid for by Maddox, with full performance completed by Benham. It also concluded that the oral contract did not conflict with the written contract, since Maddox's responsibilities to furnish "all cost estimating services" for the project related only to estimating services associated with the final design (as indicated by the title of the written subcontract—"Agreement—Final Design"), as opposed to the duties of Benham relative to developing biddable plans for the preliminary design.

Benham also argued that, even if the oral agreement was a separate agreement from the written contract, Benham had not warranted the accuracy of the bidding information and, therefore, Maddox could not recover for breach of contract. The court disagreed, finding that under Missouri law, Benham had impliedly warranted the accuracy of the bidding information by repeatedly assuring Maddox and the owner that it was qualified to do the work.

A 1997 case repeats the *Maddox* theme of implied warranty liability, although in the context of an at-risk construction management project. In *Shidmore, Owings & Merrill v. Intrawest I Limited Partnership*,[69] Intrawest, as developer, hired Skidmore, Owings & Merrill (SOM) to design a high-rise complex known as the *Newmark Building*. SOM represented to Intrawest that it could meet Intrawest's tight budget and exacting schedule. SDL Corporation, the at-risk construction manager, was hired to establish an acceptable guaranteed maximum price (GMP) based on the design completed by SOM.

During the course of construction, major defects were found in SOM's design requiring substantial changes. Because of these defects, the project cost more than anticipated. When SOM sued Intrawest to recover its outstanding fees for services, Intrawest counterclaimed for, among other things, the costs associated with the change orders required for the extra work omitted from SOM's drawings.

Although SOM had alleged that the drawings were 90 percent complete, evidence presented at trial showed that in fact the GMP drawings furnished by SOM were actually 50 to 65 percent complete. SOM conceded that its drawings were incomplete and that later design changes required additional work. However, SOM argued that the work to incorporate these design omissions would have been priced in SDL's GMP if the design had been complete. By awarding Intrawest monies for these design omissions, Intrawest would be in

[69]1997 Wash. App. LEXIS 1505, at *1 (Sept. 8, 1997) discussed in Loulakis, *Lessons Learned: How 1997's Design-Build Cases May Affect You,* 1997.

a better position than if SOM had provided a perfect design at the time the GMP was developed—in effect, meaning that Intrawest would not have even paid once for such work.

The jury rejected SOM's argument and awarded Intrawest $820,372, which included monies for the design omissions. On appeal, the court acknowledged that under normal circumstances an owner cannot recover from an A/E the full cost of the correction of a design defect based on the "betterment" principle. Under this principle, the owner can recover only the additional costs incurred as a result of the design defect (such as tearing out and rebuilding). However, the court found that, because of the "special circumstances" in this case, SOM's liability should not be limited by betterment:

> Intrawest bargained for complete designs that would allow it to establish the project's GMP.... SOM knew that Intrawest had a tight budget so that significant design changes after the GMP was set would threaten the project's feasibility. The record shows that Intrawest would not have undertaken the project had it known the true extent of its cost.[70]

SOM further argued that, because it had not guaranteed the construction price, it could not be liable. Despite the fact that the court conceded that SOM had never made such a guarantee, the court nevertheless rejected SOM's position, stating that

> [b]ecause SOM agreed that the drawings would be substantially complete for GMP purposes, it guaranteed the information that SDL used to establish the total construction costs.... SOM is more like the fixed-price contractor than the engineer.[71]

As with *Maddox,* the *SOM* court found an implied warranty, stating that "where a person holds himself out as qualified to furnish, and does furnish, specifications and plans for a construction project, he thereby impliedly warrants their sufficiency for the purpose in view."[72]

There are other cases that address an A/E's exposure for express and implied warranties for design liability. However, the three cases referenced above are particularly significant because of the manner in which damages are calculated. In *Arkansas Rice Growers,* the A/E was ultimately liable for the entire cost of the rice hull combustion facility, since the facility was basically inoperable and of no value to the owner. In *Benham* and *SOM,* the damages that were awarded were far beyond those that would be imposed under traditional theories, given the rejection of the "betterment" principle. Nevertheless, given the reliance on the A/E's work product by the plaintiffs in each of these cases, one can conclude that the courts' rejection of betterment was appropriate.

---

[70]*Skidmore,* 1997 Wash. App. LEXIS 1505, at *10.

[71]Idem, at *12.

[72]*Skidmore,* 1997 Wash. App. LEXIS 1505, at *12 [quoting *Prier v. Refrigeration Engineering Co.,* 74 Wash. 2d 25, 29, 442 P.2d 621 (1968)].

16.6   Contract Formation and Interpretation

There are a remarkably high number of the disputes in design-build settings that involve issues arising from the formation of the design-build relationship. There appear to be several general reasons for this, including

- *Difficulty in completely defining scope of work.* Unlike traditional design-bid-build projects, the drawings and specifications defining the work for design-build projects are not finalized until well after performance is commenced; thus, parties often discover that their respective understandings of the scope of work were vastly different.

- *Inexperience.* The relative inexperience of owners, contractors, and A/Es in design-build procurement.

- *Sloppiness.* Sloppiness in contract drafting and administration, with parties not taking the contracting process seriously.

Note that these problems occur at both the owner/design-builder contracting phase as well as at the design-builder/subcontractor contracting phase. The cases discussed below illustrate many of the common problems that can arise in this area, underscoring the importance of formally finalizing the contract terms and identifying the various performance standards that apply.

10.0.1   Owner/design-builder issues

A particular case, *Marshall Contractors, Inc. v. Brown University*,[73] provides an illustration of some of the difficulties in contract formation under design-build. In early 1986, Brown University ("Brown") requested design and construction proposals from Marshall Contractors ("Marshall") and three other design-builders for a new university sports facility. Brown selected Marshall's proposal of $4,627,134 and authorized Marshall to proceed with formal design development, sitework, and relocation of utilities. This initial work was done without a formal contract in order to accommodate the fast-track schedule.

By late 1987, Marshall completed final construction documents. After reviewing the documents, Brown sent Marshall a letter of intent to enter into a contract to construct the sports facility for a price of $7,157,051. The letter of intent noted that a formal contract would be forthcoming and that any changes from the scope of work identified in the letter of intent would become a change order. The parties, however, never entered into a written contract because of their continuing disagreement over what was included in the baseline scope of work and Brown's budget expectations.

---

[73]692 A.2d 665 (R.I. 1997) discussed in Loulakis, *Lessons Learned: How 1997's Design-Build Cases May Affect You,* 4, 1997.

By the time the project was complete, the final cost was well over the letter-of-intent price. Marshall specifically claimed that it incurred costs of almost $900,000 for items that Marshall believed were for extras, but which Brown claimed were within the original scope of work. Marshall ultimately filed a lawsuit against Brown for these costs. Brown defended the claim on the basis that the parties had entered into a de facto contract for the $7,157,051 lump-sum price and argued that Marshall was entitled to no more than that amount.

In attempting to demonstrate that there was no contract, Marshall introduced into evidence a memorandum authored by Brown's president near the end of the project, which acknowledged that major areas of disagreement remained between Brown's original request for proposals (RFP) and Marshall's proposal. The memo noted that while Brown wanted both the RFP and proposal included as contract documents, Marshall only wanted to include its proposal, since this established Marshall's baseline for pricing purposes. The memo also stated that Marshall had some major cost overruns that it believed were outside the scope of its work, while Brown believed that Marshall should bear these costs. Marshall claimed that this memo, among other documents, demonstrated that the parties remained in a negotiating position and had not consummated a deal. Brown alleged that because the parties had mutually agreed on all material terms of the contract, there was, in fact, a contract—notwithstanding that some terms may have been left open.

The lower court agreed with Brown and found that, considering the conduct of the parties, an implied-in-fact contract existed as of the date of Brown's letter of intent. This ruling, which was rendered early in the case, had a detrimental effect on Marshall's ability to prove its damages, since the existence of a contract prevented Marshall from demonstrating the reasonable value of its extra work. When the jury found in favor of Brown, Marshall appealed.

The Supreme Court of Rhode Island disagreed with the lower court's finding, stating that an implied-in-fact contract could not have existed because the parties never agreed on the most basic of the essential terms: the scope of work. It stated that

> [t]he inability of the parties to ever reach mutual agreement on what the scope of the project was intended to include certainly constituted the very heart and vital essence of their ongoing contract negotiations and prevented the emergence and existence of any implied-in-fact contract.[74]

The court also found ample evidence of a continuous stream of rejected offers and counteroffers well after the date of Brown's letter of intent. It is also interesting to note the court's reaction to the case:

> [i]t is inconceivable that the parties, experienced as they were in their respective roles, would ever bind themselves to a project of such enormity and cost, without first agreeing upon what Marshall was to build in return for Brown's $7,157,051.[75]

---

[74]Idem, at 669.

[75]Idem, at 670.

In light of the lower court's error, the Supreme Court of Rhode Island remanded the case for a new trial.

Aside from scope of work issues, contract formation can also affect the substantive rights of the parties relative to insurance, indemnity, and dispute resolution. Representative of this is *United Structures of America, Inc. v. Coleman-Roth Construction, Inc.*[76] Coleman-Roth retained United Structures to design and build a steel building. Dissatisfied with United Structures' work, Coleman-Roth withheld final payment and sued for breach of contract. United Structures counterclaimed for full payment. United Structures sought to compel arbitration on the basis of arbitration language in a quotation that it claimed it faxed to Coleman-Roth on April 26, 1995 during contract negotiations. The fax quotation contained three pages of project specifications and one page of "terms and conditions" which included an arbitration clause. The terms and conditions concluded with a statement that "this proposal is not a contract until it is signed by the Buyer and accepted by an agent of the Seller."

Coleman-Roth opposed the motion to compel arbitration and claimed that it never received the fourth page of the fax. Therefore, it argued it was not bound by the arbitration clause. Coleman-Roth also disputed the significance of the fax, alleging that because it had issued a purchase order to United Structures for the work, the fax was a takeoff and not an offer of contract. United Structures introduced evidence to prove that all four pages of the fax had been sent. It also argued that Coleman-Roth's purchase order specifically stated that the work was to be done in accordance with the April 26, 1995 quotation. According to United Structures, the quotation showed that this was more than a takeoff and imposed on Coleman-Roth a duty to inquire as to the terms and conditions associated with the quotation. Coleman-Roth rebutted this argument by testimony that the April 26 reference in the purchase order was to a telephone quote, not to the written quote set forth in the fax.

The lower court agreed with Coleman-Roth's version of the facts, adding that the signatures on the fourth page were required to render the arbitration clause binding. The Court of Appeals of Texas affirmed the lower court's ruling, finding that there was sufficient evidence to support its position that Coleman-Roth did not receive the fourth page of the fax and did not agree to the arbitration clause in the contract.

Contract formation problems are also created when parties do not contemplate what might happen if the project is changed or abandoned. For example, in *Construction Development Marketing, Inc. v. Bally Entertainment Corp.*,[77] Construction Development Marketing (CDM), a commercial developer, brought an action against Bally Entertainment Corp. ("Bally") to recover 5 percent of the costs of an aborted hotel-arena in Louisiana. Bally and CDM

---

[76]1997 Texas App. LEXIS 3744 (July 17, 1997) discussed in Loulakis, *Lessons Learned: How 1997's Design-Build Cases May Affect You*, 6, 1997.

[77]1996 U.S. Dist. LEXIS 7154 (E.D. La. May 22, 1996) discussed in Loulakis, *Lessons Learned: How 1996's Design-Build Cases May Affect You*, 6, 1996.

entered into a contract whereunder CDM was to act as a facilitating developer for the project to develop all site improvements for the project. The site improvements were to be performed on a design-build basis by Brown & Root Building Company (Brown & Root). CDM's compensation included a 5 percent fee, which was to be paid out of Brown & Root's 11 percent overhead and profit on its construction contract.

When the project fell through, CDM filed a lawsuit for its 5 percent fee, claiming that Bally wrongfully abandoned the project. In turn, Bally brought an action against Brown & Root for indemnity and contribution, contending that any monies due CDM were the responsibility of Brown & Root. It relied not only on CDM's contract (which had the 5 percent fee paid from Brown & Root's contract) but also on the fact that Brown & Root had paid CDM for certain services performed under Brown & Root's preconstruction services contract with Bally. A construction contract was never actually entered into by Brown & Root on the project.

The court rejected each of Bally's positions. The indemnity theory was rejected because the only contract between Bally and Brown & Root was a preconstruction agreement that limited Brown & Root's indemnity to personal injury and property damage. CDM's claims were for neither. Further, there could be no claim for indemnity based on unjust enrichment, since Brown & Root had nothing to do with the project abandonment and would not be "enriched" if Bally were required to pay CDM for damages caused by Bally's wrongful conduct. In fact, because of Bally's actions, Brown & Root itself lost the opportunity to profit from constructing the project. Finally, the contribution theory was rejected because Bally could not show any fault on the part of Brown & Root that would establish Brown & Root's liability to CDM.

16.6.2   Design-builder/subcontractor issues

The problem with contract formation is not limited to issues involving the owner. The cases discussed in Sec. 16.5, particularly *Maddox v. Benham,* address what can happen if a design-builder and A/E do not adequately establish their relationship early in the design-build process. However, perhaps the most challenging task of administering and executing a design-build contract is for the design-builder to effectively procure trade contractors with design documents that are not complete. This requires the design-builder to establish a meaningful relationship with trade contractors earlier in the process than it might do as a general contractor on a design-bid-build project.

Consider, for example, *Collins Electric Co. v. Simplex Time Recorder Co.*[78] This case involved disputes between Collins, an electrical subcontractor, and Simplex, a fire alarm sub-subcontractor, on an office building that was deliv-

---

[78]1997 Wash. App. LEXIS 708 (Wash. Ct. App. May 5, 1997) discussed in Loulakis, *Lessons Learned: How 1997's Design-Build Cases May Affect You,* 5, 1997.

ered on a design-build basis. Prior to bid, Simplex had, at the request of Collins, met with the fire marshal regarding code requirements. As a result of this meeting, Simplex based its bid and drawings on the assumption that the city code, rather than the county code, applied. During construction, it was revealed that Simplex's assumption was erroneous. Disputes arose between Collins and Simplex as to who was responsible for the costs associated with rectifying the conflict in the codes.

The court analyzed this issue by looking carefully at how the parties entered into their contractual arrangement. These facts demonstrated that after meeting with the fire marshal, Simplex provided Collins with a written quotation that provided prices, as well as disclaimers, related to certain warranties and consequential losses. The written quotation stated that "approval of the system design by the authority having jurisdiction is not guaranteed." At a later time, Simplex sent drawings to Collins for the project, which Collins used in its contract with the design-builder. Collins ultimately sent Simplex a purchase order which stated that the work was to meet the city and county requirements for the project and that Simplex was not relieved from responsibility for a complete, accepted system—even though pricing was based on Simplex's drawings. Shortly after receiving Collins purchase order, Simplex revised its drawings and sent Collins an acknowledgment of the Collins purchase order—but included the same disclaimers that had been set forth on its original quotation. The parties proceeded without ever addressing the apparent conflict in the purchase order and "acknowledgment" terms.

After a 5-week trial, the court found that Simplex had breached the express warranties in the Collins purchase order and found it liable for over $500,000 in damages. The Washington Court of Appeals upheld this determination, concluding that the Collins purchase order was a counteroffer that Simplex accepted by its conduct. The appellate court relied heavily on the fact that Simplex provided drawings in response to the purchase order. It supported this position by noting that evidence demonstrated that Simplex's home office and key project representative had approved the Collins purchase order without objecting to its terms. As a result, the court held that the acknowledgment sent by Simplex was not a part of the contract, and Simplex's accompanying disclaimers were ineffective.

The general practice of using purchase orders to contract for design-build trade contractors is widespread. This is not, in and of itself, a problem, since short forms of a contract can be an effective way of starting on a fast-track basis. However, as reflected in the *Collins* case, for example, this practice may ultimately create some ambiguity as to precisely what terms the parties have agreed upon.[79]

---

[79]See also *Roberts & Schaefer Company v. Merit Contracting, Inc.,* 99 F.3d 248 (7th Cir. 1996) (addressing this point in a design-build setting) discussed in Loulakis, *Lessons Learned: How 1996's Design-Build Cases May Affect You,* 5, 1996.

16.7   Good-Faith Dealing

Many critics of design-build state that the absence of "checks and balances" is the primary drawback of the system, claiming that the owner is at the mercy of the design-builder. This argument ignores the fact that because of the single point of responsibility liability, a design-builder is incentivized to ensure that it provides a high-quality product to the owner. It also fails to recognize that many owners will use either in-house or external professional assistance in administering the design-build contract and monitoring the design-builder's performance.

Notwithstanding that there are, in fact, appropriate checks and balances in the design-build contract, the successful design-build project is characterized by a high degree of trust and partnering between the owner and the design-builder, as well as among the members of the design-build team. This requires the design-builder to treat the owner openly and fairly, and to act in the best interests of the owner. Several courts, confronted with situations where the design-builder failed to do so, did not hesitate to find the design-builder liable for breaching this duty.

One of the most interesting cases dealing with this issue is *Combustion Engineering, Inc. v. Miller Hydro Group.*[80] In this case, the court found that the design-builder had breached its contract with the owner by designing and constructing a hydroelectric facility capable of producing *more* electricity than the contract specified. The contract enabled the design-builder to earn a sliding-scale bonus for efficiency to the extent that the facility produced power in excess of a specified level. There was also a corresponding penalty which would reduce compensation if the facility were less efficient than the stated minimum output.

The design-builder built a plant that was larger and had a capacity greater than that set forth in the specifications. The utility claimed that this was done solely for the purpose of manipulating the bonus provision and that it was not aware of the increased size until it was too late to modify the plant. The increased capacity, which cost the design-builder only $1 million to implement, resulted in an $8 million bonus to the design-builder. When the utility refused to accept the facility or make the final payment, the contractor sued. The utility counterclaimed for breach of contract and fraud, alleging that it was potentially liable for penalties for building a facility larger than its license permitted and that it could be required to reconstruct fish-protection facilities that were keyed to the originally planned, smaller plant.

The court stated that no contractor "has any right to *any compensation* for services rendered under a special contract, unless he has at least *attempted,* in good faith, to perform all its conditions."[81] The court agreed with the utility,

---

[80]*Combustion Engineering, Inc. v. Miller Hydro Group,* 812 F.Supp 260 (D.Me. 1992) affirmed, 13 F.3d 437 (1st Cir. 1993).

[81]*Combustion Engineering,* 812 F.Supp., p. 263, quoting *Veazie v. City of Bangor,* 51 Me., pp. 509, 512–513 (1863) (citations omitted).

finding that the design-builder had intentionally breached the contract, thereby barring the design-builder's claims for unjust enrichment and the reasonable value of its services.

Another case involving this type of issue was *Aiken County v. BSP Division of Envirotech Corp.*[82] Aiken County contracted with an A/E to develop the design for a wastewater treatment plant. A portion of the design called for the general contractor to design and supply the thermal sludge conditioning system, including heat treatment and other related items. The general contractor entered into a lump-sum subcontract with Envirotech to perform this work on a design-build basis.

The A/E's specifications permitted either a wet-air oxidation or sludge-to-water technology for the heat-treatment system. After repeatedly representing to the A/E that it intended to bid its standard, proven sludge-to-water system, Envirotech instead bid and furnished a less-expensive sludge-to-sludge system. During design and construction, Envirotech continuously assured Aiken County and its A/E that the sludge-to-sludge system had been tested. On this basis, the A/E approved a change order enabling Envirotech to use it.

The facts ultimately demonstrated that Envirotech's sludge-to-sludge system was new and, while it had been installed elsewhere, had never been successfully implemented. On this project, the experience was no better. The plant was not able to achieve design production rates, since heated sludge tended to plug the spacers, requiring frequent cleaning and creating inefficiency in the heat transfer. The contract required that "systems furnished...shall be placed in operation ready to operate on a 24-hour per day basis with not more than 15% of total time required for maintenance and repairs."[83] The actual repair and replacement time consistently exceeded these levels, ranging between 36 and 42 percent of the total operating time over a 3-month period.

On the basis of these findings, Aiken County sued Envirotech directly for breach of warranty, breach of contract, and fraud. The district court did not hesitate to find liability against Envirotech, with damages amounting to the amount originally bid by Envirotech's competitor for a competing system, as well as the cost of checking design submittals and inspecting the installation of replacement equipment. The lower court also assessed $1 million in punitive damages for Envirotech's fraud. Although all the damages were remanded by the appellate court for reconsideration (including the punitive damages), the underlying liability of Envirotech was not overturned.

The duty to act in good faith is a two-way street, where owners have the same obligation to the design-builder. Consider, for example, *Building Structures, Inc. v. Young,*[84] where BSI contracted to design-build a commercial project for

---

[82]657 F. Supp. 1339 (D.S.C. 1986), affirmed in part and reversed in part, 866 F.2d 661 (4th Cir. 1989).

[83]Idem, p. 1356.

[84]131 Or. App. 88, 883 P.2d 1308 (Or. Ct. App. 1994).

Young. The one-page contract between the parties required BSI to retain necessary design professionals to develop a design, with the construction price based on a negotiated cost-plus, GMP format.

After receiving preliminary drawings, Young began negotiating construction pricing with other contractors, without the knowledge of BSI. BSI's preliminary pricing was approximately $250,000, while one of the other contractors had provided pricing of $200,000. Within a month after the city approved BSI's design, Young entered into a construction contract with another party and advised BSI—although BSI had discovered another contractor working on the site before being notified by Young. BSI sued for fraud and breach of contract, alleging that Young had only wanted it to perform design work and had no intention of using it for construction.

A jury found Young liable for both breach of contract and fraud, and awarded punitive damages. This was upheld by the appellate court, which concluded that, on the basis of the evidence produced at trial, the jury could conclude that Young had not intended to honor its agreement. The court cited evidence that Young was aware that BSI was interested in doing the project only if it was retained for both design and construction and that it was not interested in bidding on the construction. The one-page agreement, although very simple, clearly stated that the work was to proceed on a negotiated cost-plus basis. Although not expressly in the opinion, it is likely that the jury was impacted by the failure of Young to even attempt to negotiate mutually acceptable terms. Had Young been open about what it was doing, the decision might have been different.

16.8   Cancellation of the Procurement

While the number of proposers on design-build projects is normally small, resulting in reduced competition, the cost associated with the preparation of design-build proposals can be substantial, particularly if the RFP requires significant preliminary design work. As a result, design-builders are particularly sensitive to the circumstances under which an owner can choose to abandon the project. Two cases illustrate the reasons for this concern.

Case 1: SC Testing Technology, Inc.   In *SC Testing Technology, Inc. v. Department of Environmental Protection,*[85] an RFP was issued by the Maine Department of Environmental Protection (DEP) for the design, construction, and operation of emission inspection stations throughout the state of Maine. This program was being implemented through legislation authorized to comply with certain federal Clean Air Act requirements.

---

[85]688 A.2d 421 (Me. 1996) discussed in Loulakis, *Lessons Learned: How 1997's Design-Build Cases May Affect You,* 6, 1997.

After receiving the RFP, several bidders expressed concerns regarding the devastating impact on their businesses if the legislature repealed the emissions testing program after contract award. Notwithstanding these concerns, DEP issued an amended RFP that included the following language in Section 5.N:

> [i]n the event the Maine Legislature repeals all or part of the program, the Department and the State of Maine shall bear no responsibility to compensate the Contractor.[86]

During contract negotiations with the successful proposer (SCI), DEP repeatedly rejected attempts to include language that would partially compensate SCI for its investment in the event of a legislative repeal of the program.

The contract ultimately negotiated between the parties appears to have consisted of a signature page attached to a 70-page document called "Rider A." Rider A's language closely tracked the amended RFP. Some provisions of the amended RFP were explicitly incorporated by reference into Rider A, as was SCI's 1000-page proposal. Section 5.N, however, was neither included expressly into Rider A nor in SCI's proposal. In fact, Rider A (and hence the contract) did not specifically address the repeal issue in any way.

Nearly one year after SCI commenced its performance, the Maine Legislature repealed the emissions control program and DEP terminated the services of SCI for convenience. SCI filed a complaint against DEP, alleging a breach of contract and a breach of the implied duty of good faith and fair dealing. The trial court ruled in DEP's favor on a motion for summary judgment, concluding that, as a matter of law, the risk of repeal was the responsibility of SCI. It based its opinion on the language of Section AA of Article I of Rider A, entitled "Conflicts," which reads as follows:

> [t]his Contract shall control in the event of any conflict between the provisions hereof and the provisions of either the RFP or the Proposal. Furthermore, only as between the RFP and the Proposal, the RFP shall control in the event of any conflict between the provisions of the RFP and the provisions of the Proposal.[87]

The court reasoned that this language expressed a clear intention by the parties to incorporate into Rider A those provisions of the amended RFP that were not in Rider A. Because Rider A was silent as to risk of loss, and because nothing in Rider A conflicted with the risk of loss provision of Section 5.N of the amended RFP, Section 5.N was considered part of the contract.

The Maine Supreme Court agreed with the lower court, concluding that in construing a contract, one should avoid an interpretation that renders any of its terms meaningless. The court also held that when a party enters into a con-

---

[86]Idem, at 424.

[87]Idem.

tract with a state agency, it does so with the understanding that the legislature could at some future time take action that nullifies the subject matter of the contract and, necessarily, the respective performance obligations of the parties.

One of the justices, however, strongly dissented. He argued that the majority opinion inappropriately transformed the Conflicts clause of Rider A into a clause that defined not only the contract documents but also what documents would be incorporated by reference into the contract. If this was what the parties intended, the dissenting judge felt that it should have been clearly stated in the agreement—particularly given the major discussions that had taken place during contract negotiations regarding the risk of legislative repeal. The dissent indicated that this problem could have been resolved if the State had drafted a clear contract. With this view, the judge believed that the correct result would have been to find the contract ambiguous and to send the case back to the lower court to determine what the parties intended regarding the risk of loss.

Case 2: American Recycling Company, Inc.   The lawsuit *American Recycling Company, Inc. v. County of Manatee*[88] involved a similar issue over an RFP issued by the County of Manatee, Florida, for the design, construction, and operation of a waste composting facility. After determining that Americycle submitted the most advantageous proposal to the County, the Board of Commissioners authorized contract negotiations to proceed. Negotiations took place for nearly 18 months, but the parties were unable to reach an agreement. As a result, the Board terminated the negotiation process and rejected all proposals. This prompted Americycle to file a lawsuit against the County for breach of contract and violation of its constitutional rights.

Americycle argued that the Board had awarded it the contract—largely on the basis of the Board's statements at several public meetings that Americycle was a responsive bidder and that negotiations would be going forward. It looked to the language of both the RFP and the County procurement code, which stated that "award shall be made to the responsible offeror whose proposal is determined…to be the most advantageous to the County." Americycle believed that the Board's actions constituted an "award of contract." The County argued that the term "award" as used in the RFP and code meant an award of the exclusive right to attempt to *negotiate* a contract, and *not* the *actual award* of a contract. Both the trial and appellate courts agreed with the Board and rejected Americycle's positions.

In both cases, the courts looked carefully at the language of the procurement code and concluded that there was a distinction between a *bid* situation, where an award is automatically made to the lowest bidder, and a *proposal* situation, where it is anticipated that discussions will ensue with the highest ranked

---

[88]963 F. Supp. 1572 (M.D. Fla. 1997) discussed in Loulakis, *Lessons Learned: How 1997's Design-Build Cases May Affect You*, 7, 1997.

offeror. In this case, the Board made it clear that it awarded Americycle a number one ranking and the exclusive right to negotiate, but that it did not award a contract. In fact, the actions of Americycle confirmed the Board's position, since several SEC filings of Americycle's parent stated that negotiations with the County were continuing and that "there can be no assurance that [Americycle] can successfully consummate the contract" with the County. Thus, because the County never accepted the Americycle offer, no contract claim could be raised.

Americycle also argued that it had a protectable property interest under the 14th Amendment of the U.S. Constitution. It had expended millions of dollars in detrimental reliance on the RFP and procurement code, and believed that the failure of the County to award the contract was a violation of that interest. In rejecting this argument, the court found that Americycle should not have expected to receive the contract because the RFP gave the County broad discretion to cancel the procurement and not award a contract to anyone.

Both *SC Testing* and *County of Manatee* involved high-stake design-build projects, where operation and financing were important components of the procurement. Although not clear from the decisions, there is an indication that both owners were vague about what they wanted and allowed the competing design-build firms to expend substantial monies at risk. However, each design-builder should have been concerned that the language of the contract did not clearly identify what would happen if the program were canceled. In fact, the contract language in both cases suggested that the proposers were proceeding at their own risks and that cancellation was a possibility.

16.9   Differing Site Conditions

As discussed in Chap. 13, *differing site conditions clauses* are contract clauses used to establish which party bears the risk of certain unanticipated site conditions.[89] Parties should not assume that because a design-builder contracts to both design and construct the project, the design-builder will automatically bear the risk of differing site conditions. In fact, many owners prefer to retain this risk, just as they would in the traditional contracting process. The owner will then only pay for the actual cost of an actual unanticipated condition, rather than the contingency included in the contractor's bid for a problem that may never arise.

Combined with the issue of differing site conditions are two other legal principles that can affect responsibility for unexpected site conditions: (1) a contractor who makes a claim for differing site conditions will be bound to what a reasonable site investigation would reveal,[90] and (2) the owner has a

---

[89]See M. C. Loulakis, "Differing Site Conditions," in *Construction Claims Deskbook, Management, Documentation and Presentation of Claims*, R. S. Brams and C. Lerner, eds., Wiley, New York, 1996, p. 131.

[90]Idem, at Section 11.14.

duty to disclose information that could be relevant to the contractor's performance, under what is commonly known as the *Helene Curtis doctrine.*[91] All of these issues were addressed within the context of a design-build project in a case before the Armed Services Board of Contract Appeals, *Pitt–Des Moines, Inc.*[92]

In *Pitt–Des Moines,* the design-build contractor entered into a firm fixed-price contract with the Government to design, develop, fabricate, and install an approximately 900,000-gal large acoustical tank in an existing building, known as Building 5, at the Naval Research Laboratory NRL in Washington, D.C. The work required, among other things (1) removal and resupport of existing structures, (2) excavating the interior of the building to a depth of approximately 50 ft, (3) dewatering of the excavation, (4) providing all tank foundations, (5) underpinning, and (6) modifying and resupporting the roof to accommodate the tank. The design-builder proposed a design that used a 28-auger cast pile system to underpin and support the existing walls during construction, as well as ring beams with timber lagging to support the excavation.

After receiving notice to proceed, the design-builder learned from newly discovered government drawings that the existing building walls were actually much thicker than previously represented in the government's RFP. The design-builder had previously attempted to obtain these as-built drawings, but was informed that no such drawings could be located. On the basis of the new information, the design-builder changed its design from a 28-pile system to a 48-pile system in order to accommodate the greater building weight.

Subsequent to the support system change, the government informed the design-builder that the existing wood piles supporting an adjacent building (Building 43) were severely deteriorated. It is not clear when the government learned about the problems of deteriorating piles at Building 43, but it is clear that prior to its issuance of the RFP for Building 5, the government had two reports in its possession regarding problems with the structural integrity of Building 43. The government did not furnish the reports to the design-builder.

In light of the deteriorating piles, the design-builder evaluated the likely effect that its construction on Building 5 would have on Building 43 and determined that its proposed dewatering would lower the water table, likely causing severe settlements to Building 43. Because the design-builder was fearful of being held responsible for the loss of structural integrity of Building 43, it proposed a different method to support and excavate the Building 5 work. The design-builder proposed a ground-freeze method in which it would inject refrigerant through a piping system placed in the ground and thus create a frozen earth cofferdam that would support the building and prevent water infiltration into the excavation. The ground-freeze method eliminated the

---

[91]See generally, Cibinic and Nash, *Administration of Government Contracts* (see fn. 26, above), pp. 368–407.

[92]ASBCA 42838, 43514, 43666, 96-1 BCA, Paragraph 27,941 (1996).

need for dewatering and thus extinguished the fear of the adjacent Building 43 settling.

Although the design-builder was able to overcome the problems associated with both the unanticipated additional weight of Building 5 and the potential settlement problems with Building 43, it lost several months in its planned schedule and incurred substantial cost increases as a result of the need to employ different means and methods of construction. It asserted a claim for differing site conditions due to the unanticipated increase in the weight of Building 5 and the deteriorated condition of Building 43. The government, relying in part on a contract clause which required the design-builder to investigate the site, argued that the design-builder had taken a "narrow view of its design responsibilities and failed to investigate adequately the conditions potentially affecting the planned work."[93]

The Board concluded that the design-builder was entitled to an equitable adjustment due to the differing site conditions. The Board stated that, as a general rule, a bidder is "not required to conduct a costly or time-consuming technical investigation to determine the accuracy of the Government drawings."[94] The Board held that this general rule applies even in a design-build context. The Board further made it clear that the government should have disclosed all the information concerning site conditions at the time of bidding or negotiation.

16.10   Ownership of Documents

As discussed in Sec. 12.8, the ownership of documents developed by the design-builder can be a critical issue in the success of the design-build project. Three of the most important considerations relative to ownership of documents are

- What happens if the design-builder defaults on the project and the owner is required to reprocure others to perform design and/or construction services?

- What happens if the owner terminates the design-builder for convenience and uses the design-builder's design for the completion of the project?

- What happens if the owner executes the design-builder's design on another project with a different design-build team?

The answer to these questions has been decided largely by how the parties contract for such risks. However, several cases shed some light on how ownership of the contract documents affects the administration of the design-build project.

---

[93]Idem, at 139, 573. See *Investigation and Conditions Affecting the Work,* FAR 52.236-3, April 1984.

[94]Idem, at 139, 575.

In *Johnson v. Jones,*[95] Johnson, an architect, was retained by Jones to undertake the design of the renovation of Jones' million-dollar home for a fee of $3\frac{1}{2}$ percent. Shortly after being retained, Johnson also agreed to manage the construction of the renovation for a 14 percent fee. After agreeing to the business deal, the parties began negotiating a written design-build contract. Johnson began performance in advance of contract execution, obtaining construction estimates, preparing general demolition and construction plans, and completing several drafts of preliminary design drawings. He also prepared and submitted plans for permitting. Jones approved the plans and made payments to Johnson.

Approximately 3 months into their relationship, Jones terminated Johnson and contracted with another architect and a construction management firm. These firms used the work product that had been developed by Johnson, prompting Johnson to bring an action for copyright infringement against Jones, the successor architect, and the construction management firm.

Jones argued that this was not a copyright case, but simply a dispute over the failure of Jones to pay a contract balance to Johnson. The court disagreed. Looking at the entire course of dealings between Jones and Johnson, the court concluded that Johnson never intended to relinquish control of his drawings, either when he agreed to prepare the plans or when he gave copies of the incomplete plans to Jones.

The court focused on the contract negotiations, where Johnson had submitted to Jones several contract drafts using AIA Document B141 as a baseline. These contracts vested ownership of the documents with Johnson, as evidenced by the following language:

> [t]he Drawings, Specifications and other documents prepared by the Architect for this Project are instruments of the Architect's service for use solely with respect to this Project, and the Architect shall be deemed the author of these documents and shall retain a common law, statutory and other reserved rights, including the copyright.... The Architect's Drawings, Specifications and other documents shall not be used by the Owner or others on other projects, for additions to this Project or for completion of this Project by others, unless the Architect is adjudged to be in default under this Agreement, except by agreement in writing and with appropriate compensation to the Architect.

The court noted that during contract negotiations Johnson repeatedly objected to language that would make Jones the owner of Johnson's work product. The court concluded that Johnson intended to grant a license to Jones to use the plans for renovation of her home only if the plans were completed by Johnson—not by someone else after Johnson had been terminated. Finding that this constituted copyright infringement, an award was rendered not only against Jones for unjust enrichment but also against the successor architect

---

[95] 885 F. Supp. 1008 (E.D. Mich. 1995).

and construction manager—who were required to pay Johnson their collective profit on the project.[96]

I.A.E. Incorporated v. Shaver[97] involved the design-build of a cargo/hanger building at the Gary, Indiana Regional Airport. The design-builder subcontracted with Shaver, an architect with extensive experience in designing airport facilities. The letter agreement between the parties provided that Shaver would develop the schematic drawings for $10,000. Although Shaver believed that he would execute additional written contracts for other design phases after schematics, nothing to this effect was stated in the letter.

After the Airport approved one of Shaver's schematic drawings—which contained a notice of copyright—another architect was retained by the design-builder to perform the remaining design work. When Shaver learned that another architect had been hired, he wrote to the airport and noted that, while he was no longer in a position to participate in the project, he trusted that his ideas as set forth in the schematics would assist the airport.

Subsequent to Shaver's letter, Shaver's lawyer sent a letter seeking, among other things, a $7000 payment for the assignment of Shaver's copyright on the schematic drawings. When the parties were unable to resolve the issue, the design-builder filed an action for a declaratory judgment that it did not infringe any copyright owned by Shaver and it had the right to use Shaver's drawings. Shaver counterclaimed for copyright infringement. His argument was based on the expectation that he would finish the design of the project.

The district court's ruling was affirmed by the Seventh Circuit Court of Appeals, which held that there was no copyright infringement and concluded that Shaver had provided the design-builder with an implied license to use the drawings. The court reached its conclusion by carefully looking at the contract drafted by Shaver, which unambiguously described his role on the project as the designer of the schematic drawings. The contract did not mention anything that would suggest that Shaver was to continue on as architect after this work was finished. The court held that Shaver created the preliminary architectural drawings and found Shaver was respectively paid for this work.

Shaver's actions confirmed that the design-builder had an implied nonexclusive license to use the drawings on the project. The Seventh Circuit noted that Shaver had delivered the copyrighted designs without any warning that their further use would be a copyright infringement, and acknowledged, in his letter to the airport, that he was no longer a contributor to the project. Neither of Shaver's actions supported his argument that he needed to remain the architect of record as a condition precedent to the use of the schematics by the design-builder.

Although the facts of the *Johnson* case are similar to the facts of *Shaver*, the courts reached totally opposite conclusions. The court in *Johnson* rejected the

---

[96] *Johnson v. Jones*, 921 F. Supp. 1573 (E.D. Mich 1996).

[97] 74 F.3d 768 (7th Cir. 1996).

notion that there was an implied license for the owner and new architect-builder to use the work product of the terminated design-builder. The different result can be justified by the facts of *Johnson,* wherein the design-builder made its position very clear during contract negotiations and performance that it would not release its ownership rights in the drawings and work product. The court was persuaded that Johnson had been retained to design and construct the *whole* project and that the owner had taken advantage of the situation in terminating the relationship. To the contrary, in the *Shaver* case the Seventh Circuit was convinced that the design-builder had done nothing wrong and the contract, drafted by Shaver, reflected that Shaver was only to perform limited services on the project.

16.11 Personal Injury and Property Damage

On projects delivered under a design-bid-build approach, the responsibility for job-site safety and personal injury rests largely with the general contractor, who will develop an overall safety plan and cause the trade subcontractors to follow it. Moreover, the contract will typically require that the general contractor indemnify the owner and A/E from any third-party injury caused by the negligence of the general contractor.[98] Notwithstanding this attempt to shift liability for third-party injuries to the general contractor, owners, A/Es, and construction managers (CMs) are frequently joined as parties to the personal injury litigation—principally to overcome workers' compensation limitations.[99]

The contract language relative to personal injury and property damage under design-build contracts is typically quite similar to that cited above under design-bid-build projects. While there is little case law to date on these issues, courts appear to be evaluating liability in the overall context of single-point responsibility, finding the design-builder, not the owner, to be responsi-

---

[98]A typical clause is found in AIA Document A201 (1997 ed.), which states in Subparagraph 3.18.1:

> To the fullest extent permitted by law and to the extent claims, damages, losses or expenses are not covered by Project Management Protective Liability insurance purchased by the Contractor in accordance with Paragraph 11.3, the Contractor shall indemnify and hold harmless the Owner, Architect, Architect's consultants, and agents and employees of any of them from and against claims, damages, losses and expenses, including but not limited to attorney's fees, arising out of or resulting from performance of the Work, provided that such claim, damage, loss or expense is attributable to bodily injury, sickness, disease or death, or to injury to or destruction of tangible property (other than the Work itself), but only to the extent caused by the negligent acts or omissions of the Contractor, a Subcontractor, anyone directly or indirectly employed by them or anyone for whose acts they may be liable, regardless of whether or not such claim, damage loss or expense is caused in part by a party indemnified hereunder. Such obligation shall not be construed to negate, abridge, or reduce other rights or obligations of indemnity which would otherwise exist as to a party or person described in this Paragraph 3.18.

[99]M. C. Loulakis, *Trends in A/E & CM Liability for Jobsite Safety,* Construction Briefings, 2d series, Federal Publications, Sept. 1997.

ble if such an injury should occur. However, two important questions remain about personal injury and property damage:

- Can the owner's active role in contract administration impact the single point of responsibility and render the owner liable?
- Will strict liability be used as a standard for assessing the liability of design-builders and their team members for personal injury or property damage?

Each of these questions is addressed in the sections that follow.

16.11.1 Owner involvement as affecting liability

The issue of how owner involvement may affect job-site liability was discussed in the case of *Moschetto v. United States*.[100] Lehrer, McGovern and Bovis, Inc. (LMB) was the design-builder on a contract with the General Services Administration (GSA) for the construction of the new United States District Courthouse in White Plains, New York. Moschetto, an LMB employee, was injured on the project and sued the federal government under the *Federal Tort Claims Act* (FTCA), which allows private actions against the federal government if a government employee was negligent while acting within the scope of his employment. The government contended that it was not liable to Moschetto because, if any negligent acts had been committed, they were done by the design builder, LMB, who was not an "employee" under the FTCA.

Moschetto attempted to overcome this argument by pleading the substantial involvement of GSA on the project. It cited GSA's contract with LMB, which, among other things: (1) reserved the government's right to "supervise, direct, and inspect" the construction site; and (2) stated that all work was to be performed under the general direction of the contracting officer. It also noted GSA's conduct during the job, whereby GSA supervisors allegedly walked the construction site daily, supervising and directing LMB's employees on how to perform their duties. Moschetto contended that this control created a principal-agent (employer-employee) relationship between GSA and LMB, making the government liable under the FTCA.

In considering Moschetto's arguments, the court acknowledged that an agency relationship can be established if the government exercises day-to-day control over the "detailed physical performance of the contractor." However, it also noted that when "the government exercises broad supervisory powers, reserves the right to inspect, or monitors compliance with federal law, an independent contractor relationship will most often be found." The court concluded that LMB, as the design-builder, was directly responsible for the job-site and construction superintendence. It held that GSA's functions were merely supervisory. GSA's placement of supervisors on site was consistent with these broad supervisory powers. In reaching its conclusion, the court noted that

---

[100]961 F. Supp. 92 (S.D.N.Y. 1997).

[r]eservation of the power to control a contractor's compliance with the contract's specification does not make the contractor an employee.

Given this view, the court concluded that LMB was a government contractor, not a government agent, and dismissed Moschetto's FTCA claim against the government.

The *Moschetto* decision begins to provide some answers about the impact that the owner's role may have in personal injury liability. It reinforces the notion that an owner does have the right to remain actively involved on its project without affecting the single point of responsibility of the design-builder. However, by looking at the facts as carefully as it did, the court appeared to leave open the question of what would have happened if the owner exercised major control over site safety, through its own forces or through an A/E or CM. Because many personal injury cases in construction are decided on the axiom "with control, comes responsibility,"[102] an active owner may very well find itself liable for personal injuries, notwithstanding the single point of responsibility.

16.11.2 Strict liability

Strict liability is a theory that has developed over the years to provide injured consumers with a means of recovering against a seller of a product without having to prove negligence or fault on the part of the seller. Sometimes referred to as "liability without fault," this concept arose in the context of manufacturers who sold products that were "defective" and "unreasonably dangerous to the user or consumer."[103] Importantly, the doctrine as developed does not provide redress for those with whom the seller has a direct contractual relationship, but instead is meant to help those for whom there is no privity of contract.[104]

Given the single point of responsibility of the design-builder, some have argued that the strict-liability doctrine should apply to the design-builder and its team members in the event of a jobsite injury or property damage.[105] They argue that the design-builder has substantial control over the "product" (i.e.,

---

[101]*Moschetto,* 961 F. Supp at 95 [quoting *Lipka v. United States,* 369 F.2d 288, 291 (2d Cir. 1966)].

[102]M. C. Loulakis et al., *Construction Management Law and Practice,* Wiley, New York, 1995, Sections 14.1–14.22.

[103]See *Restatement (Second) of Torts* §402a (1965), which states in pertinent part: "(1) One who sells any product in a defective condition unreasonably dangerous to the user or consumer or to his property is subject to liability for physical harm thereby caused to the ultimate user or consumer, or to his property, if (a) the seller is engaged in the business of selling such a product, and (b) it is expected to and does reach the user or consumer without substantial change in the condition in which it is sold."

[104]See *Restatement (Second) of Torts* §402A, Subsection (2)(b)(1965).

[105]Miller, "Comment" (see fn. 4, above).

the completed construction project) and is in the best position to bear the risk of defects, just as is the manufacturer of a product. Although some courts have found that the design-builder is providing a "product," most courts disagree, finding that construction is a "service" industry.[106]

There is a line of cases, however, that routinely has found the design-builder to be a "seller" of a "product" for strict-liability purposes. Most of these cases involved the design and sale of residences by a mass-home builder.[107] For example, *Schiffer v. Levitt and Sons, Inc.*[108] involved a suit against Levitt, the design-builder of a home by Schiffer, who was leasing the home. The injury was created when excessively hot water from the bathroom faucet scalded Schiffer's child. The New Jersey Supreme Court found that Levitt, as a mass producer of homes, was no different from a manufacturer of automobiles or other consumer products and that the consumer (in this case Schiffer) was in no better position to protect itself than would be the user of an automobile.

A similar case, *Kriegler v. Eichler Homes, Inc.*[109] involved defects in a radiant-heating system in a home designed and constructed by Eichler Homes. The system's tubing, which was constructed of steel due to a shortage of copper, corroded as a result of improper placement in the concrete slab. Eichler had constructed over 4000 homes using this type of system. The court, relying on the same public policy concerns as found in the consumer goods manufacturing cases, believed that public interest required that Eichler be responsible for the defect and was in a better situation to bear the resulting costs.[110]

There is also at least one case that found strict liability outside of the residential home market. In *Abdul-Warith v. Arthur G. McKee and Co.*[111] the court considered strict liability in the context of the design-build of a skip bridge that was to carry materials from a stockhouse to a blast furnace. The complaint had a count for strict liability under Section 402A of the Restatement (second) of Torts on the basis that the skip bridge was defective and unreasonably dangerous. The court stated

> Where the architect or engineer simply provides the design or merely supervises, without actually participating in, the construction of a challenged product, he has

---

[106]See discussion in Sec. 16.4.2 (above) on implied warranty, which involves the same consideration of product versus service.

[107]See also Tasher, "Liabilities of California Builder Contractors & Construction Professionals: The Need for Equality in Legal Responsibilities," *Cal. W.L. Rev.* **15,** 305 (1979).

[108]44 N.J. 70, 207 A.2d 314 (1965).

[109]269 Cal. App. 224, 74 Cal. Rptr. 749 (1969).

[110]See also *Rosell v. Silver Crest Enter.* 436 P.2d 915 (Ariz. 1968); *Furney Industries, Inc. v. St. Paul Fire & Marine Ins. Co.,* 467 F.2d 588 (4th Cir. 1972).

[111]488 F. Supp. 306 (E.D. Pa. 1980).

not been held strictly liable; where, however, the professional actually assembles or erects the allegedly defective item, strict liability will attach.[112]

The design-builder in *McKee* alleged that it was not a mass producer of goods and not in the best position to distribute the costs of compensating injured consumers. The court rejected the argument, finding that the volume of a defendant's sales is irrelevant as to whether one is a "seller" under a strict-liability standard.

It is true that the merger of design and construction does heighten the responsibility of the design-builder over that of an A/E or general contractor on a traditionally delivered project. However, there are several crucial differences between the construction and manufacturing industries that appear to make application of a strict liability standard against design-builders inappropriate as a matter of policy:

- Unlike the manufacturer, the design-builder is not working in a controlled environment, with the benefit of being able to build prototypes in a laboratory and mass-produce the product. The product of a design-builder is one of a kind, and will be built on a unique site for the first time under less than controlled conditions.

- Unlike the manufacturer, who can spread its cost among the many products that it will sell, the "product" of the design-builder is unique and will effectively be sold once, on specific commercial terms. There is no way to spread the risk among the consumers of this "product."

- The owner, as the consumer of the design-builder's "product," will have some role in specifying what it is interested in receiving from the design-builder. This is far different from selling a consumer good, where the manufacturer does not know, and will likely have no direct contact with, the specific consumer of the product.

Notwithstanding these policy issues, however, there is often great sympathy by courts and juries to assist innocent and seriously injured plaintiffs. This, coupled with the difficulties in assessing how far the single point of responsibility standard should extend, may result in application of a de facto strict-liability standard under the guise of another liability theory, such as negligence or implied warranty.

16.12 Limitations of Liability

As discussed in Sec. 14.16, limitation-of-liability clauses are often one of the most significant issues addressed during the negotiations of a design-build contract. These clauses come in a variety of shapes and forms and include provisions which

---

[112]Idem, at 310–311, n. 3.

- Set a ceiling on the amount of damages for which the design-builder will be liable (i.e., 50 percent of the contract price)

- Exclude liability for consequential damages

- Establish liquidated damages for performance guarantee deficiencies

- Exclude all implied warranties

- Limit the design-builder's liability for defective work to the cost of repair or redesign

There is a substantial amount of state law addressing the enforceability of limitations of liability clauses on construction projects. The following sections address the case law interpreting these clauses, including several which arise in the context of design-build projects.

16.12.1 Enforceability—generally

The question of whether a limitation of liability is enforceable starts with a determination of whether applicable state law permits the use of limitation of liability clauses as a matter of public policy. Those states which find such clauses unenforceable on public policy grounds base their decisions on so-called anti-indemnity statutes, which prevent parties from being indemnified for damages arising out of their own negligence. An example of this type of statute is as follows.

> [a]ny covenant, promise, agreement or understanding entered into in connection with or collateral to a contract or agreement relative to the construction, alteration, repair or maintenance of any building, structure or appurtenances thereto including moving, demolition and excavating connected therewith, that purports to indemnify or hold harmless the promisee against liability for damage arising out of bodily injury to persons or damage to property caused by or resulting from the sole negligence of such promisee, his agents or employees, is against public policy and void.[113]

Another case, *City of Dillingham v. CH²M Hill Northwest, Inc.*,[114] demonstrates the affect of this type of statute on limitations of liability.[115] This case involved the action of a general contractor against the owner for a differing site conditions claim wherein the owner sued CH²M Hill, the project engineer. The agreement between the owner and CH²M Hill contained a limitation-of-

---

[113]*Conn. Gen. Stat.* (Civil Action) § 52-572K (1979).

[114]873 P.2d 1271 (Alaska 1994).

[115]There are numerous indemnification-type statutes in the following states: *Cal. Civ. Code*, Paragraphs 2782, 2782.2 (West 1985); *Conn. Gen. Stat.* § 52-572K (1979); *Ill. Rev. Stat.* Ch. 29, paragraph 1 (West 1971); *Md. Code Ann. Cts. & Jud. Proc.* § 5-401 (1997); *Mass. Ann. Laws* Ch. 153, § 7 (1909); *Minn. Stat.* §§ 337.07, 337.02 (West 1983); *N.J. Stat. Ann.* § 2A:40A-1 (West 1983); *N.Y. Gen. Oblig. Law* § 5-324 (McKinney 1965); *Or. Rev. Stat.* § 30.140 (1973); *R.I. Gen. Laws* § 6-34-1 (1976); *Va. Code Ann.* § 11-4.1 (1991); *Wash. Rev. Code Ann.* § 4.24.115 (West 1986).

liability clause whereby the owner agreed to limit liability "for the engineer's sole negligent acts, errors, or omissions" to $50,000.

In reliance on this limitation of liability provision, CH^{2}M Hill moved for partial summary judgment. The owner argued that the limitation provision did not apply to intentional or bad-faith actions of an engineer, and further argued that the provision was void because of an Alaska statute that provided that construction contracts which seek to indemnify a party for liability resulting from the party's sole negligence or willful misconduct are void and unenforceable.

The Supreme Court of Alaska found that "knowing," "bad faith," or "intentional" breaches could never be limited by agreement. Moreover, the court looked to the legislative intent of the Alaska anti-indemnification clause referenced by the owner and concluded that the legislature intended to prevent parties from escaping liability for their own negligence, whether through an indemnification clause or a contractual limitation of liability. As a result, CH^{2}M Hill could not rely on this clause and was left to defend itself on the merits.[116]

If the limitation of liability clause passes the public policy test, courts will carefully examine the clause and determine whether it is reasonable, specific, and the result of an arm's-length negotiation. *William Graham, Inc. v. City of Cave City*[117] illustrates the strict interpretation that most courts exercise in analyzing limitation-of-liability clauses. The *Graham* case was based on breach of contract action brought against the designer of a wastewater treatment plant. Both parties to the contract were aware that the late submission of the plans and specifications to the government agency would cause a reduction in the project funding. The design professional failed to meet the contractually established deadline for submitting the documents—resulting in a project funding reduction of approximately $339,000. However, the design professional's contract contained the following limitation of liability clause:

> [t]he OWNER agrees to limit the ENGINEER's liability to the OWNER and to all Construction Contractors and Subcontractors on the Project, due to the ENGINEER'S professional negligent acts, errors or omissions, such that the total aggregate liability of the ENGINEER to those named shall not exceed $50,000 or the ENGINEER'S total fee for services rendered on the project, whichever is greater.[118]

The design professional argued that its fee, approximately $99,000, was the extent of its damages.

---

[116]Several states have similar holdings. See *Valhal Corporation v. Sullivan Associates, Inc.*, 1993 U.S. Dist. LEXIS 6607 (E.D. Pa. May 17, 1993); *Graham, Inc. v. City of Cave City*, 289 Ark. 105, 709 S.W.2d 94 (1986); *Wausau Paper Mills Co. v. Main, Inc.*, 789 F. Supp. 968 (W.D. Wis. 1992); *Long Island Lighting Co. (Lilco) v. Imo Delaval, Inc.*, 668 F. Supp. 237 (S.D.N.Y. 1987). See also *Harnois v. Quannapowitt Development, Inc.*, 35 Mass. App. Ct. 286, 619 N.E.2d 351 (1993).

[117]289 Ark. 105, 709 S.W.2d 94 (1986).

[118]*Graham*, 709 S.W.2d, at 95.

The court did not dispute the validity of the limitation-of-liability clause, but refused to apply the clause to limit the damages for breach of contract. The court pointed out that the clause cited only damages arising out of professional negligent acts, errors, or omissions. No mention was made in the contract of liability based on breach of contract. The court reasoned that because the parties had the opportunity to negotiate a clause that would limit damages for breach of contract, but did not clearly do so, the limitation clause would not be applied.

16.12.2   Specific design-build issues

Although the general principles discussed above apply in design-build, it is interesting to note several design-build cases that have expressly discussed limitations of liability. In *Union Oil Company of California v. John Brown E&C,*[119] Union Oil Company of California ("Unocal") contracted with John Brown E&C, a division of John Brown, Inc. (JBI), to design and construct for Unocal a polymer plant in Kankakee, Illinois. The parties entered into a cost reimbursable contract to construct the plant, and JBI's total fee was limited to $415,000. The contract further provided that

> [JBI's] maximum aggregate liability to Unocal...shall not exceed the proceeds of the applicable insurance coverages plus eighty percent (80%) of the aggregate fee paid to [JBI].... The limitations on [JBI's] liability as specified above, shall apply whether such liability arises at contract, tort (including negligence or strict liability), or otherwise. The above notwithstanding, said limitations on liability shall not apply to all or any portion of such liability which arises out of the gross negligence, fraud, or willful misconduct of [JBI].

When project completion was delayed 8 months, Unocal asserted that the delay was caused by JBI's failures and that Unocal incurred over $8 million of additional costs associated with it.

In an effort to resolve the dispute, JBI offered to credit Unocal $332,000 (80 percent of JBI's fee), which it asserted was the maximum amount for which it could be found liable. Unocal refused the offer and filed suit against JBI, alleging breach of contract, negligence, negligent misrepresentation, gross negligence, and breach of the implied covenant of good faith and fair dealing. On motion by JBI, the court dismissed all the negligence counts for failure of Unocal to state a claim. The court ruled in favor of JBI on its motion for judgment on the breach of the implied covenant of good faith and fair dealing.

The only remaining count in the litigation was for breach of contract. JBI moved for summary judgment on that count, asserting that its liability was limited to $332,000. Unocal argued that the limitation of liability was not applicable because JBI was grossly negligent. The court noted that Unocal's assertion of gross negligence had already been addressed by the court when it

---

[119] 1995 U.S. Dist. LEXIS 13173, at *1 (N.D. Ill. Sept. 8, 1995).

was dismissed as a direct count against JBI and could not be used as a defense to the limitation of liability for a contract count. Accordingly, the court upheld the limitation of JBI's liability on the basis of the contract provision.

The court briefly responded to Unocal's assertion that it is a question of fact as to whether the limitation of liability in this case was unconscionable. Although the court found the contract to be "somewhat open-ended," it concluded that Unocal was experienced with such contracts and there was nothing to indicate that Unocal did not voluntarily acquiesce to the limitation of liability clause with JBI. The court also noted that Unocal was "a giant and sophisticated company" and that there was no evidence that it had been at a bargaining disadvantage with JBI. "In short," noted the court, "Unocal has made no showing whatsoever on the issue of unconscionability."[120]

In *Valero Energy Corp. v. M.W. Kellogg Construction Co.,*[121] an oil refinery owner contracted with Kellogg to design-build a $500 million expansion of a refinery in Corpus Christi, Texas. Several years after construction was completed, a piece of equipment failed and caused damages to the rest of the plant, with flying shrapnel damaging structural I-beams, pipe rack, and several tons of catalyst. Suit was brought for gross negligence, products liability, and violations of the Texas Deceptive Trade Practices Act. Kellogg defended on the basis of a limitation of liability clause, which stated, among other things, that Kellogg, in exchange for a 2-year guarantee and a promise to correct and reengineer any defects, would be released from claims arising out of the work. The court upheld the limitation, concluding that the owner and Kellogg were sophisticated entities, with competent counsel and equal bargaining power, that negotiated the agreement over an extended period of time.

Ebasco Services, Inc. v. Pennsylvania Power & Light Co.[122] involved disputes between Pennsylvania Power & Light Co. (PP&L), the owner of a power plant, Ebasco Services, Inc. (Ebasco), the EPC contractor for the plant and General Electric (GE), and the steam turbine subcontractor to Ebasco. Although there were various claims raised by each party, one of the most significant claims made by PP&L against GE and Ebasco was for the cost of replacement power incurred by PP&L due to delays by GE in supplying the turbines. Ebasco and GE relied, in part, on a clause in GE's subcontract which stated:

> In no event, whether as a result of breach of contract, alleged negligence, or otherwise, shall [GE] be liable for damages for loss of profits or revenue resulting therefrom, or for damages for loss of use of power system, cost of capital, cost of purchased or replacement power, or claims of customers of Purchaser for service interruptions resulting therefrom, or similar items of damages resulting therefrom.[123]

---

[120]*Union Oil,* 1995 U.S. Dist. LEXIS 13173, at *10.

[121]866 S.W.2d 252 (Tex. Ct. App. 1993).

[122]460 F. Supp. 163 (E.D. Pa. 1978).

[123]Idem, at 180.

In a lengthy opinion, the court concluded that this clause was enforceable against PP&L, even though PP&L and GE had not contracted directly, and that the cost of replacement power was not recoverable.[124]

A design-build case where the court did not enforce the limitation of liability is *Koppers Co., Inc. v. Inland Steel Co.*[125] Koppers, the design-builder, entered into a cost-plus-fee agreement with Inland Steel to design, engineer, and construct a blast furnace and a battery of 69 coke ovens to be installed at Inland's Indiana Harbor location. The agreement established target costs for each area of work as well as a provision for adjusting the target costs to reflect certain specified actual costs and deviations agreed to during the work's progress. The agreement provided the following limitation of liability provision concerning overruns:

> If both Engineering and Material Costs and Construction Costs overrun their respective Target Costs, Koppers will be obliged to refund to Inland an amount equal to (a) that part of the overrun attributable to Construction Costs which does not exceed $1,500,000 plus 30% of the remaining portion of the overrun, if any; provided that, in no event will the refund be more than $4,800,000; plus (b) the amount of overrun attributable to Engineering and Material Costs.... [This provision applies to overruns resulting from excess Construction Costs.] If the balance is an overrun resulting from excess Engineering and Material Costs, Inland will be entitled to a refund equal to such balance.[126]

The agreement further attempted to limit Koppers' liability by stating, "Engineering and design errors shall be corrected at no change in Contract Price, but Koppers shall have no other liability under the contract for such errors."[127]

The project was eventually completed over $2\frac{1}{2}$ years late and $170 million over the original budget. As a result, Inland brought suit for breach of contract and was awarded over $63 million in damages. Koppers appealed, asserting that (1) the limitation of liability provisions limiting its total liability to $4.8 million must be enforced regardless of why or how the overruns may have occurred and (2) its only liability for engineering errors was to correct the engineering drawings at no charge—with no liability for the field changes resultant from engineering errors.

The appellate court disagreed. While the court recognized the parties' freedom to agree to limit damages, it held that such limitations must be stated with great particularity in clear, direct, and unmistakable language.[128] In this

---

[124]This opinion, over 60 pages long, goes through an interesting history of why organizations like GE and Ebasco mandate the use of limitation of liability clauses in their power contracts. It also addresses numerous arguments made by PP&L to overcome the effect of this clause, including arguments under the UCC.

[125]498 N.E.2d 1247 (Ind. Ct. App. 1986).

[126]Idem, at 1249–1250.

[127]Idem, at 1254.

[128]Idem, at 1251.

case the court found that the language fell far short of the particularity requirement because the breach of contract damages allegedly arose from Koppers' negligence and lack of good faith, reasonableness, and diligence in prosecuting the work—none of which were mentioned in the limitation of liability provisions. Moreover, the court held that "to the extent that the field changes were necessary to correct errors that had been made in construction by following the erroneous engineering or to the extent that the changes required by the corrected engineering would increase the contract price to be paid by Inland, such amounts were properly recoverable."[129]

16.13 Statutes of Limitations and Repose

Most A/Es and contractors are well aware of the statutes of limitation and statutes of repose applicable to their work. These legislative creations, which limit the time within which a party can be held liable, are based on a variety of public policy considerations, including the fact that with the passage of time, key evidence can be lost. Statutes of limitation accrue from the date an improper act was committed or, in some states, the date of discovery of the improper act. Because construction defects may not be discovered until many years after the work was performed, statutes of repose were legislated to provide a specific date from completion of the project that will extinguish claims, regardless of when the defect was actually discovered. Both of these statutes vary from state to state, and statutes of repose range from 5 to 20 years.

Two interesting questions arise with respect to statutes of limitation and repose in design-build. As noted in Sec. 16.4, some of the novel theories of liability (e.g., application of the UCC or implied warranties) are based on attempts to circumvent statute of limitations and statute of repose defenses. Recall the *Smith v. Arcadian Corp.*[130] discussion in Sec. 16.4.2. The other question relates to the impact of merging design and construction. In some states, the A/E's liability for design errors has historically accrued on completion of the design and issuance of construction documents. By placing these services under a single contract, it appears that the accrual date will commence on completion of the overall project.

This result was reached in *Kishwaukee Community Health Services Center v. Hospital Building and Equipment Company, et al.,*[131] where the court evaluated a defense of statute of limitations when the design was completed more than 2 years prior to the completion of construction.[132] Had the court found that the causes of action based on design defects accrued at the time the design was completed, those claims would have been barred under the applicable statute.

---

[129]Idem, at 1254–1255.

[130]657 So.2d 464 (La. 1995).

[131]638 F. Supp. 1492 (N.D. Ill. 1986).

[132]Idem, at 1504.

The court refused to find any difference between accrual dates for the design-based claims and those for claims based on construction defects, stating

> [C]omponents of a contract cannot be wrenched out of the contract for accrual purposes, especially in construction cases.... Plaintiff alleges a unified agreement to design and build a hospital free of defects. The whole contract was not breached until delivery of the hospital, not delivery of the design.[133]

In a similar case, *Welch v. Engineering, Inc.,*[134] involving a statute of repose, the court held that where design and construction are undertaken by the same entity, it is the completion of all construction that triggers the accrual of any action based on either design errors or construction defects. The court reasoned that, "[i]t would be paradoxical to start the ten-year clock running towards repose in favor of the design/builder before the dangerous condition even existed, other than conceptually."[135]

16.14   Dispute Resolution

As noted at the outset of this chapter, parties to a design-build contract are frequently interested in using creative means of resolving their disputes. Consequently, it is not unusual for a dispute resolution clause to set forth a number of approaches to addressing conflicts—starting with a conciliatory, nonbinding process (e.g., executive negotiations, mediation, or dispute review boards) and ending with either arbitration or litigation as a last resort. It is critical for the parties to give some careful thought to this issue, since a workable disputes clause negotiated before a problem arises can be quite beneficial in soothing emotions when the conflict actually arises.

The subject of dispute resolution is very broad and not unique to design-build. The sections that follow will highlight some of the alternatives available for nonbinding and binding dispute resolution and contrast some of their attributes.

16.14.1   Nonbinding dispute processes

Nonbinding dispute resolution processes [commonly known as *alternative dispute resolution* (ADR)] have become a popular and cost-effective method to resolve construction disputes. Most ADR processes are geared to the goal of facilitating settlement between the parties, and include characteristics such as

- Mandatory involvement of top decisionmakers of the parties in the settlement proceedings
- Use of a third-party neutral guiding the discussions or presentations

---

[133]Idem, at 1505.

[134]202 N.J. Super. 387, 495 A.2d 160 (1985).

[135]Idem, at 165.

- Limited formal presentations of each party's position, followed by extensive discussions among the decision makers and neutral about the strengths and weaknesses of the case
- Nonbinding nature of the results

Many parties like the nonbinding feature of an ADR process because they are not limited to "only one bite at the apple," and each party has the opportunity to evaluate a neutral third-party's objective position of the strengths and weaknesses of both party's positions before committing the resources to place the decision into the hands of a third party. There is, however, some level of distrust that can develop if one of the parties believes that the other party does not have a good-faith intention of resolving the dispute and is interested in the ADR only to receive information that could help in the binding resolution process. The most often used ADR processes are discussed in the sections that follow.

Mediation.   The most widely used ADR process in today's construction industry is *mediation,* a negotiation process guided by a neutral facilitator. The success of a mediation depends largely on the skill and conduct of the mediator. Therefore, careful consideration must be given to selecting a mediator, and the mediation agreement should require approval of any mediator by both parties. Additionally, the parties should establish mediation procedures and guidelines that cover such matters as premediation submissions and how presentations will be made by the parties. Normally, at the initial mediation session, each party makes a short presentation followed by a private meeting with the mediator. The private meetings are confidential and intended to enable all the parties to candidly discuss their positions to facilitate settlement.

Minitrials.   Minitrials are similar to a mediation; the primary difference is that the parties are expected to put forth key components of their claims and defenses to a panel of judges. The judges typically include a principal decisionmaker for each party and a neutral third party who functions in a manner similar to that of a mediator. As with a mediation, minitrial presentations are expected to be succinct; the key feature is the negotiations taking place between the judges after the presentations have been made.

Dispute review boards.   *Dispute review boards* (DRBs) are used primarily in connection with large construction projects such as power-generation facilities, dams, and large tunneling projects. Developed at the beginning of a project before disputes arise, DRBs usually consist of a three-member panel, where each party appoints a panel member and the chairman is selected by the two party-appointed representatives. The board meets at regular intervals throughout the duration of the project so as to keep informed of the job progress and any disputes that may arise.

Individual disputes are submitted to the board for review, and a hearing is conducted, often without legal representation. The hearings tend to focus more on technical issues rather than legal arguments. After hearing all the evidence, the board issues a nonbinding recommendation. The parties may choose to accept the recommendation or challenge it in some other contractually established proceeding such as arbitration or litigation. History has shown, however, that because of the fairness of the proceeding and qualifications of the experts, few DRB recommendations are challenged.

16.14.2   Binding dispute processes

Although ADR processes could be structured to be binding, this would drastically change the character and effectiveness of the proceedings. As a result, the most common binding dispute resolution processes remain arbitration and litigation. Because of its inclusion in most standard form contracts, arbitration remains the most commonly used binding process for resolving construction disputes.

Arbitration versus litigation.   Arbitration has historically been used in construction disputes over litigation for the following reasons:

- Arbitration is perceived to be a more expedient and cost-effective method to resolve disputes.
- Arbitrators are more understanding of the technical and legal issues associated with construction than are most judges and juries.
- Arbitration avoids the expense and frustration of motions practices and extensive discovery.
- Arbitrations are less formal than litigation, making it easier to present one's case.

Since the late 1990s, however, some of these perceived benefits have been challenged, since experience is showing that (1) arbitrations have become almost as costly and time-intensive as litigation; (2) the quality of arbitration panels may not be as strong as originally perceived, particularly with respect to administering the case and moving the parties along; (3) some discovery is needed to prevent "trial by ambush"; (4) parties are becoming more formal in how they present their cases—there is now little difference between a litigated and an arbitrated case.

Parties can eliminate some of these problems by agreeing in the contract as to how the arbitration process will be governed. For example, if discovery is perceived as being an important issue, the agreement could specifically address the number of depositions which will be allowed and other affirmative obligations to produce discovery. If the parties are interested in understanding the reasoning of the arbitration panel's decision, they can require a written decision based on and supported by findings of fact and conclusions of law. Moreover, if there is a concern over the quality of the panel, the parties can engage in a selection process that will resemble that used in a DRB, where each party selects one and the third is selected by the two party-appointed arbitrators.

Overturning an arbitration award.   One of the most important reasons why parties reject or choose arbitration is that the scope of judicial review of an arbitrator's decision is extremely limited. In normal litigation, appellate review of errors of law, as well as the sufficiency of evidence, is almost always available. In arbitration, however, most jurisdictions provide that arbitration is subject to review and reversal only for activities such as blatant fraud or misconduct on the part of the arbitrator—a very difficult standard to prove.

Over the years, several cases have addressed a party's attempt to overturn an arbitration award in a design-build case. Consider, for example, the case of *Clairol, Inc. v. Enertrac Corporation,*[136] where Enertrac contracted to design-build Clairol's cogeneration plant for its manufacturing facility in Stamford, Connecticut. The court considered a motion by Clairol to vacate an arbitration award rendered in favor of Enertrac. Clairol initiated the arbitration, and Enertrac filed a counterclaim. After more than 20 days of hearings over a 2-year period, the arbitration panel generally found in favor of Enertrac. Clairol moved to vacate the award on the grounds that it was deprived of a full and fair hearing.

Clairol's argument centered on the fact that it was unable to cross-examine one of Enertrac's witnesses, Thompson, who completed his direct examination and then, after the hearings resumed some time later, became unavailable. Although Enertrac offered to have Thompson's direct testimony stricken from the record, Clairol believed that the panel should have dismissed Enertrac's counterclaim in its entirety. The panel neither struck Thompson's direct testimony nor granted Clairol's request for dismissal, stating that "we are the ultimate final decision makers with respect to relevance and materiality…and we will assign [the testimony] our own weight or relevancy or materiality in our ultimate deliberations on a decision."[137]

The trial court ruled, as a matter of law, that Clairol was deprived of its absolute right of cross-examination and that the award should be vacated. The Connecticut appellate court disagreed and reinstated the award, on the position that Clairol failed to prove that the lack of cross-examination resulted in substantial prejudice to Clairol. Its decision noted that Clairol did not provide the trial court with transcripts to demonstrate where the prejudice occurred:

> Absent [Thompson's] transcript, it was impossible for the trial court to determine whether any of his testimony was detrimental to Clairol. Moreover, the transcript of all other witnesses would have been necessary to determine whether Thompson's testimony was merely duplicative, or whether the other testimony impeached or explained his testimony, rendering it harmless to Clairol.[138]

The court also observed that Clairol did produce an additional witness to refute the testimony of Thompson and that the panel itself stated that

---

[136]690 A.2d 418 (Conn. 1997).

[137]Idem, at 420.

[138]Idem, at 422.

Thompson's testimony would have either no effect or a negative effect on Enertrac's position. In light of this, and the policies favoring arbitration, vacating the award to Enertrac was inappropriate.

16.15 Conclusion

Many other areas of liability arise in the context of design-build projects, including

- Licensing issues
- Trade subcontractor liability, particularly for those trades who also provide design services
- Disputes over extra work between the owner and the design-builder, as well as between the design-builder and its subcontractors

It is incumbent on the parties to the process to think carefully about the fact that design-build is different than other delivery systems and that liability will be judged largely on what is contained in the contract, as well as the assumptions by the trier of fact as to what the parties intended by entering into a design-build relationship.

The Use of Construction Management on Design-Build Projects

17.1 Introduction

There has been much debate within the construction industry as to whether the use of a construction manager (CM) adds value to the design-build delivery process. The answers to this question vary with the respondent. Examples are

- *Design-builders.* Design-Builders, particularly when they are directly selected by their clients and help in the programming of the project, often contend that a CM is an unnecessary project cost. They submit that the CM, as it seeks to justify its fees and preserve its professional standing, may engage in "turf protection" that will interfere with (1) the design-builder's single point of responsibility for performance of the work and (2) the close and direct partnering relationship between the owner and design-builder.

- *A/Es.* Design professionals who assist in developing the owner's design-build program and the preliminary design often feel that an independent CM will intrude into their professional relationship with the owner, marginalize their role on the project, and reduce their potential for fees. These design professionals contend that there should only be one prime consultant to the owner on a design-build project and that they should be the one to perform any needed CM services.

- *Construction managers.* Those CMs who are not design professionals believe that the owner still needs an independent professional to help manage certain areas of the project, regardless of whether the owner's design professional or design-builder will be involved. They claim that their unique management skills can improve the project's success and ensure that the owner is getting that for which it bargained.

In the authors' opinion, each of these positions has validity. CMs have skill sets that can greatly assist owners with limited in-house resources and complex design-build projects. However, just as design-build is not the best delivery choice for every owner and every project, construction management is not an appropriate management technique for many design-build projects because of the unique relationship between the owner and the design-builder. Thus, the question must be answered based on specific client and project needs.

The purpose of this chapter is to review construction management principles and the types of service that a CM can provide on design-build projects. This chapter also addresses some of the unique issues associated with the integration of a CM into the design-build relationship.[1]

17.2   Construction Management—Generally

Construction management was introduced to the private sector in the late 1960s as a tool to overcome some of the problems associated with the design-bid-build process, including project delays, claims, and litigation. Since that time, it has played a major role in the domestic and international construction industry. As observed in a 1974 *Harvard Business Review* article entitled "How to Avoid Construction Headaches," the construction management concept

> has been applied and has worked in many cases and under a wide variety of circumstances. Some of the results have been spectacular; others have been mediocre; and some have undoubtedly been failures. Obviously the use of a construction manager does not guarantee successful project management, and in some cases the costs of this approach may exceed the benefits. But the construction management approach is one of the more exciting and promising developments in the field of facilities construction, and it represents an alternative to traditional procedures of which all potential construction owners should have the right to be aware.[2]

This article, written in the mid-1970s, accurately reflects today's construction management experiences. While results have been mixed, most large and complex public construction projects delivered through design-bid-build or multiple prime contract systems use some form of construction management.[3] The widespread use of construction management has also been created because of the major downsizing of corporate facilities departments that took place in the late 1980s and early 1990s—which caused owners to look to consultants for project management services.

---

[1]M. C. Loulakis, "An Introduction to Construction Management," in *Construction Management: Law and Practice,* Wiley Law Publications, New York, 1995, Chap. 1, pp. 3–32.

[2]E. W. Davis and L. White, "How To Avoid Construction Headaches," *Harvard Business Review,* p. 93 (March/April 1973).

[3]The numbers bear this point out. *Engineering News Record*'s Top 100 CM-for-fee firms billed clients $5.12 billion in 1997 [*Engineering New-Record* p. 53 (June 15, 1998)].

17.2.1 Role of the construction manager

Experience has shown that construction management means different things to different people. Milt Lunch, the former general counsel of the National Society of Professional Engineers (NSPE) and widely recognized authority on construction industry trends, captured the essence of this problem by stating that

> The one, and probably only, area of agreement regarding construction management is that there is no consensus as to what it is, what it ought to be, and how it should be applied.[4]

Mr. Lunch's observations, which are shared by others in the industry,[5] demonstrate the importance of defining in the contract precisely what services the owner expects the CM to perform and the relationship of the CM to others on the project.

Several trade and professional associations have taken positions on the definition of construction management.[6] Although these definitions differ somewhat, they are consistent in portraying the relationship as one where the CM is to assist the owner's team on issues affecting project cost, time, and quality. These definitions also establish some overriding principles that may be helpful to an owner in deciding whether it needs a CM on its design-build project.

CMAA. The Construction Management Association of America's (CMAA) *Standard CM Services and Practice*[7] notes that construction management is the "process of professional management applied to a construction program from concept to completion for the purpose of controlling time, cost, and quality."[8] It goes on to say

> [C]onstruction management refers to the application of integrated systems and procedures by a team of professionals to achieve the owner's project goals. These systems and procedures are intended to bring each team member's expertise to the project in an effective and meaningful manner. The desired result is to achieve a greater benefit from the Team's combined expertise than that which could be realized from each individual's separate input.[9]

CMAA, organized in the early 1980s to promote the use of construction management, believes that the CM should be the leader of the owner's project consulting team[10] and that construction management is appropriate for a wide range of projects, including those using design-build.

---

[4]M. Lunch, "New Construction Methods and New Roles for Engineers," *Law Contemp. Probs.* **46,** 83 (1983), quoted by M. C. Loulakis, *Construction Management* (see fn. 1, above), p. 17.

[5]Loulakis (idem), p. 19.

[6]Idem, Chapters 1 and 2.

[7]Published by the Construction Management Association of America, McLean, VA, 2d ed., 1993.

[8]Idem, at 3.

[9]Idem.

[10]Loulakis, *Construction Management* (fn. 1, above), p. 27.

AGC.   The Associated General Contractors of America (AGC) has also been actively involved in defining construction management and instructing its members on its use. In its 1979 publication entitled *Construction Management Guidelines,* it defines construction management as

> [O]ne effective method of satisfying an owner's building needs. It treats the project planning, design and construction phases as integrated tasks within a construction system. These tasks are assigned to a construction team consisting of the owner, the Construction Manager and the Architect-Engineer. Members of the Construction Team ideally work together from project inception to project completion, with the common objective of best serving the owner's interests. Interactions between construction cost, quality and completion schedule are carefully examined by the team so that a project of maximum value to the owner is realized in the most economic time frame.[11]

Although many AGC members practice at-risk construction management, AGC does not endorse the use of construction management over any other delivery system.[12] The authors are unaware of any positions taken by AGC on the use of a CM on design-build projects.

AIA.   The *Architect's Handbook of Professional Practice* published by the American Institute of Architects (AIA) defines construction management as

> [M]anagement services provided to an Owner of a Project during the Design Phase, Construction Phase or both by a person or entity possessing requisite training and experience. Such management services may include advice on the time and cost consequences of design and construction decisions, scheduling, cost control, coordination of contract negotiation and awards, timely purchasing of critical material and long-lead items, and coordination of construction activities.[13]

Unlike CMAA, AIA views the architect as being the leader of the owner's consulting team on the project and that the CM will assist the owner and architect in certain specific project areas.

17.2.2   Agency versus at-risk construction management

Construction management is practiced through two very different contract formats: agency CM and at-risk CM. Each of these formats is discussed more fully below (Fig. 17.1).

---

[11]*Construction Management Guidelines,* The Associated General Contractors of America, AGC Publication 540, Alexandria, VA, March 1979, p. 2.

[12]Loulakis, *Construction Management,* p. 18.

[13]*Architect's Handbook of Professional Practice, Glossary of Construction Industry Terms* 8, American Institute of Architects, Washington, DC, 1991, quoted by Michael C. Loulakis, pp. 18, 19.

COMPARISON OF AGENCY AND AT-RISK CONSTRUCTION MANAGEMENT	
Agency CM	**At-Risk CM**
Acts solely as owner's agent or advisor and generally assumes no financial risk of performance of trade contractors	Assists owner and A/E in specific preconstruction activities, and then becomes "at-risk" for price and schedule performance
Paid on a fee basis or cost-plus arrangement	Fee is included in a fixed contract price or GMP
Owner contracts directly with the trade contractors on the project and bears responsibility for their performance	At-risk CM contracts directly with the trade contractors and is responsible for cost overruns, quality problems, and schedule deficiencies
Liability standard is similar to that of a design professional (standard of care and usually negligence)	Liability standard is similar to that of a general contractor (obligation to meet contract documents for a specific price and within contract time)

Figure 17.1   Agency CM versus at-risk CM.

Agency CM.   Construction managers who act solely as the owner's agent or advisor and take no direct risk for cost overruns, timeliness of performance, quality of construction, or design deficiencies are typically referred to as "agency CMs." The agency-CM approach can be used on a variety of project delivery systems, including design-bid-build, multiple prime, and design-build. It is not a delivery system in and of itself, but rather is intended to make the underlying delivery system work better.

The agency-CM approach contemplates that the owner will contract directly with those design professionals and contractors who will be performing the actual work. The services of the agency CM will be dependent on the owner's needs. They can range from providing discrete advice on conceptual scheduling or budgets to standing in the shoes of the owner in supervising and coordinating the design and construction of the entire project. However, most agency CMs assume no financial risk of construction (unless through contract incentives) and are generally paid on the basis of a fee or cost-plus arrangement.

At-risk CM.   Some owners prefer to have conceptual budgets and schedules drafted and constructability reviews conducted by the parties who will actually perform the work. This preference resulted in the development of a concept typically known as "at-risk construction management." Under this concept, the at-risk CM is placed in virtually the same legal and contractual position as a general contractor, except that it will become involved earlier in the project than a traditional general contractor and perform specific preconstruction activities.

As is the case with an agency CM, the at-risk CM can assist the owner and A/E in breaking down the project into discrete bid packages to allow the project to be fast-tracked. Unlike the agency CM, however, after the design is sufficiently complete, the at-risk CM can provide a price, in the form of either a fixed price or a guaranteed maximum price (GMP), and contract directly with the trade contractors. Cost overruns, quality problems, and schedule deficiencies will typically become the responsibility of the at-risk CM.

Liability differences.   Aside from the major contractual and contract administration differences between the agency CM and at-risk CM approaches, there is also a major difference in liability potential. The agency CM is typically held to a professional standard of care, similar to that to which an architect or engineer would be held. As a result, claims against an agency CM will often be based on negligence, with the agency CM given leeway to make errors within the industry standard. On the other hand, the at-risk CM is treated like a general contractor and must meet the terms of its contract with the owner.

17.3   Program Management—Generally

Most agency CMs are brought in on a project to perform discrete tasks related to the design and construction of the project—usually interfacing with one designer, one contractor, or one design-builder. On complex projects using multiple designers and contractors, a management concept commonly known as *program management* has become popular. Program management is a variation of agency CM that has been primarily used on large public sector infrastructure projects, although it has found some application on complex private-sector projects.

As with an agency CM, a program manager (PM) performs services which are totally within the control of the owner and based on its unique project needs. In general, however, the PM has broader responsibilities than the typical agency CM, including assisting the owner with the development of the overall project concept; assisting in the commissioning, start-up, and operation of the project; and evaluating exit strategies for the project. Additionally, many PMs operate as actual agents for the owner, retaining the designers, contractors, and others and standing in the shoes of the owner relative to the administration of the contracts. On some projects, the PM may even be in a position to determine whether to use design-build as a delivery process for some or all of the project and will administer the design-build contract accordingly.[14]

17.4   Determining Whether a Construction Manager (CM) Should Be Used

As noted at the outset of this chapter, the decision to use a CM on a design-build project should be based on the unique characteristics of the project and

---

[14]Because a PM is essentially an agency CM with greater scope and responsibility, the remaining sections of this chapter typically refer to the entity performing such services as an "agency CM."

the specific needs of the owner. To perform an objective evaluation of this issue, the owner must first determine what skills the CM should have to meet the specific project needs. With those in mind, the owner can then develop a matrix that identifies the pros and cons of using a CM. See Fig. 17.2.

Although numerous factors can be considered, some of the most common and critical issues that will likely affect this decision are described below. Note that the work associated with some of these issues—such as the skill sets needed to handle a complex project—can be performed by either the design-build team or a CM. The owner is the one who must determine whether its long-term interests are better served by having a professional manager assist in the process.

QUESTIONS TO CONSIDER IN DETERMINING WHETHER TO USE A CONSTRUCTION MANAGER ON A DESIGN-BUILD PROJECT	
1.	Does owner need assistance in developing its program?
2.	Does the owner's overall program consist of more than one project?
3.	Does the project consist of more than design and construction?
4.	Is the project highly complex?
5.	Does the owner have in-house expertise that is familiar with the type of project being designed and constructed and is available for the project?
6.	Does the owner have familiarity with the design-build process?
7.	Was the owner successful in its previous projects with its existing staff?
8.	Can the owner delegate authority and trust those with responsibilities?
9.	Will the design-builder be selected through competition?
10.	Does the owner need outside assistance in evaluating proposals and contracting?
11.	Does the owner want a direct relationship with the design-builder?
12.	Does the owner intend to use its lead design professional for oversight duties?
13.	Do lenders or other third parties require, or feel more comfortable with, CM oversight?
14.	Will the owner need third-party assistance to monitor the design-builder's performance?
15.	Is the project of such a significant size that the owner needs personnel to monitor the paperwork flow on the project?

Figure 17.2 Considerations in using a CM on a design-build project.

17.4.1 Unique project characteristics

Each design-build project has unique characteristics that may impact an owner's decision to use a CM. These characteristics may include

- The project's overall technical complexity
- Whether the project is advancing the state of the art
- The extent of predesign work, such as site investigation or environmental remediation, required to develop the project
- The owner's intent to use other contractors to execute some or all of the project
- The postconstruction work to be outsourced by the owner, including personnel move-in and relocation, leasing and exit strategies, and operation and maintenance

Generally speaking, the more complex the project, and the more work the owner requires in addition to design and construction, the greater is the likelihood that the project could benefit from a CM.

17.4.2 Owner's internal capabilities

One of the most critical factors that an owner should evaluate in determining whether to use a CM is its internal capabilities and experience in developing projects of the type being considered. Although many owners believe that they have adequate internal resources, this perception may not match reality. For example, if the owner has several projects under development, its internal resources may, as a practical matter, be unavailable for the project being considered. The owner should also objectively determine whether the type of project stretches the technical, financial, and management capabilities of its staff. This might be the case if the project is advancing the owner's state of the art or is more complicated than other projects that the owner may have developed.

There are several areas in which an owner may not have requisite expertise or sufficient staff to manage the design-build project. These include (1) establishing the program for the project, (2) understanding how design-build is different from other delivery systems, (3) determining whether and how to deviate from its existing design and construction standards, (4) monitoring the development of the design, and (5) inspecting and monitoring the construction process. The latter area, in particular, can be quite manpower intensive and is a primary reason why many owners elect to hire a CM.

Finally, even though an owner may have substantial in-house resources, it should nevertheless objectively examine its success on previous projects and determine whether it is time for a change in approach. If a previous project was viewed as unsuccessful (however defined), was this in any respect attributable to a lack of experience or capabilities of the internal staff? Some owners will periodically hire a CM or PM simply to keep their in-house personnel up-to-speed on newer and more efficient ways of managing projects.

17.4.3 Owner's corporate culture

In addition to a realistic comprehension of its internal capabilities, an owner must have an understanding of its corporate culture and management style before determining whether a CM or PM will be useful on a design-build project. A corporate culture that has a "not invented here" mentality or is overly bureaucratic may have difficulty in deriving the full benefits of a management professional. Questions that should be asked by the owner include

- Does the owner need hands-on involvement to be comfortable with the design-construction process?
- Can the owner readily delegate authority?
- Does the owner have a partnering and team-oriented philosophy?
- Can the owner easily learn to trust those to whom it has outsourced responsibility?

Depending on the answers to these "personality" questions, the owner may be wasting its money by having an independent outside professional provide it with advice.

17.4.4 Selection process for the design-builder

As noted in Chap. 8, there are several ways to select design-builders. In general, if the process is a competitive one, it is likely that the owner will need a consultant to help develop its program and define the basis on which selection will be made. Few owners have the internal resources necessary to develop criteria packages, preliminary design, or requests for proposals (RFPs). They may also need the talents of an outside professional to assist in the evaluation of proposals and the actual contracting of the design-builder. CMs and PMs are among those who can provide these services.

17.4.5 Owner's relationship with design-builder

Some owners want to have a direct relationship with the design-builder from the start through completion of the project. They find that the introduction of a third-party consultant, regardless of its skills, will jeopardize the partnering and trust relationship they are seeking to develop. In these cases, as well as on projects where the owner has a preexisting relationship with the design-builder, it is unlikely that a CM or PM will provide added value.

On the other hand, some owners want to distance themselves from the design-build process and create a reporting structure in which the design-builder will operate. This may be done for (1) internal political reasons; (2) the benefit of third parties, such as stockholders; or (3) to ensure that the design-builder will keep on its toes and not take advantage of a cordial relationship with the owner. In these cases, the use of an independent CM or PM may be justified.

17.4.6 Owner's other consultants

A CM is only one of several owner's consultants that could be used on the design-build project. Depending on the nature of the project, the owner may have (1) a design professional to develop the owner's program and preliminary design; (2) specialty professionals to deal with site acquisition, permits, leasing, operations, information technology, and other technical areas; and (3) contract specialists to help in the design-build procurement process.

Many CMs can provide these services. However, if the owner believes that other consultants are better suited to perform these services, the CM may be able to offer little aside from coordinating the overall work of these consultants and monitoring the design-builder's construction process. This may lead some owners to conclude that they will perform the management services in house and that the construction will be monitored by one or several of the other consultants, thus foregoing the need for a CM.

17.4.7 Third-party requirements

CMs are often introduced to projects because of the requirements of lenders, equity participants, or others who have a financial interest in the project. In fact, many design-build projects involving new technologies and unproven developers can move forward with financing only if a CM or PM is brought on to help manage the process, regardless of the developer's expertise or capabilities.

17.4.8 Cost/benefit ratio in using the CM

The cost of the CM is a factor the owner should consider in evaluating the need for a CM. The direct cost of employing the CM is relatively easy to determine, as are the savings that would arise if the owner did not have to use an in-house person to manage the project. Other financial benefits of using a CM may be more difficult to assess, but could include the following:

- More favorable pricing from design-builders because of the involvement of the CM and its special expertise
- A greater likelihood that the project will be administered successfully and the owner will achieve its financial goals
- The ability to use contractual incentives based on the CM's expertise and involvement

An owner who wants to perform an objective cost/benefit analysis of the use of CM should work with its financial advisor and the CM to develop a matrix of possible benefits and a range of the savings that are achievable.

17.5 Typical CM Services

As noted in the preceding sections, an owner can use a CM for a broad range of services. On a traditional project, these are generally categorized as preconstruction services and construction services. Preconstruction services typically include (1) value engineering, (2) constructability and design reviews, (3) conceptual estimating, (4) conceptual scheduling, and (5) trade package development and procurement.[15] Construction services may include[16]

- Providing on-site staff to monitor construction
- Developing and monitoring schedules
- Preparing cost controls
- Conducting periodic progress meetings
- Administering change orders and requests for information
- Establishing and monitoring a safety program
- Making preliminary inspections of the work
- Administering payment processes
- Assisting in dispute resolution

Additionally, if the CM will be performing program management duties, a variety of predesign and postconstruction activities may be required in the contract.

The reasonable range of services to be performed by the CM on a design-build project will be different from those of a design-bid-build or multiple prime contract delivery system. Generally, these services will be categorized into (1) project development, (2) procurement of the design-builder, and (3) monitoring the design-builder's performance. Note that in the discussion that follows, there is a presumption that the owner has already decided to retain a CM on the basis of factors discussed in the preceding section.

17.5.1 Project development

There is substantial front-end work associated with the conception and development of the project that must be performed by the owner prior to starting the design-build process. This includes not only the owner determining its site and the purposes underlying the project but also the development of a program that will enable the design-builder to understand its responsibilities.

---

[15]Loulakis, *Construction Management* (fn. 1, above), p. 19 and Chapters 8 and 9.

[16]Idem, Chapter 17.

The owner's program is particularly significant when the design-builder will be selected on the basis of a competitive selection.

As noted in the *Design-Build RFQ/RFP Guide for Public Sector Projects* published by the DBIA,[17] the owner's program should

> [S]pecify the owner's intentions for the character or image of the facility; the activities to occur in the facility; the functions of the individual units housed in the facility; and their interactions with each other and with visitors to the facility. Additionally, the program should specify the exact or minimum amount of usable floor areas required, and the environmental conditions (power, light, heating, cooling, ventilation, etc.) required in each programmed space. Depending on the specific needs of the project, facility programs can include design directives, and design configuration criteria developed from functional and programmatic needs.[18]

If the owner determines that a CM will develop its program, the CM should have the appropriate design professionals in house to perform such work. Even if the CM is not performing the work directly, it may be called on to assist the owner's independent design professional in the development of the overall package.

Other tasks related to the development of the design-build project that can be performed by the CM include

- Preparing a procurement and legal review to determine any constraints on the contracting process
- Interfacing with interested governmental agencies
- Obtaining permits or easements
- Determining how to manage interfaces with third parties who will be affected by the project—such as adjacent landowners and utilities

Finally, the CM can provide specific guidance to the owner about how the design-build process will be conducted and the differences between design-build and other methods of project delivery. This may include in-house educational programs as well as partnering sessions with agencies or others who are interested in the project, but who are unfamiliar with how their roles will be performed in a design-build setting.

17.5.2 Procurement of the design-builder

The CM usually plays a major role in assisting the owner with the procurement and selection of the design-builder. The CM may draft the owner's program and assist the owner in developing the overall procurement scheme for the project.

---

[17]*DBIA Manual of Practice,* Document 212 (1995).

[18]Idem, at pp. 3–4.

If the design-build selection process contemplates the owner's issuance of a request for qualifications (RFQ) and the design-builder's preparation of statements of qualifications (SOQs), the CM will generally prepare the RFQ and advise the owner on how the shortlist will be developed. It will also play a role in the evaluation of the SOQs—typically providing advice to the owner on specific areas of the submissions, such as technical competence and management plans.

If an RFP is to be issued, the CM will often be the primary party responsible for its development. Among the most important aspects of the RFP on which the CM will provide guidance are

- Establishing the weighting criteria for design-builder selection
- Providing a draft of the design-build contract, including any incentives, penalties, and other commercial terms
- Determining whether to use an honorarium for the unsuccessful proposers
- Determining the proper mix of performance and restrictive specifications to use

Many CMs use workshops to develop this information, working in conjunction with legal counsel and the owner's in-house staff.

Finally, the CM will assist in the review of responses to the RFP and answer any questions that the selection committee may have. This includes providing quick and practical advice to respond to unforeseen problems in the procurement process, such as (1) shortlisted proposers electing not to submit proposals, (2) proposed costs that exceed budgets, and (3) last-minute changes from third parties that require changes to the owner's program or the RFP.

17.5.3 Monitoring the design-builder's performance

The CM services described in Secs. 17.5.1 and 17.5.2 are most frequently performed when the CM is assisting the owner in a competitive procurement and is the lead professional on the project. These services are to be contrasted with those performed by CMs in the monitoring of the design-builder's design and construction work, services which are typically performed by a CM regardless of the selection method or role of other professionals working for the owner. In fact, the monitoring services of the CM on a design-build delivered project are quite similar to those traditionally performed by a CM on projects delivered through design-bid-build or multiple prime contracting systems.

In monitoring the design work of the design-builder, the CM is often the owner's primary quality control check point in determining whether the design complies with the contract documents and meets the owner's needs. The CM will likely receive interim design submissions, meet with the design-builder to review the submissions, and then provide guidance to the owner in approving the design.

If the contract price will be developed after execution of the design-build contract, the CM will also play a major role in determining the cost-effectiveness of the design and the proposed prices submitted by the design-builder. The CM should carefully review the design-builder's periodic estimates and the bases for any proposed GMP. It may also suggest value engineering changes, although this will be far different from what it might do in a design-bid-build system, since the design-builder does have the right to control the design and can reject value engineering suggestions.

During construction, the CM will provide the same types of services that it performs under other delivery systems, including the following:

- Review the design-builder's quality assurance/quality control (QA/QC) program

- Analyze and negotiate change orders

- Participate in progress meetings

- Monitor the design-builder's schedule

- Review and certify payment applications

- Assist the owner in obtaining owner-furnished deliverables, such as permits, utilities, and site information

Depending on the owner's internal resources and the extent to which the CM will be involved on behalf of the owner, these services may be conducted in conjunction with others (where the CM is basically providing a "body shop") or solely by the CM (with the CM reporting directly to the owner's senior project management).

17.6 Selecting the Right CM

Due to the various types of services that a CM can provide, an owner needs to select the CM who is right for the specific needs of its design-build project. This means looking not only at the skill sets of the individuals the CM proposes to do the work but also at the proposed CM's organizational structure.

It is well recognized that the types of organizations selling construction management services span the entire spectrum of the construction industry. They include

- Engineer-constructor (E/C) firms with integrated in-house design and construction capabilities

- General contractors

- Design professionals, including both architectural and engineering firms

- Pure construction management firms[19]

Although many firms claim to have construction management capabilities,

---

[19]Loulakis, *Construction Management,* pp. 22–23.

the reality is that many so-called CM or PM firms are simply practicing their traditional line of business—whether as designers or contractors—and have not focused on how the advisory role of a CM differs from this work. This lack of true CM capabilities can be particularly problematic on a design-build project, where the so-called CM may feel some professional jealousy as a result of the unique position the design-builder enjoys in having single-point responsibility for design and construction. This jealousy—which can prompt the CM to second-guess the design-builder's design, motives, and competency—is a leading reason for many owners' skepticism about the use of CM in design-build.

Given the above, the owner who concludes that it needs a CM on its design-build project should approach the CM selection process thoughtfully and objectively. Some of the factors that should be considered are discussed below.

17.6.1 General expertise of the CM

The CM should be able to demonstrate its specific expertise and qualifications in the areas where the owner will be using its services. This process by which an owner evaluates the CM's qualifications will be similar to those processes used by any owner retaining any other consultant who is selected on the basis of qualifications. However, some issues that are unique to the design-build system that should be explored, include:

- *Prior design-build experience.* The CM should have a proven track record in working on design-build projects or, at the very least, in understanding how the design-build process differs from other delivery systems.

- *Design expertise.* If the CM will be providing design advice, either in the development of the RFP or in reviewing the submissions of the design-builder, it should have in-house design expertise.

- *Familiarity with performance specifications.* Because of the importance of performance specifications in the design-build process, the CM must demonstrate its familiarity with the use of these specifications and the process of monitoring the design-builder's compliance with these specifications.

The first issue may be the most important. Many CMs and other owner consultants tend to forget that they are not controlling the design or construction and that the design-builder should be given the discretion to perform the work as it sees fit within the constraints of the contract documents. These parties find it particularly difficult to avoid being prescriptive, which can be a recipe for disaster if something goes wrong.

17.6.2 Organizational structures

As noted at the outset of this section, the owner will have its choice of a number of different organizations to serve as the CM. Depending on what services the owner is interested in having the CM perform, some organizations may be better suited for this role than others.

E/C firms and contractors.   E/C firms and general contractors are often strong choices to serve as an owner's CM because of their experience and expertise in overall trade management, coordination, and scheduling of projects—particularly on a multiple prime contracting basis. However, the strength of these organizations is often on the construction side of the process, where their experience in monitoring project resources and seeking cost-saving means and methods is important. They may not have as much experience with the design activities needed to be performed on the project—with the exception, of course, of some of the major E/C firms with large resident engineering staffs.

Design professionals.   A CM who has extensive experience in design can greatly assist the owner in the early analysis of the design. As a result, A/Es are often well situated to act as the CM on a design-build project—particularly when a competitive RFP will be assembled. The A/E can also be an excellent choice for CM when construction methods are driven by an evolving design based on field conditions—as might be the case with environmental remediation projects.

Notwithstanding the positives, many A/Es do not have the skills to effectively deal with value engineering concepts, conceptual budgeting and scheduling, and other construction-related areas of expertise. Moreover, many A/Es have diluted their in-house expertise in monitoring construction because of liability concerns. Thus, before an owner decides to use an A/E as the CM, it must look carefully at whether the services it needs are more heavily weighted to the design, procurement, or construction side of the process.

Pure construction management firms.   Many firms have developed in-house expertise to provide the types of design, construction, and management services that are expected of an agency CM. In fact, some of these firms are among the "best and brightest" in terms of meeting the needs of an owner engaged in the design-build process, since they have the requisite balance of design and construction expertise.

Owners should remember, however, that merely because the firm identifies itself as a construction management organization does not render the organization capable of performing the level of services that are expected. The firm providing the CM services is only as good as the experience of its principals and those who are responsible for the specific project. Care should be taken to ensure that the practicing CM has the requisite experience in the type of project being constructed—in the capacity of either a design professional or a construction contractor.

17.7   Special Considerations

In determining whether and how to select a CM for the design-build project, the owner should attempt to "think outside the box" and evaluate how the CM will integrate with the other players on the project. Consideration of the following points may be helpful in making this determination.

17.7.1 Using CM in negotiated design-build procurements

As noted throughout the preceding sections of this chapter, an owner that uses a competitive mode of design-build selection will often require the assistance of a consultant, who could be a CM, to develop and administer the RFQ-RFP process. However, the role of a consultant in a direct or negotiated design-build selection process is less clear. Many owners will rely on their relationship with and the integrity of the design-builder in formulating the contracting strategy and implementing the project—and will not have a CM or other third-party professional intervene in this process.

A CM can be helpful in a negotiated design-build environment, particularly if

- The owner has insufficient internal resources to monitor the execution of the design-builder's work.

- The owner has needs that are beyond the expertise of the design-builder, such as unique permitting capabilities, leasing strategies, or operation and maintenance.

- Third-party financiers require an independent professional as part of the owner's team.

Developing a comprehensive cost/benefit matrix will assist the owner in analyzing the relative value of using the CM in these noncompetitive procurements.

17.7.2 Leadership of the team

There has been a great deal of industry debate as to the proper role of the CM as contrasted with other consultants working on behalf of the owner. On design-bid-build projects, this controversy is manifested by the question as to whether the CM or the A/E should take the lead and control the selection and monitor the performance of the other.

CMAA, not surprisingly, takes the position that the CM will serve as the owner's principal agent in connection with the project.[20] AIA, on the other hand, has developed contract documents that reflect the CM's role as that of an "advisor" to the owner, with the CM's authority subordinate to the architect.[21] For example, although the AIA contracts call for the CM to review the design during the preconstruction stage, the CM is simply "assisting" the

---

[20]Article 2.3, CMAA Document A-4, *Standard Form of Agreement between Owner and Design Professional,* 1990 ed.

[21]Loulakis, *Construction Management,* Section 1.14.

architect and its role is primarily to make "suggestions" and "recommendations."[22] Ultimate responsibility for virtually every major issue rests with the architect under the AIA contract documents.

The merger of design and construction into a single entity helps alleviate some of the tension as to whether the owner's A/E or CM should be the leader. Although little has been written about this subject, experience shows that if several consultants are needed for a project, the specific project needs should generally define the role and relationship of the CM to others. Complex projects may demand that the CM be fully active from project inception to completion and that the CM control the selection of all parties. In fact, this is the theory behind program management. On the other hand, on projects where the CM will simply be providing construction-phase field monitoring, the CM should likely not have any meaningful leadership role.

17.7.3   Participation by the CM in the design-build team

Questions may arise as to whether a CM who is providing advice to the owner at an early stage of the project could ultimately become an at-risk member of the design-build team—either as the design-builder or as a subcontractor to the design-builder. This question is particularly important given the view that at-risk CM is often an attractive alternative to design-build on some projects.

One of the strong selling points behind agency CM is that the CM is acting in a fiduciary capacity with the owner, providing the owner with objective advice intended to benefit the project and not the CM. Placing the CM in a position where it could become at risk on the project may compromise the CM's objectivity to the detriment of the owner. This is particularly true if the CM's role is to assist in the development of contract language and pricing assumptions (i.e., what costs may be deemed reimbursable). If the CM is also involved as a PM early in the life of the project, the opportunity to go "at risk" may lead to inappropriate conclusions about the delivery systems that should be used for portions of the work.

If the owner and CM discuss in detail at the outset of their relationship the circumstances under which the CM could become a member of the at-risk design-build team, it should be possible for the CM to shift over to a nonfiduciary relationship at some point in time. This is similar to the approach taken by CMs that perform preconstruction services on a fee basis and then provide the owner with a GMP. However, both CMs and owners should recognize the conflict of interest that this situation creates and objectively determine what, if anything, can be done to minimize its impact.

---

[22]For example, Article 1.1.1 of AIA Document B801 requires the CM to prepare preliminary cost estimates but then assist the architect in achieving the owner's program. Similarly, the CM makes recommendations regarding constructability and value engineering issues, but does not make decisions, pursuant to Article 1.1.2 of AIA Document B801.

17.7.4 Accountability of the CM

As noted at the outset of this chapter, the agency CM is held to a standard of care consistent with the industry. Owners who use design-build are often interested in having more accountability from their entire project team, including their consultants. It may be appropriate for the parties to evaluate what incentives—including cost, time, and quality—can be used in the CM contract to achieve this accountability.

Institutional Challenges for Design-Build

Design-build and single-source responsibility contracting has been a small but enduring aspect of the American design and construction industry despite the ascendance of design-bid-build during the Industrial Revolution. Although the use of the design-build method ebbed further in the years following World War II, design-build was still the method of choice for some firms in the power, process, and industrial/distribution sectors. With design-build found only in isolated market niches, the focus of most laws and practices was the so-called "traditional" design-bid-build process. Laws, regulations, procurement procedures, customs, and educational programs all evolved around the linear, sequential process of design-bid-build.

The following 10 challenges are discussed in this chapter:

1. The cultural chasm
2. Legal and regulatory barriers
3. Complexity of the process
4. Changes in the marketplace; changes in roles
5. Availability of industry products and services for design-build
6. Identity of design-builders
7. Education and training
8. Exploding the persistent myths
9. Permitting the design-build project
10. Removing barriers to design-build

18.1   The Cultural Chasm

18.1.1   Public perception

Construction is arguably the most observed of all occupations. Everyone seems to love to watch the frenetic activity on a construction site, no matter what is being constructed: a road project, a deep foundation for a massive skyscraper, or the creation of a new reservoir and pumping station. Whatever the activity, construction stirs the imagination; as designers and construction personnel, piles of materiel and fascinating machines converge to deliver a new facility.

What is in the public's mind, though, when they think of how the products, personnel, and machines were assembled to create the final project? Do they think that the builder is responsible for the eventual design of the project? Most likely they do not. Does the public then think that the designer is building the project? Again, most likely not. Generally, the public believes that the designer creates the design and the contractor accomplishes the construction.

In viewing the project delivery process, the majority of the public believes that the design-bid-build process is how most projects are delivered. The public perception of how projects are procured, particularly by public entities, is not benign in viewing the barriers to design-build. The public generates views that affect how press coverage and how politicians determine the projects will be procured. If the public believes that it is not receiving the full benefit of its tax dollars, then both politicians and the press raise questions or clamor for investigations.

The traditional design-bid-build process offers little in the way of innovation or efficiency, but what it does offer is selection of a construction contractor based on the lowest initial cost. The public clearly draws an incorrect connection between the term *low bid* and the concept of competition. This erroneous parallel unfortunately leads to criticism of design-build in that the more appropriate way to procure design-build is through a best-value or qualifications-based process. The fact is that best-value purchasing does not generate commodities-oriented competition but rather, performance-oriented competition by attracting the most quality-driven firms to enter the competition for a project. No quality-driven company wants to enter into what amounts to an auction (i.e., a low-bid auction) for a project. How will the low bidder design and build the project if the low bid is below the cost of the project? Do taxpayers win by having that firm default on that project and having the surety complete the project?

Are these public perceptions a barrier to design-build? The notion that design-bid-build delivery and low-bid procurement are the normal way for public design and construction to be properly and legally delivered and procured serves to retard efforts by reform-minded public officials to adopt procedures to facilitate design-build. Enlightened owners understand that it is in their self-interest to implement project delivery and procurement processes that reward top performers rather than low first-cost bidders. Public perception in the postindustrial economy is also beginning to shift toward quality and value as predominate attributes over low initial cost.

18.1.2   Industry perception

The design-construction industry has historically been characterized by ease of entry into the marketplace by design and construction firms. These firms are often sole proprietorships or closely held businesses, and most commonly family-owned rather than corporately held. Typically, there are very low barriers to entering the business of construction or construction management. A specific business license is usually required, sometimes with a testing requirement, but capital requirements are very low and the skill trades can be subcontracted.

Traditionally, design has been procured separately by the owner via qualifications-based selection, with construction procured separately by the owner using the competitive low-bid method. These separate procurements have been especially prevalent for public-works projects since it (1) accommodated all those who wanted to compete for a public-works project, (2) made for fairly straightforward selection of designers through the use of qualifications, (3) selected contractors by use of only one evaluation factor—price, and (4) assured the public that the construction project went to the lowest initial bidder. Since this "traditional" system was focused on procuring services from the professions and trades that were in existence, no change from the status quo was championed by these professions and trades (which could alter the good livelihood that was being provided). The "traditional" system was institutionalized by codifying it in public procurement laws, from the federal government to the most local public jurisdiction.

With the exception of the rise in the late 1960s and early 1970s of the construction management form of project delivery, the "traditional" or design-bid-build method predominated. It was well understood. It was relatively easy to administer by public procurement officials, and as far as the public understood, it achieved the lowest price. The relatively recent advent of widespread use of design-build caught many in the design and construction industry without a full understanding of how integrated services could be applied. Such a drastic shift in market dynamics has not been seen since the open shop movement, which occurred in the late 1960s and throughout the 1970s. The open-shop movement altered business relationships, affected where skilled artisans were trained, and changed market opportunities depending on a company's union or nonunion orientation.

The growth of design-build has stimulated similar market adjustments. The reaction by those who have watched their traditional market segments erode was predictable. They opposed it through opposition to procurement reform initiatives that would enable design-build; or they have advocated revisions to the design-build process that shackle design-build's full advantages. A powerful advantage for design-build, however, is that this market phenomenon is within the design and construction industry. It's an industry fraught with entrepreneurs who recognize market opportunities and are risk takers. Many designers and constructors are weighing the risk, determining the risk to be reasonable, and are succeeding by assuming that risk. Risk/reward tradeoffs are being made daily as owners increasingly demand design-build solutions from the marketplace.[1]

18.2   Legal and Regulatory Barriers

18.2.1   Licensure

Licensing laws and regulations present a special challenge to design-build. Licensing laws for designers are to protect the public in the interest of health, safety, and welfare in the design of structures and improvements to property. No one wants or could possibly support the practice of design by those who have not been fully educated in technical design and the building arts. Licensing laws protect the public from the unlawful practice of design. However, some states have licensing laws going beyond the determination of the ability of who can practice (the area of individual professional licensing) and extend to how practice can be applied (the arena of business and employment). These expanded licensing laws have been put into place not only to protect the public but also to protect the profession from intrusions into the profession's markets.

In many states, the licensing laws provide that only licensed professional designers are in the position to provide design services and that design services cannot be practiced or even offered through another entity. This approach may prevent the practice of design-build in those instances where a construction-oriented firm (or any party other than a licensed designer) leads the design-build consortium. Virtually every state has case law that bars unlicensed persons from recovering monies otherwise due under a contract.

Individuals who are appointed to the licensing boards may have a bias in favor of "traditional" practice, where the designer contracts directly with the owner, and in opposition to anything that may move away from that practice. Since design-build does not have the constituency that the traditional practice has, the members of a licensing board may be more prone to stall meaningful reform than to approve an amendment in the licensing regulations to permit design-build. One common remedy is to allow the practice of architecture and engineering by licensed professionals within business entities that are not majority-owned design firms. Explicitly, where the responsible-in-charge designer is a registered professional (such as the lead architect within a design-build firm), the state licensing board would extend its gaze beyond the business entity and recognize the individual employee(s) as meeting the letter and spirit of the licensing statute.

---

[1]ASCE, *Design-Build in the Federal Sector,* Washington, DC, 1991.

18.2.2 Procurement

The intent of most public procurement laws have been to provide a framework for buying goods from the private sector while allowing all competitors an equal opportunity to provide offers or bids to the buyer. These laws were enacted during the rise of the design-bid-build method and focus largely on the procurement of professions and trades to facilitate design and construction, rather than to facilitate overall project delivery.

Public and private owners are demanding more from their goods and services providers. The realization that low first cost may be an incomplete indicator of project performance is becoming apparent to most facilities buyers. Since 1990, the federal government has made past performance a required measure during the procurement cycle. Use of past performance and other measures (such as the energy-savings potential, life-cycle costing and sustainable design) will slowly wean contracting officials away from low-bid purchasing. A combination of objective and subjective procurement metrics is gaining acceptance, and this new procurement philosophy will eventually lead to multiple-attribute decisions about what solution represents the best value for the owner. Until these systems are more fully implemented, however, the knee-jerk reaction in support of low bid will continue to cause problems for innovative project delivery and procurement systems.

18.2.3 Public-sector funding

The manner in which planning and design services and construction services are separately funded on federal projects (through the appropriations and administrative cycles), coupled with the way funds are earmarked for varying points in the project development process, results in less than optimum cost and time savings when design-build is employed. Appropriations for military construction projects are provided through various accounts, depending on the purpose of the expenditure. The principal accounts for real property improvements are the planning-and-design (P&D) account and the construction account. It is generally accepted that P&D funds must be used only for those purposes. A similar rule applies to construction. Agencies may not mix or transfer these funds among accounts without the approval of Congress (this process is known as *reprogramming*).

This strict constraint on funding creates problems when a military agency uses design-build. Design funds that should be attached to a design-build project cannot be transferred by an agency to the construction account because of legislative prohibitions. Currently, with the legislative prohibition on transfer of funds, agencies often proceed with design-build projects using only the funding from the construction account (the exception is use of P&D funds for design criteria or preliminary design for design-build projects). Obviously, without design funds (which may represent 6 to 9 percent of construction costs), a project that uses design-build may be reduced in scope by not including this funding in the project's development.

Federal procurement officials recognize that government agencies receive better prices for major acquisitions if all the funds for the project are fully appropriated at the beginning of project development. With fully integrated funding for design-build on federal projects, the delivery process would be facilitated by having federal agencies in control of the funding stream. Long-standing institutional biases may prove substantial in an attempt to alter the fragmented appropriations and agency funding processes. Nevertheless, greater use of best-value selection and fixed-budget, best-technical-response procurement will stimulate further interest and discussion leading to regulatory and administrative reform.

18.3   Complexity of the Process

The primary reason why owners choose design-build has varied over the years. In the early 1990s, owners began to utilize design-build because of the reported cost and time savings. The old adage of "time is money" was true in this case, and owners who had projects dependent on generating early cash flow moved to design-build. Some owners also began to realize the power of single-source responsibility with its attendant reduction of claims and ease of administration.

During the mid-1990s, owners gravitated to design-build in order to obtain savings, but they soon found that the cost expectations under traditional process were not directly translatable to design-build. A few owners did not understand that their requirements should be set in terms of performance as opposed to needs stated through prescriptive specifications. The need to make decisions quickly also confounded owners who were not used to the rapid pace of design-build.

Design-build requires a different outlook for owners and other members of the building team. There is a transition from a command-and-control mentality to significant up-front identification of facility requirements, the building in of contractual "hooks" for quality, and a lessening of the control and monitoring during design and construction. An owner must answer the question of "How active do I want to be in this process?" The owner may want to engage a design criteria professional or an independent PM/CM to draft the performance specifications and RFP and to monitor the work during its design-build phases. Each of these stages demands new and different approaches, tools, and techniques to be adopted by traditional practitioners.

A design-builder accepts the dual role of designer of record and constructor of record. The team leader role requires the design-builder to be a collaborative decisionmaker where design is a primary issue, but concern for design does not supersede the owner's goals for balancing quality, cost, and schedule. The greatest challenge for integrated services delivery remains the unwillingness on the part of staunchly traditional designers and constructors to subordinate a portion of their self-interest for the good of the project and the owner.

18.4   Changes in the Marketplace; Changes in Roles

A study of public and private owners' attitudes toward design-build, which was released in 1996 by two professors at the University of Colorado,[2] cited the decrease in overall design and construction time to be the primary reason why both public and private owners were gravitating to design-build. Other top seven issues for owners (in rank order) included the ability to establish firm costs early in the project, the potential for reducing overall project costs, potential for innovation in process or product, capability in establishing firm schedules early in the project, reduction in claims, and applicability to large or complex projects. As speed to market becomes increasingly important in the age of the Internet, techniques that squeeze inefficiency out of the project delivery process will enjoy greater use.

Still, there are those who oppose any deviation from the project delivery status quo. Design-build does require practicing professionals to comport themselves differently, to assume greater or different responsibilities, and to work at understanding more of the entire design and construction process. Understanding what motivates designers and constructors within the design-build process holds keys to overcoming barriers to design-build:

Designers (architects and engineers) within the design-build process are seeking

- Contact with and access to the owner

- An opportunity to provide the design solution

- Recognition for design contributions

- Ability to land the work and earn a fee

- Ways to achieve quality, safety, and reliability

- Limits on risks (most prefer the traditional design standard of care or negligence standard)

Designers in a design-build relationship may have to forgo their independent consulting practice (on the subject project) depending on contracts among the parties. Designers may also need to look beyond their business management techniques (based on professional advice for fee) to adopting a cost-of-final-product approach. These cultural changes present surmountable, but nonetheless challenging, obstacles for traditional design firms.

Constructors with the design-build process are seeking

- Control over the cost and schedule aspects of the process

- Increased volume of work and/or more satisfying working relationships

- Satisfaction of managing the entire design and construction process (the same satisfaction can accrue to the designers in designer-led design-build)

---

[2] Professors Anthony Songer and Keith Molenaar.

- Profit margins commensurate with the risks

Constructors in constructor-led design-build arrangements must adopt the mantle of collaboration and responsibility for design. The design-build entity is both the designer of record and the constructor of record, and the perpetuation of a brusque low-bid contractor mentality is not appropriate (and not welcomed or expected).

Industry complaints about the design-build process have included issues such as bundling of contracts and use of subjective evaluation requirements (which can be just as routinely applied to design-bid-build as they can to design-build). Among the other objections to design-build are that subjective evaluation criteria are often used to select the design-builder. Some have complained that developers are commingling public monies that may be given to advance an economic enterprise. The project is steered toward design-build, say the objectors, and then the public-private ventures use public funds to choose contractors using subjective evaluation factors. In reality, the Competition in Contracting Act of 1984 placed competitive negotiation on a par with sealed bidding, ushering in the beginning of widespread use of subjective evaluation criteria well before the large growth in design-build.

Experienced design-build practitioners recognize that both home office and job-site personnel must become in tune with design-build culture and procedures. Not every employee within the design-construction industry who has been schooled in the design-bid-build process can make the transition to design-build and its adherence to shared goals and performance attainment. In fact, management in most construction firms that collaborate on design-build projects quickly identify the personnel who can make the transition. Those who cannot think and act collaboratively, and with equal care and concern for design and construction issues, are relegated to staffing traditional design-bid-build projects.

18.5   Availability of Industry Services and Products for Design-Build

A barrier to the full and widespread application of design-build is the paucity of services and products to assist the design-builder. During the 1990s' surge of design-build growth, surety and insurance products targeted for design-build use were in very short supply. Professional liability insurance coverage for general contractors leading a design-build project was not generally available from the top providers until 1996, and one of the largest professional liability insurers persistently refuses to write errors and omissions insurance for any entity unless the firm is a professional design firm.

Surety availability for medium- and small-volume design-builders remains a problem in some markets. In 1998, a spokesperson for a major bonding company wrote about options for obtaining surety for design-build and included a recommendation for separating the design contract from the construction contract, thereby negating the very premise of design-build. Traditional performance bonds are third-party guarantees that the work will be completed according to the plans and specifications. Design-build contracts, on the other hand, include the total amount of the contract including design, which causes the surety to be concerned about additional obligations and exposures that extend over a longer period of time. Some of the more progressive bonding companies are providing additional training to their producers to assist them in identifying risk mitigation measures (such as adequate insurance for design errors and omissions) as a condition for underwriting.

Similarly, building product manufacturers have been slow to adopt a building systems approach to developing and marketing their products. Among the manufacturers that have made the transition are companies that produce heating and cooling equipment, curtainwall manufacturers, low-voltage control systems producers, and roofing manufacturers. Still, firms that produce single product lines such as lighting fixtures, gypsum wallboard, or brick have not necessarily discovered the demand for integrated solutions: that is, how their products fit within an assembly or element that contributes to the overall performance of the building or a functional part of a building. Information and technology is expected to drive demand for systems-based products that can be quickly analyzed, costed, selected, and implemented by the integrated design-build team.

18.6   Identity of Design-Builders

Other than the historic appellation of "master builder," there is no single contemporary term that adequately describes the role of "design-builder" (the Design-Build Institute of America uses the hyphenated version; the National Association of Professional Engineers uses the term "design/build," and the American Institute of Architects simply contracts the two words into one: "designbuild"). Before 1995, it was nearly impossible to find design-builders listed as service providers in any construction guide (such as the *Blue Book*) or listed in the Yellow Pages of the telephone book. How were/are owners able to locate design-builders? By going to an architect? An engineer? A general contractor? Were or are all of these entities capable of performing design-build services? Some, certainly yes; others, emphatically no. Owners are beginning to recognize that there is a difference between those that "do" design-build and those that have crossed the cultural divide and "are" design-builders.

Although more than $70 billion of design-build work is done annually in the United States, the U.S. Census Bureau does not yet have a category for design-builders. Beginning in 1994, the Design-Build Institute of America began working with the U.S. Department of Commerce to create a Standard Industrial Classification (SIC) code for design-build establishments.[3] In 1998,

---

[3]DBIA, *Overcoming Fragmentation*, report about the industry compiled by Chwat & Assoc., Washington, DC, 1994.

three countries—Canada, Mexico, and the United States—combined their statistical resource base to develop the North American Industrial Classification System (NAICS). It is likely that the effort to include design-build firms in the census of construction will be successful by the NAICS revisions of 2007. Once design-build receives a code from the U.S. Economic Classification and Policy Committee, establishments will be able to self-select (by coding on the census form) whether their business should be classified as architecture, engineering, general contracting, or design-build.

18.7 Education and Training

Fragmentation of the design-construction industry is replicated by college and university programs, where individual schools of architecture, engineering, or construction have usually not embraced cross-disciplinary study. At present there is increased interest, but thus far there are few undergraduate or graduate schools granting degrees in design-build. Architectural schools maintain the design studio as the central component of their programs; engineering schools tend to emphasize theory and applied technology; and schools of construction typically emphasize project planning, budgeting, scheduling, and control.

How fragmented are the disciplinary departments within the schools? An Associated Schools of Construction student design-build competition was challenged when an *architectural* student was part of a competitor school's team (all of the other teams were populated by construction management students). The ideal student team, of course, would reflect real-world team members; at minimum, an architectural student, an engineering student, and a construction sciences student would make up the multidisciplinary teams, and careful consideration could be given to engaging business or management students who may fulfill their career goals while working for a facility owner or practitioner.

A number of major universities offer design-build courses (see sample list of courses in Fig. 18.1) within their architecture, engineering, or construction management programs. As of spring 2000, Georgia Institute of Technology and Washington State University have completed planning for Masters Programs in Design-Build. These programs will draw from the pool of graduates who possess first professional degrees in design or construction, and then tailor the individual program to emphasize management of integrated design and construction services.

During the time of the Roman Empire, Vitruvius wrote:

> The master builder should be equipped with the knowledge of many branches of study and varied kinds of reasoning, for it is by his judgment that all work done by others is put to the test. This knowledge is the child of practice and theory. Practice is the continuous and regular exercise of employment where manual work is done with any necessary material, according to the design of a drawing. Theory, on the other hand, is the ability to demonstrate and explain the productions of dexterity on the principles of proportion.... Master builders who have aimed at

BUSINESS MODEL – INTEGRATED SERVICES FOR CAPITAL FACILITIES

CONTINUING EDUCATION COURSES FOR EACH STEP IN THE CYCLE
- ○ Existing Course (offered by DBIA as of 2001)
- ➤ Potential Course

I. FACILITY OPERATION/FACILITY NEEDS IDENTIFICATION
- ➤ Facility Management – A Systems Approach
- ➤ Useability and Serviceability Assessments (ASTM Facility Standards)

II. ACQUISITION PLANNING/FEASIBILITY
- ➤ Project Development Market and Feasibility Studies
- ➤ Concepts in Acquisition Planning – Lease, Purchase, Construct or Forgo?
- ➤ Site/Right-of-Way Analysis; Preliminary Environmental Studies

III. FINANCE/FUNDING/RISK ASSESSMENT
- ➤ Basic Project Finance
- ➤ Advanced Project Finance
- ○ Public/Private Partnerships and Privatization Approaches
- ○ Bonding/Insurance Products for Design and Construction

IV. PROJECT DELIVERY SELECTION
- ○ Comparative Project Delivery Systems
- ○ Design-Build Studies; Research and Findings
- ○ Design-Build Variations and Options
- ➤ Roles in Design-Build—Disciplinary and Multidisciplinary Thinking

V. CONTRACTS FOR FACILITY DELIVERY: COMMERCIAL TERMS AND CLAUSES
- ➤ Engaging and Managing a Design Criteria Professional
- ○ Comparative Design-Build Contracts
- ○ Commercial Terms of D-B Contracts: Cost Plus; GMP; Fixed Price; Other and Their Effects on Projects and People
- ○ Teaming: To Pursue D-B Projects; To Implement D-B Projects
- ○ Alternative Dispute Resolution—Applied to Design-Build
- ○ Structural Variations of Design-Build Consortia: JV; Constructor Prime; A/E Prime; Integrated Firm; Developer-Led; Other

Figure 18.1   Potential courses for integrated delivery by Jeffrey Beard, Design-Build Institute.

acquiring manual skill without scholarship have never been able to reach a position of authority to correspond with their pains; while those who relied only on theories and scholarship were obviously hunting the shadow, not the substance.[4]

----

[4]Morgan, M. H., *Vitruvius—The Ten Books on Architecture,* Harvard Univ. Press, London, 1926, p. 5.

VI. PROCUREMENT/PURCHASING OPTIONS (SOLICITATIONS)

- ○ Using RFQs, RFPs and IFBs in Design-Build
- ➤ Purchasing Options and Selection Methodologies for Public Owners
- ○ Best Value Procurement – Advanced Source Selection
- ➤ When to Procure a Design-Build Advisor or CM – Guidelines for Owners
- ➤ Gaining Sole Source or Strategic Alliance Status in Private Sector Procurement

VII. PROGRAMMING/STUDY AND REPORT; TECHNICAL REQUIREMENTS (CRITERIA FOR DESIGN)

- ➤ Programming for Design-Build
- ○ Performance Specifying (Using a Systems and Assemblies Approach)
- ➤ Advanced Design and Construction Classification Systems (Uniformat; CADD; Object-Oriented XML; Etc.)
- ➤ Environmental Impact Studies and Assessment for D-B Projects

VIII. DESIGN AND CONSTRUCTION

- ○ Marketing Design-Build Services
- V Conceptual Design for Design-Build
- ○ Conceptual Estimating and Scheduling
- ➤ Innovative Methods and Materials in D-B
- ➤ Value Analysis and Life Cycle Costing in D-B
- ➤ Project Control by Design-Build Teams
- ➤ Safety Programs in Design-Build
- ➤ Quality Assurance/Quality Control in D-B
- ➤ Human Resource Issues in D-B
- ➤ Electronic Project Management (Web; CADD; Cost Systems Linkages and Related Technology)

IX. COMMISSIONING/OPERATION

- ➤ Facility Commissioning by Design-Build Teams
- ➤ Design-Build-Operate
- ➤ Electronic Documentation of Facility and Equipment

X. FACILITY MANAGEMENT/DISPOSITION

- ➤ The Military Base Realignment and Closure Act Model for Facility Disposition
- ➤ [[See Courses Listed in "I" Above]]

Figure 18.1 *(Continued)*

The growth of Web-based learning will break down the barriers between discipline-based schools operating within universities. Self-directed bachelor's and master's programs will offer an alternative to traditional on-campus profession-based offerings. In response to industry needs for design-build project managers, courses from at least three distribution areas will be amalgamated to educate the future manager of the multidisciplinary process. A healthy dose of theory and practice, of shadow and substance, of philosophy and business, and of design and construction will form the curricula for training the designer-builder.

18.8   Exploding the Persistent Myths

A myth is "a traditional story that originated in a preliterate society consisting of a recurrent theme that becomes embedded in the culture of a people." Until a brave Genoan captain proved otherwise, the notion that the earth was flat was a widely held, but fictitious, myth. Design-build has slowly but persistently overcome some of the mythological traditions espoused by those who have held onto dogma despite evidence or facts to the contrary.

A recurrent theme by design consultants who are in opposition to design-build is that the "fox is guarding the henhouse," a reference to the design-builder's direct contract for both design and construction with the owner, without requiring a separate initial design contract. In truth, the design-build entity is both the designer of record and the constructor of record, and owes the owner duties of agent (for health and safety, advocating for the project and advocating for the owner) and of vendor (providing a functioning facility according to the performance requirements of the contract). The sum of integrated services is greater than its separate parts because the combined standard exceeds that of the ordinary negligence standard. A design-build entity is probably more akin to the collaborative efforts of the sheepdog and the shepherd who better the odds, by working together, of herding the entire flock to the shearing pen.

There is also a perception among those who are unfamiliar with the process that design-build is suitable only for simple, straightforward, or repetitious projects. In reality, design-build has been used liberally for many of the world's largest and most technically difficult projects. The General Services Administration (GSA), as stated earlier in this book, concluded that design-build is applicable to all types of projects, *regardless of size or complexity,* as long as the requirements were relatively known and stable.[5]

Confronted with the erroneous ideology that it was difficult, if not impossible, to achieve design excellence with design-build, the Design-Build Professional Interest Area (PIA) of the American Institute of Architects undertook a study to determine whether some award-winning designs since 1975 had been produced through design-build. The resultant study and guidelines show how design-build competitions, when properly planned and conducted, can deliver outstanding design. Better yet, design proposals submitted for design-build competitions are usually accompanied by cost and constructability information, increasing the likelihood that the graphically represented solution will become architecture, rather than merely art.

---

[5]GSA, *Design-Build Assessment,* internal report of the General Services Administration, Washington, DC, 1992.

18.9   Permitting the Design-Build Project

A hallmark of successful design-build is predicting, with accuracy, the costs and schedule requirements at the signing of the contract. One vexing issue in design-build project management is the unfamiliarity of some permitting agencies with the design-build process. The phased design-construction approach, which is necessary to accelerate project completion, often runs counter to permit reviewers' expectations (most are more comfortable with 100 percent complete plans and specifications). A challenge for the design-build industry will be to conduct educational programs describing integrated project delivery variations for permitting officials.

18.10   Removing Barriers to Design-Build

The challenge for owners and practitioners who want to gain the advantages of design-build delivery is to overcome both the internal and external objections to its application. The barriers fall within five areas:

Cultural barriers

- There is a concern among all the parties about gaining responsibility/liability versus giving up control.
- For either the public or the industry, there is a lack of recognition of design-build as a profession or discipline.
- There may be a loss of independence when working in concert with designer or constructor.
- Design-build may change the designer's or the constructor's legal relationship with the owner.
- Recognition for contributions to the project may be submerged (design, subcontracting, specialty consulting).
- Fear of change; fear of the learning curve by traditional owners and practitioners.

Legal barriers

- Inflexible procurement laws that do not allow purchasing of integrated services, or that preclude use of other than low-first-cost-based procurement.
- Licensing statutes that go beyond individual practice competency and protection of health and safety issues to affect business practices.
- Building, zoning and fire code processes that are suited to traditional design-bid-build, but have not been updated to accommodate performance-based standards and codes.
- Permitting procedures that are based on the linear design-bid-build process and do not accommodate phased or systems-oriented project management.

Educational barriers

- Schools of architecture, engineering, and construction within universities and colleges are usually focused on single-discipline training; few place emphasis on cross-discipline training.

- Single-discipline programs may not teach the overall process of project delivery or facility life cycle, but instead concentrate on the contribution of the single discipline.

- Student activity groups (e.g., AIA, AGC student chapters) may perpetuate the cult of the single discipline, rather than emphasizing the need for collaborative teams.

- Continuing education programs for practicing professionals are reaching only a small percentage of the industry.

Technical barriers

- Most producers of products and assemblies have not changed their informational approach for design-build, that is, by publishing complete performance data in accordance with up-to-date standards, or showing how their products fit within systems or assemblies to meet the owner's needs.

- Existing formats and classification systems do not meet the predesign needs of integrated services delivery (one exception is the CSI-DBIA performance specifying product called PerSpective that is based on UNIFORMAT).

- Different approaches to quality and innovation are possible with design-build delivery, but A/Es are unsure how to stimulate (through design criteria, incentives/disincentives, and other measures) increased quality and innovation.

Business barriers

- Bonding and insurance companies are still in the process of creating products for design-build; some claim that additional loss data is necessary before creating fully reliable underwriting guidelines.

- Lending institutions reviewing small and midsized projects may be unfamiliar with the design-build process and may require construction drawings before approval of financing.

- Funding of federal and state construction projects may hinge on approval of design before appropriation of construction monies.

- The standard industrial classification system (now NAICS) was developed to fairly represent business establishments in America. There is currently no NAICS code for design-builders to mark when encoding the census of construction industry form.

Design-Build Plus and the Future of Integrated Services

19.1 Design-Build Plus

Basic design-build services are characterized by the simple combination of design and construction under a single source contract. In 1989, design-build was considered to be a revolutionary concept for the delivery of commercial and institutional projects. The mere combining of design with construction was enough of a philosophical change to cause the lobbying arms of major design and construction trade groups to battle its acceptance as recently as 1995. But as design-build has gained currency in the twenty-first-century marketplace, the notion of integrated services on an at-risk basis continues to blossom. *Design-build plus* is the generic name for project delivery methodologies that can incorporate additional services such as business planning, feasibility studies, site acquisition, programming, finance, operation, maintenance, asset management, and other deliverables to meet the market.

The need for flexibility and adaptability, the presence of global competition, and the driving force for efficiency in capital economies all point to the increased use of a delivery system that can be tailored to meet a customer's goals. Four trends will spur the increased use of integrated delivery approaches that are anchored at their core by design-build:

- *Facility obsolescence.* Change and churn plus growth and capacity shifts are rapidly outstripping the functionality and serviceability of facilities. A bank building formerly was designed to exude permanence and to last 50 years as an anchor within an American city. Today's urban bank may be a tenant, perhaps a major tenant, but a tenant nonetheless, in a developer's building under a 5- or 10-year lease. City avenues are regularly ripped up and reconstructed because of new water or gas lines, subway systems, or fiberoptic cabling. The need for coordination among the many disciplines working in a collaborative manner is heightened. Facilities that can be expanded, adapted, and swiftly altered (or razed and rebuilt) will be in demand, paralleling the short cycle time between initial design of a new service or manufactured product and its introduction into the marketplace.

■ *Reorganization and reinvention.* Firms and organizations are adapting to external change by altering their philosophies and modifying their internal operating procedures. As competitors encroach on what was formerly protected turf, organizations are confronted with a choice of either reinventing their core business and operating strategy or going into decline. Organizations are modifying their internal procedures with horizontally sweeping control or management systems that break down the departmental fiefdoms. Greater automation and downsizing within facility owner organizations has created a demand for service providers who can be the single source for facility finance, delivery, and operation of capital facilities.

■ *Proliferation of technology.* Use of technology in facilities design, delivery, and management is still at a relatively immature stage. Excellent new software products are running on hardware platforms that aid estimating, scheduling, cost control, and computer-aided design and drafting (CADD). But each of these applications is a standalone system without the ability or ease of communicating across databases. Advances in microchips, bandwidth, and electronic imaging are about to change the discipline-by-discipline availability of software tools, with programs that can cross-communicate, checking for consistency and accuracy. The new cross-cutting software tools, available through the Internet and intranets, will more closely mirror the multidisciplinary model of design-build. Information will be entered, integrated, hubbed and spoked, and progressively manipulated, providing the electronic project manipulators with a competitive advantage.

■ *Striving for efficiency and quality.* Traditional discipline-based hierarchies in owner organizations (and in the design and construction industry) are slowly being replaced with systems-based problem solving, where the power of ideas and solutions overtakes the status quo. *Quality* is a mantra that is now repeated at all levels, not just at the designer level. To achieve the owner's desired quality level, a quality strategy should be implemented through clear communication with all participants, training, tactical implementation, and measurement of results. Owners are interested in a defined level of quality and cost. An integrated services provider will trade off levels of quality against levels of cost to attain the appropriate level of value. A seasoned design-builder understands that an owner may need assistance in defining the best-value facility—that is, what level of quality is cost-effective given the multiple goals of the owner. With the speed of change and unrelenting competition in the marketplace, the firms that develop the most efficient ways of conducting business, while maintaining or increasing service and product quality, will survive and thrive. Efficiency in design and construction can be attained through some of the newer processes available under the design-build umbrella.

19.2   Design-Build Finance

The addition of project finance to integrated design and construction provides an enormous benefit to owners who may not have access to other than annually budgeted funds. Financing, when combined with project delivery, makes sense when the projected revenues over the economic life of the facility are adequate to meet or exceed the pro forma projections for repayments, debt service, and return on investment.

Project financing will vary according to facility type, perception of risk about the project, and economic climate. For equity investors, the focus is largely on the potential for return of principal plus substantive dividends. Debt financiers, on the other hand, will scrutinize the credit risks and will pass judgment on the fiscal viability of the venture. Sources of loans for capital facilities may include traditional banks, pension funds, insurance companies, and real estate investment trusts. Lenders will ask the design-build-finance team to provide annual balance sheet projections showing the project's performance over time.

Design-build finance opportunities are concentrated largely in three sectors today:

1. The buildings market utilizes design-build finance for commercial and institutional structures. The General Services Administration (GSA) has employed design-build finance on a number of facilities, with creditable success, as have many commercial real estate developers.

2. The water utility sector of the civil infrastructure market is rapidly adopting design-build finance for major projects. The guaranteed revenue streams for potable-water treatment and wastewater treatment facilities are easy to quantify, and these utilities are regarded as relatively safe risk ventures for financing sources.

3. The transportation sector of the civil infrastructure market has been slowly stepping into design-build finance where toll revenues are regarded as sufficient to warrant proceeding with the project. Because of the voluntary nature of transportation choices, lenders and investors scrutinize projected traffic counts or ridership for these projects to gauge whether the pro forma revenue projections are realistic.

Is the project financeable at a rate that is attractive to the market? A number of ratios and metrics provide insight into a project's chance for success:

- *Payback period.* How long before the project has allowed the sponsors to regain their initial investment? Usually measured in terms of total returns against total initial cost, but usually excluding the costs of design, construction, and interest, the payback period should be as short as possible to satisfy lenders and investors.

- *Return on investment (ROI)*. What is the internal rate of return for the project's cash flow? To be financeable, a project should have an ROI of at least 8 points above the prime lending rate; 10 to 20 points (above prime) would be an attractive range.

- *Return on equity.* What is the internal rate of return for the leveraged cash flows? Investors and lenders will be looking for a rate of at least 10 to 12 points above the prime rate; 12 to 25 points above prime would be a healthy range for a successful business venture.

- *Coverage of debt service.* Is the project's revenue from operations sufficient to cover principal and interest payments as well as to throw off additional funds for operating expenses? A design-build finance project should have a debt coverage ratio of at least 1.25:1, with higher ratios preferable for project financial success.

A number of financing instruments are available for capital facilities, from partnership equity to shares of stock and from unsecured loans to capital market instruments; but the financial stalwart for design-construction projects is lending from commercial banks. Using strict underwriting procedures, banks will lend funds for projects with no recourse or limited recourse to the project shareholders. The lending institution will often vary the terms and interest rate depending on the completion status or operating status of the facility. For example, many banks regard the period of construction to be the most risky part of the project's life. The second most risky phase is the period between start of operation and the point at which cash flow meets the *cash flow adequacy test* (usually a minimum of 125 percent of debt service payments).

Design-build finance delivery is likely to grow in the foreseeable future, as a method for jump-starting worthy projects that are artificially stymied by rigid or inflexible funding sources. Design-build companies and financing companies are beginning to see the advantages of business alignments that draw on the expertise of each party. Design-builders are seldom capitalized sufficiently to want to tie up large portions of their equity in financed projects; instead, many would rather win the design-construction contract, execute the work, and move on to the next design-build project. On the other hand, banks, pure investors, and developers, all of whom are well capitalized, prefer to invest for long-term returns. Their business modeling lends a needed fiscal discipline to the process and paves the way for life-cycle thinking in design and construction.

19.3 Design-Build Operate/Maintain

The powerful combination of design and construction with operation and maintenance responsibilities is a realistic answer to owners' concerns about facility quality and long-term performance. In the owner's mind, the project is an asset that must perform in a satisfactory manner over time. Its success or failure is not tied to its functionality or serviceability on day 1 or even after the first year (during the term of the standard one year contractor's warranty), but in terms of multiyear performance.

A local governmental unit that is charged with providing wastewater treatment for a population of 30,000 may opt for a design-build operate/maintain contract including a term of 20 years to ensure adequate treatment and compliance with federal and state environmental regulations and at a reasonable cost to rate payers. During operation, the design-build entity may be confronted with latent equipment or construction defects. As operators of the plant, the team must maintain the operating efficiency of the treatment system, plus provide ongoing training to ensure safety and mitigate against breakdowns.

Design-build operate/maintain is a way to incentivize the services provider to incorporate quality measures and operating vigilance that will prevent technical failure or obsolescence of the facility during the contract term. Value analysis studies have shown that there are three major areas for efficiency improvements: planning/design, construction, and maintenance and operations. The area that is least integrated into facility life-cycle models is operation/maintenance, but it is the area that can have the greatest impact on life-cycle costs and improved facility value for the owner.

19.4   Program or Portfolio Management

Generally, program management is not considered to be a project delivery system. A program manager is normally an agent for the owner on a fee basis, acting as a surrogate "extension of staff" for the owner in the facilities delivery contracting process. An exception to the rule is when the program manager is at risk for the total amount of the design and construction costs, and is required to complete the facilities delivery on a defined timetable with a required level of quality; then the program manager at risk becomes a design-builder for a portfolio of owner projects.

Program managers are tasked with providing extensive facilities services to their owner clients. Services at the front end of program management may include feasibility studies, site analysis, financial feasibility studies, and site acquisition. The program manager, acting for the owner, may have to guide the owner through a rezoning application and hearing, and navigate the environmental permitting processes for the owner. In addition to securing the facility site, the program manager must understand the facility programming and study and report services in order to convert the owner's needs to A/E/C problems and goals that will be solved by design and construction professionals. After the facility is completed, the program manager may be retained to commission the facility, assist in start-up of operations, provide training to facility managers/operators, or conduct ongoing management or maintenance services for the facility.

Program management is especially well suited to those projects or portfolios in which the contract is for an extended duration and the cumulative mass of the contract is larger or more complex than a traditional design-construction contract. Owners who employ program management are looking for services beyond what is traditionally provided by architects, engineers, and constructors. These services may include business process analysis, planning and zoning law, computerized value analysis, and comprehensive facility management. When done on an at-risk basis, program management becomes a design-build-plus variation.

19.5   Other Variations (BOTs, BOOs, DBOT, etc.)

Integrated project delivery variations have proliferated since 1980, as owners in government and in the private sector have sought creative ways to allow substantial projects to move forward. Essentially, there has been a return to the nineteenth-century model of financing, designing, and building by private enterprise in a modern era of capital markets and national regulation. Three ingredients are necessary for megascale integrated project delivery: (1) permission by government jurisdictions to undertake the project and to charge fees for cash flow to repay lenders and investors, (2) a project that has an acceptable risk profile to enable completion and cash generation within a reasonable period, and (3) lenders and investors willing to fund the project coupled with designers and constructors willing to bring the project to life. Fundamentally, medium- or large-scale integrated delivery is about the creation and implementation of a good business model—a scheme that can reward, through profits, those willing to take calculated risks.

The more popular design-build plus delivery methods currently in use (in each of the following examples, the letter D for design is implicitly included; therefore, BOT becomes DBOT, BOO is really DBOO, etc.) include the following:

- *Build-operate-transfer (BOT)*. This delivery method has been used worldwide in projects where the owner grants the right to build and operate a facility that will generate cash flows and profit. The facilities are typically (but not exclusively) infrastructure-related, such as power plants, toll roads, water-treatment plants, or ferries. At the end of the term of the operating contract, the operator agrees to turn the facility over to the owner. When using the BOT approach, the project sponsors often conclude a turnkey contract with a design-build entity. The design-builder may also be one of the project's sponsors, especially where the design-build entity has been providing professional services since the inception of the project and is tying part of its future economic fortune to the BOT project.

- *Build-own-operate-transfer (BOOT)*. In addition to the permission to operate a project for a period of years, the BOOT variation enables the design-build-plus consortium to have ownership of the property during the operating period. Development rights may provide opportunities to maximize returns for investors and equity participants. For example, a toll bridge may have a scenic approach with an ideal location for a rest stop with restaurants, fuel, and other revenue-generating amenities. A BOOT agreement incentivizes the consortium to find creative ways to maximize the facility and environs, by taking advantage of business opportunities available through the facility and its user base.

- *Build-own-operate* (*BOO*). The BOO arrangement incorporates the financing, designing, constructing, and operating of a facility but without the requirement for transferring full ownership over to the owner at the end of the agreement term. Some governments do not sanction any BOO arrangements because much of the control over the services that the facility provides is ceded to the consortium. On the other hand, a BOO is attractive to the sponsors since it allows syndication or public offering of stock in the venture.

- *Lease-sale-transfer variations* (*LST*). Innovative delivery methods occasionally incorporate leasehold rights that may be granted to the developer, to the investors, to the owner, or to the design-build team. The design-build-operate team may begin the project with ownership rights allowing the project to attract favorable financing and rates. After a stipulated period of time, the facility ownership rights transfer to the owner, who leases the facility back to the team for operation. The concept also may be applied by private owners who issue long-term property rights to developers and/or financial consortia, where the owner retains the right to lease a portion of the completed facility for their corporate use. The owner benefits by acquiring needed facilities without huge investments, and can therefore place capital at work for their core business or in other more liquid investment vehicles (other than buildings or physical plant).

The days of relatively straightforward design-build projects are here and now. Tomorrow's design-build-plus variations will include various forms of finance, ownership, and operation of stand-alone facilities and portfolios of facilities. The flexibility design-build-finance delivery systems in adapting to multiple funding sources, diffusion of risks and responsibilities, and complex ownership and shareholder schemes is proven. What remains to be seen is the potential for exciting new mutations of design-build-finance through the changes in government controls over creative ownership (e.g., public-private partnerships) and alternative sources of government, nonprofit, and private-sector funding.

19.6   Total-Facility Systems

The changing landscape of twenty-first-century design and construction is bringing new players to an industry where (1) the large number of transactions (the total nonresidential and residential building industry encompassed over $600 billion of activity in 1999, according to the Department of Commerce) and (2) the relative inefficiency of the process present enormous opportunity to managers and entrepreneurs. One of the outcomes has been the globalization of the design-construction market, bringing competitors from outside countries to nearly every locale with sizable project opportunities. A second outcome has been manifested by the business management disciplines, who have seen, through quantitative analysis and operations research, the possibilities of better long-range facilities strategies for corporate and governmental owners.

The new competitive forces chipping away at the traditional design and construction industry are able to broaden the scope of services offered to facility owners. Business management consultants are engaged by corporations and government agencies to examine core processes and products. The management studies offer keen insights into the organizations' strategic focus. These business advisors are in the position to help shape the direction of their customers, and with powerful knowledge about the goals and aspirations of their clients, are able to devise business approaches that perpetuate their business relationship.

The business management consultant is already inside the door when the customer decides to embark on a capital facilities expansion. What does the onset of this activity portend for the design and construction practitioner? The competitive response will be to develop strategic alliances with these out-of-industry business entities, or to broaden the professional services offered by the design-build firm to encompass management studies, outsourced business services, facility operations, and other nontraditional offerings. The design-construction industry currently has better knowledge and understanding about life-cycle consequences of owner facility-planning decisions; and it is a market advantage that could be further exploited.

A total-facility delivery system will cast a wider net; encircling business planning and modeling; facility feasibility and user analysis; performance definition and programming; structuring and placement of finance; public outreach and involvement; design and construction; commissioning; maintenance; operation and facility upgrade (as standards change); multiyear financial management of the asset; and finally, disposal or revitalization of the asset by renovation, sale or transfer, or demolition. During the extended life of the project, the total-facility deliverer may also provide services that are intimately tied to the facility, such as resort management services for an island resort or academic instruction for a charter school system. The willingness to understand the facility owner's needs and to take risks beyond traditional design, construction, and finance will characterize the entities that assume total-facility delivery responsibilities.

19.7   Technology

Information technology tools are beginning to have an enormous impact on the design-construction industry. The power of automation is allowing owners and practitioners to organize and track the many tasks, transitions, and activities that constitute project and product management. Product management is included here because it is no longer sufficient to think simply of the narrow phases of design and construction—the management of these two phases does not necessarily focus on the owner's goals. Project delivery is in some ways a short-term view, whereas product (i.e., facility asset) management is concerned with the success of the venture over time. Newer software tools are beginning to incorporate this broader concept of long-term acquisition and operation of facility assets.

Mapping of the facility delivery-operations process would provide a chronological chart of the key activities in the life of a facility (see hypothetical timeline in Fig. 19.1). Included on this map would be the various phases of facility's gestation, birth, and productive life: needs identification, financing, performance

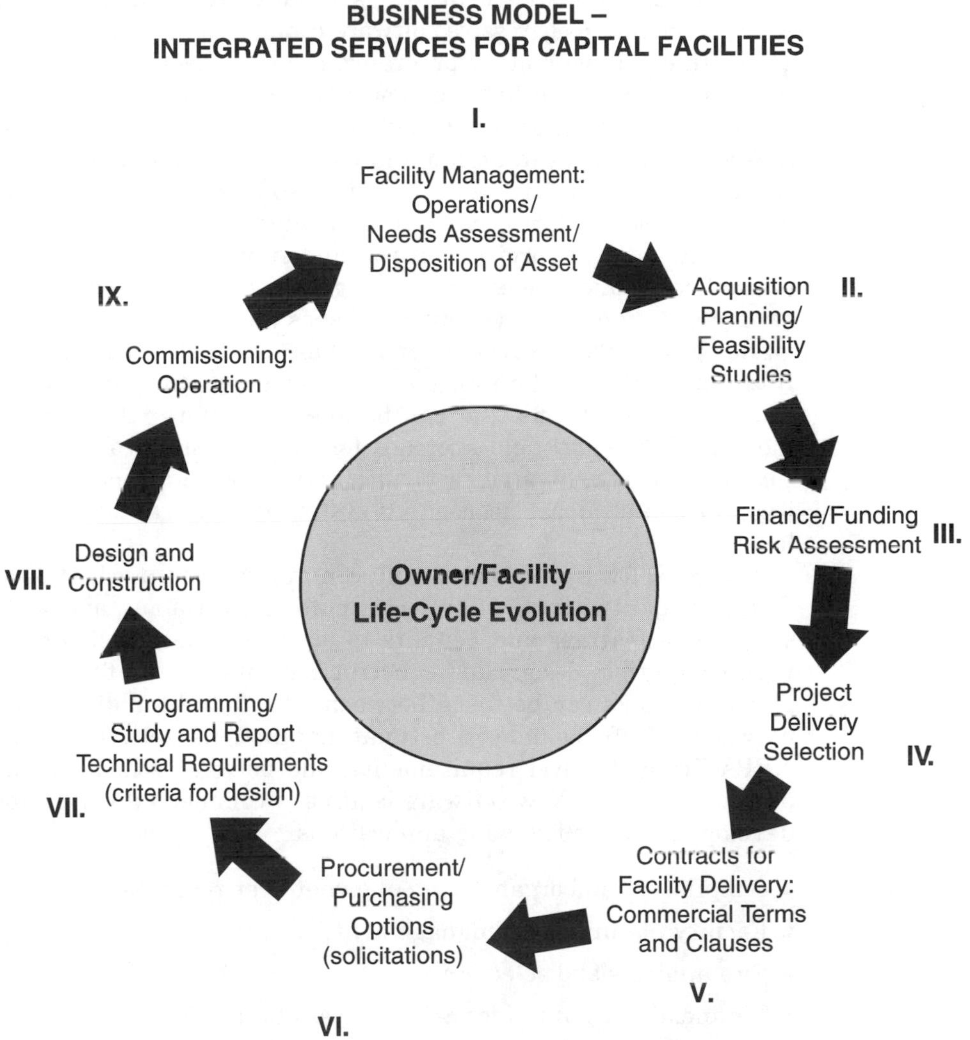

BUSINESS MODEL –
INTEGRATED SERVICES FOR CAPITAL FACILITIES

I.

Facility Management:
Operations/
Needs Assessment/
Disposition of Asset

Acquisition
Planning/
Feasibility
Studies   **II.**

IX.

Commissioning:
Operation

Finance/Funding
Risk Assessment **III.**

**Owner/Facility
Life-Cycle Evolution**

Design and
VIII. Construction

Project
Delivery
Selection   **IV.**

Programming/
Study and Report
Technical Requirements
(criteria for design)
VII.

Procurement/
Purchasing
Options
(solicitations)

Contracts for
Facility Delivery:
Commercial Terms
and Clauses

V.

VI.

Figure 19.1 Facility life-cycle chart.

requirements setting, scope definition, conceptual estimating and design, design development, construction documentation and construction, commissioning, operation, maintenance, and ongoing business management until revitalization, transfer or disposal.

Current software applications tend to focus on individual phases in the facility life cycle and ignore other phases. Indeed, most software tools are currently incompatible with other tools because of different standards and protocols exacerbated by the limited focus of the program. At the start of the twenty-first century, most of the software available for design and construction applications is limited to organization (e.g., tracking of change orders or schedule activities), graphic representation [CADD and three-dimensional (3D) modeling], and reporting (dissemination of drawings or project information to the project team). A few of the more aggressive software companies are attempting to create programs that link with other programs, or are developing software that incorporates analysis or built-in "intelligence" features.

There are readily available software packages for many of the communication and data needs in design and construction. A number of firms have concentrated in computer-aided design software, such as Intergraph and AutoCAD. Other firms have concentrated on cost estimating, scheduling, materials tracking, project control, and submittal updating. A few entrepreneurial companies are experimenting with systems in which the operator can generate design, and the software keeps pace with materials quantities and guide pricing. With proven technology for establishment of local networks and Internet availability for projects, owners and practitioners are looking forward to advanced programs that can handle large materials inventories, procurement processes, 3D field sketches for rapid on-site problem solving, ongoing comparisons of estimated costs with actual costs for each element of the facility, and use of global positioning system data for automated surveying and layout.

Future software products will contain increased capabilities to answer "What if" questions about project inputs and variables; and will move beyond standalone features and benefits to form artificial-intelligence (AI) linkages with compatible design and construction information technologies. Valuable electronic links can be fused between CADD and building code information; between specifying and cost estimating functions, between 3D simulation and NFPA fire exit travel requirements; and between scheduling and labor productivity metrics. New software is already commercially available (or is being developed) for the following applications:

- compilation and organization of owner/user requirements
- Facility planning and management
- Site analysis and selection
- Financial feasibility for various types of facilities
- Sourcing and combinations of financing (customization for project sponsors)

- Labor and productivity (e.g., research done by the Lean Construction Institute)

- Project delivery selection or, alternatively, design-build variation selection

- Sourcing materials-equipment systems and analyzing their performance and availability for the project

- Facility programming

- Life-cycle prediction tool for assemblies

- CADD to fabrication or CADD to construction

- Projection of construction schedules

- Simulated construction sequencing (adjustable for workforce resources or types of materials, etc.)

Software for the design-construction industry may be suitable for specific applications, or it may be used to link vast amounts of related data for use by the project team. These applications, many of which will be available through the Internet, should be configured to produce customized reports for specific audiences. The programs should also accept and retrieve organized information, analyze data and develop reports, and link with other software. One of the most exciting new software products in the 1990s—the 3D CAD—was a significant tool during the Gulf War. The allied forces used the software to virtually deconstruct (rather than to virtually construct) Iraqi military facilities satisfying the concern for guiding smart bombs to the right spaces within buildings.

A "beyond the silos" overview of electronic tools for facility delivery and management may lead owners and practitioners to a new definitional model. The model would include essential variables of performance, quality attributes, dimensional geometries, and costs and values that would be scalable from an entire portfolio of facilities down to an individual piece of material. The process map would enable a virtual portfolio, project, or piece of equipment to evolve over time, adjusting for its predicted performance at any point within the pro forma life cycle. This tool would be useful for owners or designer-builders who make countless tradeoffs between quality and cost every day while trying to satisfy competing goals. A simplified matrix of the major components of the model is shown in Fig. 19.2.

Advancing technologies in the design-construction industry will favor the integrators. The integration will not occur solely within the confines of a single corporation, but will often be a "paraenterprise" or strategic alliance that is customizing its services in response to a real and immediate marketplace need. The service-oriented design-construction team will gain a competitive advantage over the traditional nonintegrated design-construction practitioners because they will be focused on their customers' goals and needs together, and will have team underpinnings (through their business alliance) of trust, openness, and financial sharing. By contrast, traditional independent designers and

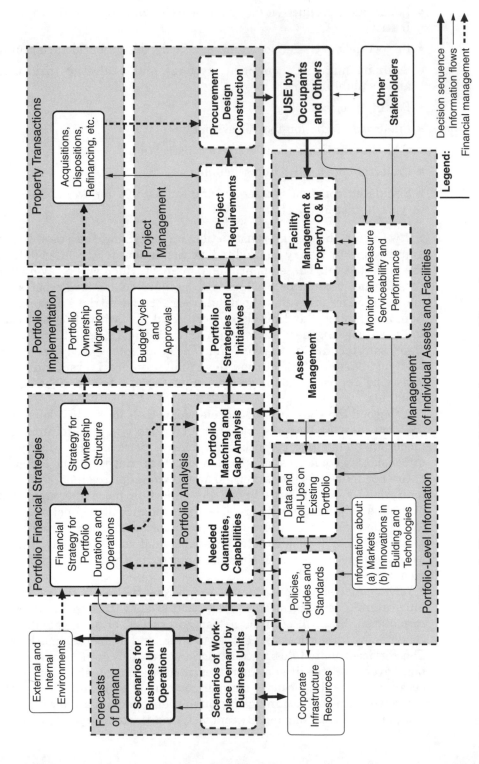

Figure 19.2 Integrated portfolio cycle (*Source: G. Davis and F. Szigeti, International Centre for Facilities.*)

constructors will continue to have a tendency to second-guess each other, will lose vital speed, and will begin to erode customer satisfaction.

19.8   Prognosis

According to an article in *The Economist,* General Electric's Web-based links to its services and goods suppliers have cut procurement cycles in half, processing costs by one-third, and have lowered costs (depending on the product or service) by 5 to 50 percent. Its partners have become more efficient, thereby making GE more efficient and profitable. Vertically based companies with linear supply chains are no longer as competitive. In the current economy, it is the collaborative entity—mirroring the flexibility of Web-based relationships and yet inextricably tied together for mutual business purposes—that gains the customers.

Design-build represents a range of possibilities. It is a profound business strategy that undergirds collaborative partners and allows them to use all the tools and resources at their disposal in the most efficient manner possible. Although the days of direct selling of individual goods and services will never disappear, it is the integrator within a multidisciplinary, multiple-trade, and multiple-product world, who can add significant value to the process.

Index

ABOUT THE AUTHORS

JEFFREY L. BEARD is the President & CEO of the Design-Build Institute of America (DBIA). The DBIA is a 501(c)6 nonprofit association dedicated to promoting the process of design-build project delivery and to supporting design-build entities that practice design-build throughout North America. DBIA is a broad-based, member-driven organization comprised of architects, engineers, constructors, specialty contractors, owners, and other professionals who are committed to best practices for integrated design-build-finance.

Mr. Beard has over 25 years of management and policy experience in the design and construction industry, including project management, contract administration, governmental relations, issue analysis, product development, and program implementation. His experience embraces both the design and construction sectors, including responsible-in-charge positions with industry organizations and with trade/professional associations serving the A/E/C community.

During his earlier tenure as Manager, Federal Programs and Regulatory Affairs with the American Society of Civil Engineers, Mr. Beard authored the seminal report and procurement survey entitled "Design-Build in the Federal Sector." Earlier, he worked as Director of Construction Administration for a Baltimore-area architectural firm; as Associate Director of the Heavy-Industrial Division, Associated General Contractors of America; as Features Editor of *CONSTRUCTOR* magazine; as District Engineer with H.H. Robertson Company; and as Project Manager for a Maryland-based general contractor.

Jeffrey Beard serves on the Board of Directors of the Lean Construction Institute; the McGraw-Hill Design-Build Magazine Editorial Advisory Board; the CSI-DBIA Joint Venture Board; and the Design-Build Institute of America (ex officio). During 1994, in conjunction with the Georgia Institute of Technology, Mr. Beard created the annual Professional Design-Build Conference, which has become the largest conference dedicated to integrated facilities delivery in the world.

Mr. Beard holds a bachelor's degree in communications and political science from Rutgers University; a master's degree in planning (concentration in facilities planning) from the University of Virginia School of Architecture; and has completed coursework in architecture at the University of Maryland. In his rare spare time, he is restoring a streamline Chris Craft express cruiser (a 1939 design originally done for the New York World's Fair).

MICHAEL C. LOULAKIS, ESQ., is a senior shareholder in the Vienna, Virginia office of Wickwire Gavin, P.C., a national law firm which specializes in the representation of domestic and international clients on construction-related matters. Since joining Wickwire Gavin in 1979, Mr. Loulakis has represented clients on projects around the world in the power, telecommunications, healthcare, public facility and transportation sectors of the construction industry. His role on these projects ranges from advising owners about project delivery systems to contract drafting, claims administration, and dispute resolution.

Mr. Loulakis is widely known for his expertise in the design-build process. He has been the lead counsel in the drafting and negotiating of design-build contracts on numerous high visibility projects around the country, as well as for projects in Mexico, Ireland, and Pakistan. In addition to his contract work, Mr. Loulakis has had substantial involvement in assisting public owners such as the Federal Bureau of Prisons and Bay Area Rapid Transit (BART) in developing best practices for their design-build programs and projects. He also represents numerous contractors and design professionals around the country by providing strategic planning, training, contract drafting, and conflict resolution advice on their design-build programs.

Mr. Loulakis has been an active member of the Design-Build Institute of America (DBIA) since its founding in 1993 and is currently a member of its Board of Directors. As Chairman of DBIA's Manual of Practice Task Force, he played a lead role in the development and drafting of the DBIA model design-build contracts. Mr. Loulakis was also the author of DBIA's Design-Build Contract Guide, which defines DBIA's perspective on best design-build contracting practices.

A frequent and highly regarded public speaker and author on design-build, Mr. Loulakis has written over 100 publications on project delivery systems and construction law, including the "Legal Trends" column in *Civil Engineering* magazine, which he has written for almost 20 years. Particularly noteworthy are two books that he has co-authored for Aspen Law Publications, *The Design-Build Contracting Handbook* and *Construction Management: Law and Practice.* One of his most recent and innovative works is the development of an interactive CD-ROM program entitled *Construction Project Delivery Systems: Evaluating the Owner's Alternatives,* produced and distributed by A/E/C Training Technologies. This critically acclaimed program provides a comprehensive review of all major project delivery systems and has been used by organizations around the country as a tool for design-build training. He also is the author and publisher of *Design-Build Lessons Learned,* an annual publication that reviews design-build case law from around the country.

Mr. Loulakis graduated with honors from Tufts University with a B.S. degree in civil engineering. Before earning a Juris Doctorate from Boston University School of Law, he practiced as a geotechnical engineer with a consulting firm in the greater Boston area. He can be reached at *mloulakis@wickwire.com.*

EDWARD C. WUNDRAM received a B.S. degree in architecture from the Georgia Institute of Technology. He has over 35 years of professional practice in design and procurement of major public facilities, including 25 years utilizing performance based specifications for the acquisition of building systems and complete facilities. Mr. Wundram has considerable experience managing large public projects as the design coordinator of an international architectural, engineering and project management firm. He was the administrator for a system-building program for the Portland Public Schools from 1970 to 1972 and was responsible for the start-up of an Anglo-American construction management firm (Heery-Farrow, Ltd.) in London during 1976 and 1977.

Since 1980, Mr. Wundram has been a sole proprietor and principal of his own firm (The Design Build Consulting Group, Beaverton, Oregon), limiting his professional practice to project management and the administration of design and design-build competitions for public facilities

As an advisor to the Board of Directors of the Design-Build Institute of America (DBIA), and Co-Chairman of their Manual of Practice Task Force, he authored one of their publications: *Design-Build RFQ/RFP Guide for Public Sector Projects,* and a similar guide for small projects. In 1993, Mr. Wundram received one of DBIA's first Leadership Awards. He currently serves as Vice-Chair of DBIA's Policy Committee. Mr. Wundram is also a member of the American Institute of Architects and the Construction Specifications Institute. He maintains his professional architectural licenses in Oregon, Washington, California and Georgia.

He has served as a consultant to the Government of Canada and their Department of Foreign Affairs and International Trade to review their design-build procurement procedures for Canada House in Trafalgar Square, London. Mr. Wundram has also provided a similar service to the Canadian Department of National Defence where he was responsible for recommending procedures for design-build proposal solicitations for their Infrastructure Reduction Program (base relocations). Mr. Wundram was also a member of the consulting team in Washington D.C., advising the Federal Bureau of Prisons on procedures and contract forms for their recently initiated design-build procurement program for major federal correctional facilities. He has also developed design-build procurement documents for the University of California at Berkeley, for a 920-unit graduate student family housing project at University Village, Albany, California.

Mr. Wundram sits on the Editorial Advisory Board of *The Construction Specifier,* the journal of the Construction Specifications Institute. His responsibilities include advising the editorial staff on matters pertaining to the design-build project delivery method, and to occasionally author articles on related subjects. Recently, he has also been appointed to the six-member, Year 2000 Advisory Group for the American Institute of America's DesignBuild Professional Interest Area (D/B PIA). The D/B PIA advises member firms on aspects of single-responsibility design-build project delivery and designer-led design-build in particular.

In addition to his professional practice, Mr. Wundram often lectures at universities and colleges on the design-build procurement methodology. Over the last several years, he has lectured at Georgia Tech, Arizona State University, University of Iowa, University of Wisconsin-Madison, and the University of Montreal. In 1996 and 1997, the Construction Specifications Institute (CSI) and the Construction Information Group (CIG) of The McGraw-Hill Companies engaged him to deliver a series of six lectures on design-build procurement for owners, designers, builders and manufacturers' representatives throughout the country. His most recent publications include articles in *The Construction Specifier,* "Improving Your Chances of Winning Design-Build Competitions" (February 2000), and "The Right Way to Write a RFQ" (August 2000). He can be reached at *wundram@msn.com.*